Nesrin Horzum, Mustafa M. Demir, Rafael Muñoz-Espí, Daniel Crespy (Eds.)

Green Electrospinning

Also of interest

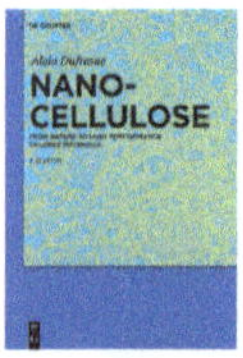

Nanocellulose
From Nature to High Performance Tailored Materials
Dufresne, 2017
ISBN 978-3-11-047848-8, e-ISBN 978-3-11-048041-2

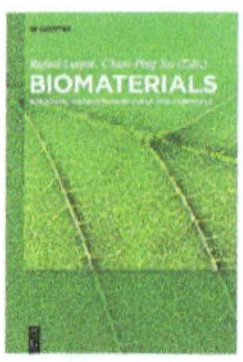

Biomaterials
Biological Production of Fuels and Chemicals
Luque, Xu, 2016
ISBN 978-3-11-034230-7, e-ISBN 978-3-11-034242-0

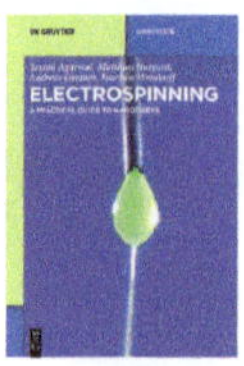

Electrospinning
A Practical Guide to Nanofibers
Agarwal, Burgard, Greiner, Wendorff, 2016
ISBN 978-3-11-057808-9, e-ISBN 978-3-11-057909-3

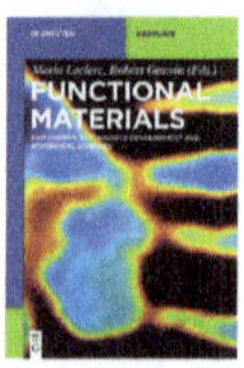

Functional Materials
For Energy, Sustainable Development and Biomedical Sciences
Lerclerc, Gauvin (Eds.), 2014
ISBN 978-3-11-030781-8, e-ISBN 978-3-11-030782-5

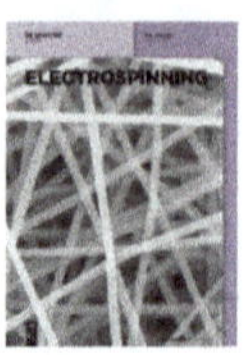

Electrospinning
ISSN 2391-7407

Green Electrospinning

Edited by
Nesrin Horzum, Mustafa M. Demir,
Rafael Muñoz-Espí, Daniel Crespy

DE GRUYTER

Editors
Dr. Nesrin Horzum
Izmir Katip Celebi University
Department of Engineering Sciences
35620 Izmir
Turkey
nesrin.horzum.polat@ikc.edu.tr

Prof. Dr. Mustafa M. Demir
Izmir Institute of Technology
Department of Materials Science and Engineering
35430 Izmir
Turkey
mdemir@iyte.edu.tr

Dr. Rafael Muñoz-Espí
Universitat de València
Institute of Materials Science (ICMUV)
C/ Catedràtic José Beltrán 2
46980 Paterna, Valencia
Spain
rafael.munoz@uv.es

Assoc. Prof. Dr. Daniel Crespy
Vidyasirimedhi Institute of Science and Technology (VISTEC)
Dept. of Mat. Science and Eng.
Colloid Organic Materials Lab.
Rayong 21210
Thailand
daniel.crespy@vistec.ac.th

ISBN 978-3-11-056180-7
e-ISBN (PDF) 978-3-11-058139-3
e-ISBN (EPUB) 978-3-11-058863-7

Library of Congress Control Number: 2019937555

Bibliographic information published by the Deutsche Nationalbibliothek
The Deutsche Nationalbibliothek lists this publication in the Deutsche Nationalbibliografie; detailed bibliographic data are available on the Internet at http://dnb.dnb.de.

Typesetting: Integra Software Services Pvt. Ltd.
Printing and binding: CPI books GmbH, Leck
Cover image: whitehoune / iStock / Getty Images Plus

www.degruyter.com

Preface

Over the last two decades, a great deal of effort has been put in to understand the fundamental laws behind the electrospinning process and the potential applications that result from its product, fibrous mats. Unfortunately, the electrospinning process has to date mainly used toxic solvents for the dissolution of polymeric chains that are spun into fibers. Considering the low concentration of polymers in precursor solutions and the entire volume of solution for bulk production, a huge amount of organic solvent is consumed, and so the process itself creates a large environmental issue, that is, handling the waste organic solvents. Electrospinning of polymers that avoid the use of harmful organic solvents by spinning either from aqueous solutions/suspensions or from the molten state (skipping the dissolution or dispersing steps) is defined as "green electrospinning." You will see some important examples of both approaches in this book. In addition, electrospinning from the molten state using waste polymeric products, which are then used for functional applications, presents a promising, greener approach.

Green Electrospinning summarizes the current state of knowledge in the field of electrospinning performed by employing environmentally benign chemicals and processes for the development of sustainable technologies. In this book, we have highlighted the key features that focus on (1) fundamental principles of electrospinning, (2) electrospun mats that are exemplified including the ones produced from solution and melt electrospinning, and (3) potential applications. We hope that this book will provide readers a valuable insight into an important area of green electrospinning. The interdisciplinary nature of the topics in this book will help students and researchers from diverse backgrounds. No doubt, the main credit of this book goes to the contributors, whose expertise in the field of electrospinning has been comprehensively written in their respective chapters.

https://doi.org/10.1515/9783110581393-201

Contents

List of contributors

Valeria Annibaldi
Avectas
Maynooth, Ireland
vannibaldi@avectas.com

Razan Badran
Department of Chemical and Petroleum Engineering
American University of Beirut, PO Box 11-236, Beirut 1107-2020, Lebanon
rab29@mail.aub.edu

Funda Cengiz Çallioğlu
Textile Engineering Department
Engineering Faculty
Süleyman Demirel University
Isparta, Turkey
fundacengiz@sdu.edu.tr

Daniel Crespy
Department of Materials Science and Engineering
School of Molecular Science and Engineering
Vidyasirimedhi Institute of Science and Technology
Rayong 21210, Thailand
daniel.crespy@vistec.ac.th

Mustafa M. Demir
Department of Materials Science and Engineering
İzmir Institute of Technology
35430 İzmir, Turkey
mdemir@iyte.edu.tr

Ryan Gharios
Department of Chemical and Petroleum Engineering
American University of Beirut, PO Box 11-236, Beirut 1107-2020, Lebanon
rtg03@mail.aub.edu

Motonobu Goto
Department of Materials Process Engineering
Nagoya University, Furo–cho, Chikusa–ku
Nagoya 464-8603, Japan
goto.motonobu@material.nagoya-u.ac.jp

Matthew A. Hood
Max Planck Institute for Polymer Research
Ackermannweg 10
55128 Mainz, Germany
matthewahood@gmail.com

Nesrin Horzum
Department of Engineering Sciences
İzmir Katip Çelebi University
İzmir 35620, Turkey
horzumnesrin@gmail.com

Tuğba Isık
Department of Materials Science and Engineering
İzmir Institute of Technology
İzmir 35430, Turkey
tugbaisikedu@gmail.com.

Hideki Kanda
Department of Materials Process Engineering
Nagoya University, Furo–cho, Chikusa–ku
Nagoya 464-8603, Japan
kanda.hideki@material.nagoya-u.ac.jp

Hülya Kesici Güler
Textile Engineering Department
Engineering Faculty
Süleyman Demirel University
Isparta, Turkey
kesicihulya@gmail.com

Karin Kogermann
Institute of Pharmacy
University of Tartu
Tartu, Estonia
kkogermann@gmail.com

Kyung Jin Lee
Department of Chemical Engineering and Applied Chemistry
College of Engineering
Chungnam National University
99 Daehak-ro (st), Yuseong-gu
Daejeon 305-764, South Korea
kjlee@cnu.ac.kr

https://doi.org/10.1515/9783110581393-202

Shani L. Levit
Chemical and Life Science Engineering
Virginia Commonwealth University
Richmond, VA 23284, USA
levitsl@mymail.vcu.edu

Siti Machmudah
Department of Chemical Engineering
Institut Teknologi Sepuluh Nopember
Kampus ITS Sukolilo
Surabaya 60111, Indonesia
machmudah@chem–eng.its.ac.id

Seongjun Moon
Department of Chemical Engineering and Applied Chemistry
College of Engineering
Chungnam National University
99 Daehak-ro (st), Yuseong-gu
Daejeon 305-764, South Korea
Seongjun@cnu.ac.kr

Rafael Muñoz-Espí
Institute of Materials Science (ICMUV)
Universitat de València
C/Catedràtic José Beltrán 2
Paterna 46980, Spain
rafael.munoz@uv.es

Ezgi Özliseli
Pharmaceutical Sciences Laboratory
Åbo Akademi University
Turku, Finland
eozlisel@abo.fi

Mirja Palo
Pharmaceutical Sciences Laboratory
Åbo Akademi University
Turku, Finland
mirja_palo@hotmail.com

Amanda L. Pham
Chemical and Life Science Engineering
Virginia Commonwealth University
Richmond, VA 23284, USA
phamal2@mymail.vcu.edu

Robin D. Rogers
525 Solutions, Inc.
720 2nd Street
Tuscaloosa, AL 35403, USA
Robin.Rogers@525Solutions.com

Didem Sen Karaman
Pharmaceutical Sciences Laboratory
Åbo Akademi University
Turku, Finland
didem.sen.karaman@ikc.edu.tr

Julia L. Shamshina
Mari Signum, Ltd.
3204 Tower Oaks Boulevard
Rockville, MD 20852, USA
Julia.Shamshina@MariSignum.com

Christina Tang
Chemical and Life Science Engineering
Virginia Commonwealth University
Richmond, VA 23284, USA
ctang2@vcu.edu

Ali R. Tehrani-Bagha
Department of Chemical and Petroleum Engineering
American University of Beirut, PO Box 11-236
Beirut 1107-2020, Lebanon
at711@aub.edu.lb

Wahyudiono
Department of Materials Process Engineering
Nagoya University, Furo–cho, Chikusa–ku
Nagoya 464-8603, Japan
wahyudiono@b.mbox.nagoya-u.ac.jp

Rebecca C. Walker
Chemical and Life Science Engineering
Virginia Commonwealth University
Richmond, VA 23284, USA
walkerrc4@mymail.vcu.edu

Nesrin Horzum, Rafael Muñoz-Espí, Matthew A. Hood,
Mustafa M. Demir and Daniel Crespy

1 Green electrospinning

Abstract: In the last two decades, electrospinning has grown in popularity; however, the majority of the setups are based on solution processing from toxic organic solvents. As green processing and environmental stewardship have also become important in recent years, for political and economic reasons, the subsequent increase in demand for the scaling up of electrospinning requires that an environmentally toxin-free process be championed. This book comprehensively addresses clean and safe electrospinning for the fabrication of green nanofibers and evaluates their potential applications.

Keywords: clean electrospinning, environmentally friendly electrospinning, green fibers, safe electrospinning, solvents

1.1 Introduction

Electrospinning is an efficient method that has been recognized for the fabrication of continuous polymer fibers with diameters down to a few nanometers. When the diameter of fibers decreases from micrometer to nanometer, several characteristics show up, such as a large surface area to volume ratio, flexibility, and mechanical performance. Initially, only polymers were accepted as fiber-forming materials. As time went on, ceramics and glasses were also considered as fiber precursors. By virtue of being a very simple and easily controlled technique, the utilization of electrospinning has spread into all fields of work, including biomedicine, pharmaceutics, filtration, energy, sensor, cosmetics, and food packaging.

In a typical electrospinning setup, there are three basic components necessary to the process: a high-voltage power supply, a capillary tube with a pipette or needle of small diameter, and a metal collecting screen. During the process, the polymer solution or melt is pumped through the capillary tube and a high voltage is

Nesrin Horzum, Department of Engineering Sciences, İzmir Katip Çelebi University, İzmir, Turkey
Rafael Muñoz-Espí, Institute of Materials Science (ICMUV), Universitat de València, Paterna, Spain
Matthew A. Hood, Max Planck Institute for Polymer Research, Ackermannweg, Mainz, Germany
Mustafa M. Demir, Department of Materials Science and Engineering, İzmir Institute of Technology, İzmir, Turkey
Daniel Crespy, Department of Materials Science and Engineering, School of Molecular Science and Engineering, Vidyasirimedhi Institute of Science and Technology, Rayong, Thailand

https://doi.org/10.1515/9783110581393-001

used to create an electrically charged jet of polymer out of the tube. The nozzle of the capillary tube serves as an electrode and a 100–500 kV m^{-1} electric field is applied. The solvent evaporates and the polymer solidifies while it is transported through the electric field. An interconnected web of fibers is therefore formed on the collecting screen. Considering that the solvent residuals are still present in the resulting nanofibrous mats, they may need to be post-modified to remove or neutralize the undesired solvent traces. The most common solvents for electrospinning are normally toxic organic solvents and, therefore, the presence of solvent residues will limit the use of obtained electrospun mats for any biomedical applications. It could also cause the operator to work under unhealthy conditions. Not only the toxicity of the electrospun scaffolds but also environmental concerns of the evaporating solvent have led to the concept of "Green Electrospinning" as a current issue (Figure 1.1).

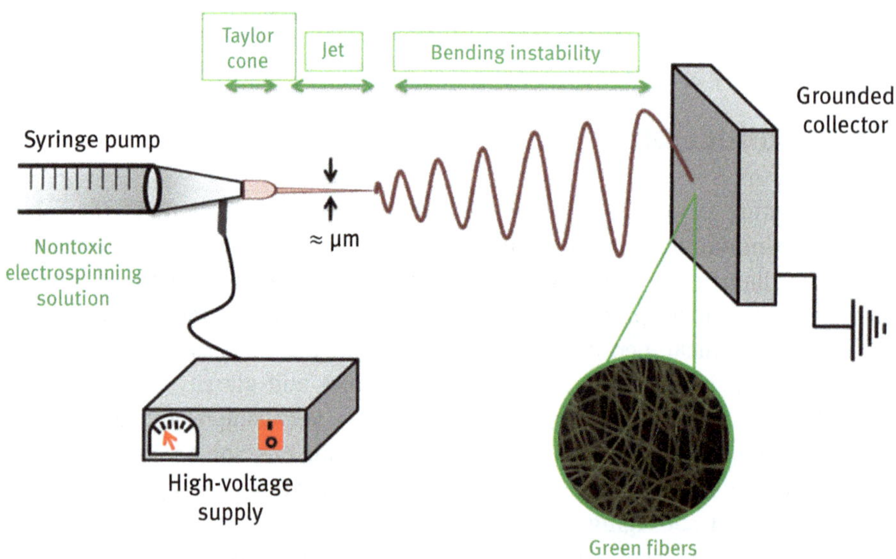

Figure 1.1: Schematic representation of green electrospinning setup.

Traditionally, electrospinning has relied heavily on the use of volatile organic solvents (VOCs) [1]. This is because solvents for electrospinning must not only be capable of dissolving the polymer chain but also be capable of evaporating within the short distance that separates the nozzle from the collector. Unfortunately, VOCs can be expensive, toxic, and hard to remove from the environment [2].

The number of journal articles using the term "green electrospinning" has been growing over the last 8 years. Green electrospinning is essentially the same or a similar procedure to traditional electrospinning and follows, therefore, the same overall physical principles that control the fiber diameter. Alterations in the procedure are attempted in order to remove the VOCs from the process and perhaps improve the recyclability and reusability of the various products and starting materials [1].

There are three main hazards in traditional electrospinning that arise due to the use of VOCs:

1) There are risks to the product user, particularly in the area of biomedicine, where materials will be implanted along with living tissue and cannot come in contact with toxic solvents.
2) Risks to the product producer: many VOCs are quickly brought into the ambient environment around electrospun fibers. If a chamber equipped with vacuum is a part of the electrospinning setup, the solvent exposure is minimized. However, trace amounts left in the porous fiber mats can overtime be dangerous to those working with the mats.
3) Risks to the environment: with the clear risk of VOCs and synthetic polymers accumulating within the air, waste that must be handled in particularly ways, with stray fibers that slowly accumulate in the soils (an unfortunate side effect of even washing of synthetic fiber clothes), and VOCs evaporating into the air effect our water, air, and soil quality [3–5].

Some particularly useful polymers for electrospinning, like polyimides, polyurethanes, and polyamides, require solvents such as dimethylformamide (DMF) or dimethylacetamide to be dissolved, which unfortunately are not volatile [6]. Other common harmful organic solvents may include strong acids such as formic acid, dichloromethane, ether, and chloroform. [7].

Green electrospinning would aim not only to reduce the use of VOCs and other toxic solvents, but to select suitable materials that produce more eco-friendly products. Some more common polymers due to their solubility in more aqueous solutions include polycaprolactone, poly(ethylene oxide) (PEO), and poly(vinyl alcohol), all of which are more easily degraded than polyolefins [1].

1.2 "Greener" electrospinning: on the way to a "truly green" electrospinning

As with many things in life, whether a glass is half empty or half full depends on the perspective of the observer. To consider a certain process "green" or not is one of these cases in which some people see the glass half full and others see the glass half empty. How "green" should an electrospinning process be to be called "green"?

Any person reading this book is certainly aware of the popularity that the adjective "green" has attained in the last few years. The title of the book is the best demonstration. However, in these lines, we have no intention to open sterile controversies, but rather to act somehow as devil's advocate in our own field and incite in the reader a certain degree of self-criticism when confronted with the so-called – or "self-called" – green strategies. A rapid view of the literature will immediately show us that the concept of green fibers is not univocal and depends very much on the authors. For instance, when the resulting materials are used in biomedical applications, the biocompatibility of the final product is put in the foreground and then the label "green" seems to be almost self-explanatory. But were all chemicals and solvents used during the synthesis of the final biomaterial "truly green"? How green were the processing steps? It is obvious that such a "quantification" is far from trivial. Herein, we would like to point out a few concerns that should be kept in mind, regardless of whether the process is or not called "green":

a) The electrospinning of natural polymers, such as gelatin or chitosan, is commonly considered as "green," because the products can be considered as such. However, most of the solvents used in these processes are fluorinated (e.g., hexafluoro-2-propanol, tetrafluoroethylene, and trifluoroacetic acid). Similarly, there is a general trend to speak about "green electrospinning" when herbal extracts or essential oils are used in the process, no matter whether synthetic polymers or organic solvents are used.
b) In some of the processes classified as "green electrospinning," toxic reagents (e.g., cross-linkers or reducing agents) are used during the preparation of the fibers.
c) Some of the so-called green electrospinning processes involve high temperatures and high energy consumption. This situation applies to the preparation of most of the inorganic fibers, which are produced by electrospinning a template polymer and a metal precursor solution, followed by heat treatment steps. The energy consumption and the sustainability of the process are typically neglected when assigning the adjective "green."
d) Finally, there are cases in which water or so-called benign solvents (ethanol, limonene, acetic acid, etc.) are employed, but the final electrospun polymer is a conventional synthetic polymer. Can we then really speak about "green"?

The real fact is that most authors label electrospinning processes as "green" as soon as at least one of the "green" requisites is fulfilled, for instance, the use of water or a benign solvent, the presence of natural products or natural polymers, or biomedical applications and biocompatibility. A large majority of the examples presented in this book will fall within one of these cases. Processes being "truly green," in which none of the concerns stated above can be made, are rare and rather the exception. It is for this reason that, even if we are in the right track toward an environmentally friendly electrospinning, there is still plenty of room for optimization to achieve “greener” processes. This is the context in which this book should be placed and understood.

1.3 History of green electrospinning

It has been possible for many years to fabricate nanofibers via processes including drawing. However, none of these processes has allowed for the amount of control over the size of the fiber with as much consistency as found by using electrospinning. Electrospinning is a versatile technique that could be scaled up and could repeatedly yield reproducible results. While there have been patents filed for various electrospinning setups at the beginning of the twentieth century, it is only in the last two decades that industry and academia have been looking into using electrospinning to fabricate various nanofibrous mats for application in many fields. Scheme 1.1 presents an abbreviated version of the history of electrospinning with a more complete history having already been well documented in a number of literature including by Agarwal and Tucker [8, 9].

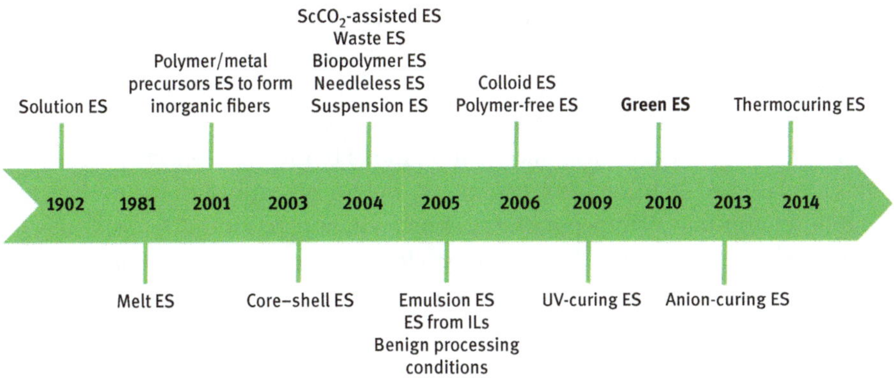

Scheme 1.1: Timeline of electrospinning.

From the eighteenth to the twentieth century, a great number of researchers had discovered the physical understanding of hydrodynamics and electrodynamics that makes up the basis for the electrospinning technique. Soon, a basic electrospinning apparatus was patented by Cooley and Morton between 1900 and 1902 [10, 11]. Approximately 30 years later, Anton Formhals patented several setups that produced yarns made out of electrospun fibers that were the progenitors of modern-day devices [12]. In 1936, C.L. Norton patented a device for the use of electrospinning from a melt rather than a solution and created the field of melt electrospinning [13]. A huge step was then made in the understanding of the electrospinning process: between 1964 and 1969, Sir Geoffrey Taylor contributed to the theoretical principles of electrospinning by mathematically modeling the shape of the cone formed by the fluid droplet under the effect of an electric field, now known as the Taylor cone [14]. However, due to the slower fiber formation compared to conventional industrial fiber production techniques, the textiles industry did not fully take up electrospun fibers. During the latter quarter of the nineteenth century and the first two decades of the twenty-first century, a great number of advancements have been made in the theory and application of electrospinning. The parameters of solution electrospinning were defined for PEO solution in water [15] and PU in DMF [16].

In 2003, Huang et al. reported that almost 100 different polymers were transformed into ultrafine fibers using solution or melt electrospinning [17]. The processing of core–shell nano-/mesofibers by coelectrospinning of two components was described, providing the entrainment of a nonspinnable core material by a spinnable outer shell [18]. Li and Xia noticed that not only polymer solutions/melts but also liquid crystals, suspension of solid particles, and emulsions can be electrospun into nanofibers [19]. Yang and coworkers observed electrospinning of poly(*m*-phenylene isophthalamide) from its solution in ionic liquid for the first time [20]. In a different approach, by mixing colloidal silica particles with polymer solutions, followed by electrospinning, organic/inorganic fibers can be obtained and applied as photonic devices and microfluidic control devices [21]. In general, polymers have been taken as a structural framework for the fabrication of nanofibers through the entanglement and overlapping of the polymer chains. However, McKee successfully fabricated polymer-free nanofibers from phospholipids because of the cylindrical micelle formations in their concentrated solutions [22]. Compared to solution electrospinning, only a few groups have been interested in solvent-free electrospinning such as melt, supercritical CO_2-assisted, UV-curing, anion-curing, and thermocuring electrospinning, which eliminate the usage of toxic solvents and provide effective use of precursors [23]. Sun et al. [24] reported that the term "Green Electrospinning" was first used in 14th European Conference on Composite Materials (ECCM14) in 2010 by Ramakrishna and his coworkers. "Green/Safe/Clean Electrospinning," which can be described as an approach to toxicology, safety, and environmental issues, can be accomplished by electrospinning from environmentally friendly solvents instead of toxic solvents, and from biodegradable natural polymers instead of synthetic polymers. Scheme 1.2 shows the classification of electrospinning with all processes that fall into the green category. Very commonly, green electrospinning has been related to emulsion and suspension electrospinning [1].

In this book, we extend the terminology to six groups: polymer-free, solvent-free, solution, and colloid electrospinning can be considered as *green processes*, while electrospinning from natural polymers and blends can be named as *green fibers*.

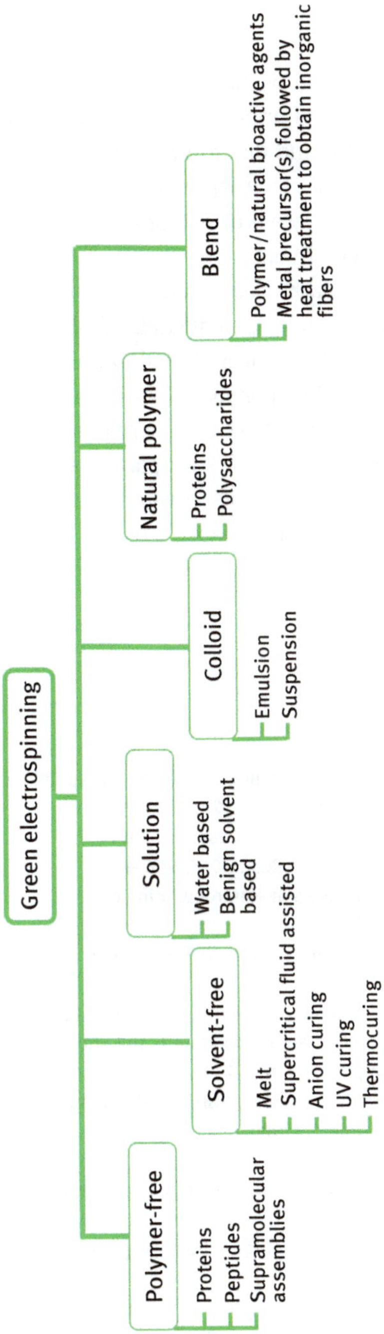

Scheme 1.2: Classification of Electrospinning referred to as 'Green'.

1.4 Structure of the book

After our succinct introduction to green electrospinning, we have discussed to what extent green electrospinning is truly green and touched upon the reality that some methods of green electrospinning yield better health and environmental outcomes than others. We have also given a brief the history of electrospinning and the advent and uses of green electrospinning. In Chapter 2, we will elaborate the green electrospinning in terms of green processes and green fibers.

We then discuss the different methods that are harnessed in green electrospinning. Chapters 3, 4, and 5 focus on polymer free, melt, and supercritical fluid–assisted electrospinning, respectively. In Chapters 6, 7, and 8, the focus is shifted to the use of water-based electrospinning and the use of natural and biopolymers (particularly cellulose and silk), respectively. In Chapter 9, we go over needless electrospinning. It is important to note that throughout the literature, some techniques for particular electrospun materials might take advantage of a number of the topics covered by various chapters, such as melt coaxial electrospinning [25]. Finally, Chapters 10, 11, and 12 deal with the applications of green electrospun fibers, including the formation of filters, biomedical materials, and batteries.

References

[1] Agarwal, S, & Greiner, A. On the way to clean and safe electrospinning – green electrospinning: emulsion and suspension electrospinning, Polym. Adv. Technol., 2011, 22, 372–378.

[2] Chen, X., Ma, L., & Zhao, M. Pollution characteristics and health risk assessment of atmospheric VOCs in the pharmaceutical enterprises, Biomed. Res., 2017, 666–672.

[3] Teo, W-E., Inai, R., & Ramakrishna, S. Technological advances in electrospinning of nanofibers, Sci. Technol. Adv. Mater., 2011, 12, 013002.

[4] Bläsing, M., & Amelung, W. Plastics in soil: Analytical methods and possible sources, Sci. Total Environ., 2018, 612, 422–435.

[5] Horton, AA., Walton, A., Spurgeon, DJ., Lahive, E., & Svendsen, C. Microplastics in freshwater and terrestrial environments: evaluating the current understanding to identify the knowledge gaps and future research priorities, Sci. Total Environ., 2017, 586, 127–141.

[6] Jiang, S., Hou, H., Agarwal, S., & Greiner, A. Polyimide nanofibers by "Green" electrospinning via aqueous solution for filtration applications, ACS Sustain. Chem. Eng., 2016, 4, 4797–4804.

[7] Kamal, MS., Razzak, SA., & Hossain, MM. Catalytic oxidation of volatile organic compounds (VOCs)–A review, Atmos. Environ., 2016, 140, 117–134.

[8] Agarwal, S., Burgard, M., Greiner, A., & Wendorff, J. Electrospinning: A Practical guide to nanofibers, Walter de Gruyter GmbH & Co KG, Germany, 2016.

[9] Tucker, N., Stanger, J., Staiger, M., Razzaq, H., & Hofman, K. The history of the science and technology of electrospinning from 1600 to 1995, J. Eng. Fiber Fabr., 2012, 7, 63–73.

[10] Cooley, J. Improved methods of and apparatus for electrically separating the relatively volatile liquid component from the component of relatively fixed substances of composite fluids, United Kingdom Patent, 1900, 6385, 19.

[11] Morton, WJ. Method of dispersing fluids. Google Patents; 1902.
[12] Anton, F. Process and apparatus for preparing artificial threads. Google Patents; 1934.
[13] Norton, CL. Method of and apparatus for producing fibrous or filamentary material. Google Patents; 1936.
[14] Taylor, GI. Disintegration of water drops in an electric field, Proc. R Soc. Lond. A, 1964, 280, 383–397.
[15] Doshi, J., & Reneker, DH. Electrospinning process and applications of electrospun fibers, J. Electrostat., 1995, 35, 151–160.
[16] Demir, MM., Yilgor, I., Yilgor, E., & Erman, B. Electrospinning of polyurethane fibers, Polymer, 2002, 43, 3303–3309.
[17] Huang, ZM., Zhang, YZ., Kotaki, M., & Ramakrishna, S. A review on polymer nanofibers by electrospinning and their applications in nanocomposites, Comp. Sci. Technol., 2003, 63, 2223–2253.
[18] Sun, ZC., Zussman, E., Yarin, AL., Wendorff, JH., & Greiner, A. Compound core-shell polymer nanofibers by co-electrospinning, Adv. Mater., 2003, 15, 1929–1932.
[19] Li, D., & Xia, YN. Electrospinning of nanofibers: Reinventing the wheel?, Adv. Mater., 2004, 16, 1151–1170.
[20] Yang, W., Yu, H., Zhu, MF. et al. Poly(m-Phenylene Isophthalamide) ultrafine fibers from an ionic liquid solution by dry-jet-wet electrospinning, J. Macromol. Sci. Phys., 2006, 45, 573–579.
[21] Lim, JM., Moon, JH., Yi, GR., Heo, CJ., & Yang, SM. Fabrication of one-dimensional colloidal assemblies from electrospun nanofibers, Langmuir, 2006, 22, 3445–3449.
[22] McKee, MG., Layman, JM., Cashion, MP., & Long, TE. Phospholipid nonwoven electrospun membranes, Science, 2006, 311, 353–355.
[23] Zhang, B., Yan, X., He, HW., Yu, M., Ning, X., & Long, YZ. Solvent-free electrospinning: opportunities and challenges, Polym. Chem., 2017, 8, 333–352.
[24] Sun, JY., Bubel, K., Chen, F., Kissel, T., Agarwal, S., & Greiner, A. Nanofibers by green electrospinning of aqueous suspensions of biodegradable block copolyesters for applications in medicine, pharmacy and agriculture, Macromol. Rapid Commun., 2010, 31, 2077–2083.
[25] Street, RM., Huseynova, T., Xu, X. et al. Variable piezoelectricity of electrospun chitin, Carbohydr. Polym., 2018, 195, 218–224.

Nesrin Horzum, Rafael Muñoz-Espí, Matthew A. Hood, Mustafa M. Demir and Daniel Crespy

2 Green processes and green fibers

Abstract: “Green Electrospinning” not from only non-toxic solvents but also from biopolymer solutions has become popular in recent years. Green fibers are particularly interesting for biomedical applications such as tissue engineering, drug delivery, biocompatible scaffolds, biosensors, and for photovoltaics, supercapacitors, fuel cells, battery components as energy fields, and for filtration membranes as environmental applications. In this chapter, we classified green electrospinning into two groups: (i) green processes as polymer free, solvent free, solution, and colloid electrospinning, (ii) green fibers from natural polymers and blends.

Keywords: benign solvent, bioactive agents, clean electrospinning, colloid electrospinning, natural polymer, polymer-free, solvent-free

2.1 Green processes

2.1.1 Polymer-free electrospinning

Typically, electrospinning studies are carried out using high-molecular-weight polymers and high solution concentrations because of chain entanglements and the continuous stretching of the charged jet. Long et al. [1] reported that high-molecular-weight polymers are not the only the requirement, but the presence of sufficient intermolecular interactions can also act as chain entanglements for the continuous fiber formation. Therefore, besides easily electrospinnable polymers, globular proteins and low-molecular-weight compounds [such as Gemini surfactants, phospholipids, diphenylalanine peptides, and cyclodextrins (CDs)] have also been electrospun into fibers, since they exhibit a similar behavior to polymers in solution [2–5]. Electrospinnability of globular proteins (namely bovine serum albumin) was attributed to disruption of the tertiary structure and reduction of intramolecular disulfide bonds, allowing the

Nesrin Horzum, Department of Engineering Sciences, İzmir Katip Çelebi University, İzmir, Turkey
Rafael Muñoz-Espí, Institute of Materials Science (ICMUV), Universitat de València, Paterna, Spain
Matthew A. Hood, Max Planck Institute for Polymer Research, Mainz, Germany
Mustafa M. Demir, Department of Materials Science and Engineering, İzmir Institute of Technology, İzmir, Turkey
Daniel Crespy, Department of Materials Science and Engineering, School of Molecular Science and Engineering, Vidyasirimedhi Institute of Science and Technology, Rayong, Thailand

https://doi.org/10.1515/9783110581393-002

reformation of intra- and intermolecular disulfide bonds [6]. Increasing the concentration of Gemini surfactants and phospholipids in the appropriate solvent results in the formation of entangled cylindrical or worm-like micelles. Furthermore, diphenylalanine peptides that self-aggregate into nanotubes were transformed into nanofibers due to the presence of π–π interactions. A cyclic oligosaccharide, CD, self-assembles in their concentrated solutions to form aggregates through intermolecular hydrogen bonding, which makes possible to obtain electrospun nanofibers. Polymer-free nanofibers of α-, β-, and γ-CDs have been fabricated by the selection of suitable solvents and concentrations ensuring sufficient viscosity and conductivity [7–10]. The morphology and the diameter of the resulting fibers are affected by not only the type of solvents but also the type of CDs. CD derivatives, such as hydroxypropyl-β-cyclodextrin (HPβCD) [11, 12], hydroxypropyl-γ-cyclodextrin (HPγCD) [11, 13], and methyl-β-cyclodextrin (MβCD) [11], have been obtained using water, *N,N*-dimethyl formamide (DMF), and dimethyl acetamide (DMAc) as solvents. Among them, DMF is found to produce bead-free fibers with the three derivatives [12].

Compared to other small molecules, CDs are advantageous because they can form host–guest inclusion complexes (ICs) with different compounds, such as drugs, volatile compounds, food or cosmetic additives, and antibacterial agents. The CD inclusion complexation enhances thermal stability and water solubility of the hydrophobic guest molecules, which provides a promising platform for drug delivery applications. Nanofibers of CD–ICs containing 4-amino benzene [14], spironolactone [15], triclosan [16], diclofenac [17], geraniol [18], vanillin, limonene [19], sulfobutyl ether$_7$-β-CD [20], vitamin E [21], camphor [22], and linalool [23] have successfully been produced without using a polymeric matrix. The most commonly used solvents for CD–ICs are water, ethanol, aqueous sodium hydroxide, DMF, DMAc, dimethyl sulfoxide, and ionic liquids (ILs, e.g., 1-ethyl-3-methyl imidazolium acetate). To obtain nanofibers as reconstitutable solids for drug-release applications, electrospinning IC is also used as an alternative to the freeze-drying process for the preparation of fast-dissolving CD-based solid complexes containing limited soluble drugs [17].

On the one hand, because of their hydrophobic cavity and the hydrophilic surface, CDs can form the host–guest ICs with various bioactive compounds to be used for the fast-dissolving, prolonged release, and long shelf-life of the active component, enhanced thermal stability, and water solubility. On the other hand, CDs are used as both reducing and stabilizing agents for the green synthesis of gold nanoparticles [24].

Apart from CDs, Allais et al. [25] enlarged the list of possible small molecules used for electrospinning using tannic acid without the addition of any polymer. They also reported the cross-linking of tannic acid nanofibrous membranes by the oxidation of galloyl groups with sodium iodate and ferric ions to obtain mechanical integrity.

2.1.2 Solvent-free electrospinning

Techniques associated with solvent-free electrospinning neither have a risk of residual solvents being present for biomedical applications, nor is there any solvent that may evaporate into the air [26]. To remove solvents, special conditions are needed, because electrospinning requires that the polymer chains are able to flow and extend in an electric field, so that the fibers are formed, which will be covered in detail in Chapters 4 and 5.

The most common techniques for solvent-free electrospinning are electrospinning from the melt state, supercritical carbon dioxide (CO_2)-assisted anion-curing, UV-curing, and thermocuring [27]. Two methods that we will focus on are the use of supercritical CO_2 as a "solvent" and melt electrospinning. In both systems, there is no traditional solvent used to dissolve the polymers. Supercritical CO_2 uses the semiliquid semigaseous properties of CO_2 under high pressure and temperatures to aid in the flow of polymer chains. Electrospinning of polymer melts, on the other hand, eliminates the use of toxic solvents by heating semicrystalline polymers or glassy polymers above their melting temperature (T_m) and glass transition temperature (T_g), so that a viscous solution in which polymer chains are capable of flowing can be formed.

Supercritical carbon dioxide

CO_2 is a gas at standard temperature and ambient pressure. When temperature and pressure are increased above a critical point, CO_2 behaves somewhat like both a gas and a liquid [28]. Electrospinning in the presence of supercritical CO_2 is similar to the one at ambient temperature, but the supercritical CO_2 liquid is used to alter the viscosity of the polymer, much like a solvent or a plasticizer. CO_2 is used because it is relatively nontoxic, nonflammable, inexpensive, easily available, odorless, tasteless, and relatively environmentally friendly. In addition, CO_2 evaporates at ambient conditions and is therefore easily released from the products [29].

The solvating power of supercritical CO_2 is connected to its density, which in turn depends on the pressure and the temperature [30]. The solubility of the polymer in supercritical CO_2 therefore may be controlled greatly by pressure and temperature, as a higher density generally translates to higher solubility.

Melt Electrospinning

In melt electrospinning, polymers are processed by using heat to melt the polymer. Unlike traditional electrospun fibers, which form due to the precipitation out of solution of the polymer as the solvent is evaporated, fibers are formed in melt electrospinning by the cooling down of the polymer melt while being collected [31]. Melt electrospinning is like an extrusion technique with the addition of a high voltage to further stretch the fibers [32]. The greatest problem with melt electrospinning is that the absence of solvent greatly reduces the surface charge density of polymer melts, which results in instabilities of the fiber jet [33].

2.1.3 Solution electrospinning

In a classical electrospinning process, nanofibers are fabricated by dissolving the polymer in an appropriate solvent. However, except water, most of the electrospinning solvents used to dissolve polymers are toxic and harmful to the environment and human health. Thus, the use of green solvents is significant to reduce environmental and health impacts. The main alternatives to traditional organic solvents are (i) water, (ii) mild solvents, and (iii) ILs.

Water as a Solvent

Water is certainly a good solvent for water-soluble polymers. However, mechanical strength of nanofibers electrospun from water-based polymers is low, limiting their use in aqueous systems. To enhance their mechanical properties and make them water-insoluble, additional modifications such as cross-linking, UV, or plasma treatment are required. On the other hand, frequently used cross-linkers like gluteraldehyde, diethylene glycol, glyoxal, epichlorohydrin, and 1-ethyl-3-(3-dimethylaminopropyl) carbodiimide together with *N*-hydroxysuccinimide are toxic. New trends are directed toward the use of nontoxic reagents such as citric acid, proanthocyanidin, epigallocatechin-3-gallate (EGCG), genipin to reduce the use of the aforementioned toxic cross-linkers. Table 2.1 reports a summary of nanofibers from aqueous precursors by green electrospinning, including the details about fiber diameters, cross-linking and/or reducing agents, and applications.

A different water-based electrospinning approach, which is also considered as green, is used to fabricate inorganic fibers. The production of inorganic fibers is based on the simultaneous electrospinning of water-soluble polymer and metal precursor(s) blends followed by calcination. Table 2.2 lists works on metal oxide fibers electrospun from aqueous polymer/metals salt(s) solutions, heat treatment parameters, as well as their potential applications. The resulting metal oxide nanofibers have been mostly used in energy applications such as solar cells, electrodes for batteries, fuel cells, and supercapacitors.

By taking Tables 2.1 and 2.2 into consideration, except the studies involving only fabrication and characterization, these water-based electrospun fibers find applications mainly in biomedical and energy fields.

Table 2.1: A summary of green nanofibers from aqueous precursors including information about fiber diameters, cross-linking and/or reducing agents, and applications.

Polymer or blend	Fiber diameter (nm)	Reducing/stabilizing agent	Cross-linking agent	Application	Reference
PEO/PHA-*b*-PEO	350–550	–	–	–	34
PVA/soy protein isolate	600–4,500	–	–	–	35
PEG/polyindole	720	–	–	An electrode for supercapacitor	36
PEO/silk fibroin/honey	480–2,230	–	–	Wound dressing	37
PVA/zinc acetate	~500	–	–	NH_3 sensor	38
PEO/silk fibroin/grape seed extract	~420	–	–	Skin care, skin regeneration, wound healing	39
PEO/chitin nanofibrils/biolignin	<150	–	–	Tissue engineering	40
PVA/AgNPs	~150	–	–	Antiseptic dressing	41
Silk fibroin/vitamin E	384–722	–	–	Skin care	42
PVA/soy protein isolate	200–900	–	–	Biodegradation	43
Folic acid/dextran; folic acid/PVP; folic acid/ODA-MMT	304–379	–	Reactive electrospinning and thermal treatment	–	44
PVA/ODA-MMT/poly(maleic acid-*alt*-acrylic acid)	50–300	–	Thermal cross-linking	–	45

(continued)

Table 2.1 (continued)

Polymer or blend	Fiber diameter (nm)	Reducing/stabilizing agent	Cross-linking agent	Application	Reference
PVA/citric acid	125–255	–	Thermal cross-linking with citric acid	–	46
PVA/sodium alginate	140–350	–	Physical cross-linking with $CaCl_2$ in ethanol	–	47
PEO/PHA-*b*-MPEG PEO/PCL-*b*-MPEG	340–600 275–720	–	Physical cross-linking	–	48
PVA/PEO/ulvan	75–109	–	Physical cross-linking	–	49
Silk fibroin/vitamin B5	625	–	Ethanol vapor	Skin care	50
PVA/PAA blend	225–408	–	Thermal cross-linking	Antimicrobial activity	51
PVA/ODA-MMT/folic acid	~100	–	Reactive electrospinning	Anticancer activity	52
Silk fibroin	623–805	–	Ethanol vapor	Tissue engineering	53
PVA/CMC/$AgNO_3$	100–150	CMC	–	–	54
PVA/WPU+ PVA/$AgNO_3$	255–405	PVA	–	–	55
PVA/$AgNO_3$/cyclodextrin	290–500	PVA and HPβCD	–	–	56
PEO/CMC	50–200	Immersed in $AgNO_3$ solution reduced by UV irradiation	–	Antimicrobial dressing	57

PVA/Au nanoparticles	50–450	*Couroupita guianensis* leaves extract	–	Antibacterial wound dressing	58
PVA/$AgNO_3$ (PVA-Ag) PVA/$AgNO_3$ (PVA-Ag@heat) PVA/$AgNO_3$ (PVA-Ag@UV)	300–500 150–350 150–350	PVA Heating UV irradiation	–	Biocidal activity and cytotoxicity	59
PVA/PAA/SiO_2 nanoparticle/ $AgNO_3$	200–800	UV reduction	Thermal cross-linking	Antibacterial air filtration	60
PVA/*O*-carboxymethyl chitosan/ $AgNO_3$	100–200	Ascorbic acid	Glutaraldehyde vapor	Antimicrobial activity	61
PVA/Triton X-100/$AgNO_3$	150–250	PVA	Thermal cross-linking	Catalytic hydrogenation	62
PVA/chitosan/$AgNO_3$	150	Glucose	Glutaraldehyde	Antimicrobial wound dressing	63
PVA/$AgNO_3$/EGCG PVA/PEI	630 508	EGCG	Glutaraldehyde vapor	Biosensor	64
PVA/$HAuCl_4$ PVA/$AgNO_3$ PVA/H_2PtCl_6	275–355 295–385 260–360	EGCG	Glutaraldehyde vapor	Biosensor for H_2O_2 detection	65

CMC, carboxymethyl cellulose; HPβCD, hydroxypropyl-β-cyclodextrin; ODA-MMT, octadecyl amine-montromorillonite; EGCG, epigallocatechin-3-gallate; AgNP, silver nanoparticles.

Table 2.2: Metal oxide nanofibers fabricated by electrospinning an aqueous solution containing metal salt(s) or metal alkoxide and a template polymer.

Electrospinning solution	Heat treatment parameters	Metal oxide(s) nanofibers	Application	Reference
PVP/TTIP/EtOH/AA/ $Zn(NO_3)_2{\cdot}6H_2O$	500 °C for 2 h	Zn-doped TiO_2 hollow fibers	Dye-sensitized solar cells	[66]
PVA/polymerized sucrose	350 °C (stabilization) 650 and 950 °C (carbonization in Ar/H_2)	Porous carbon nanofibers	Electrode for supercapacitor	[67]
PVP/$Cu(CH_3COO)_2{\cdot}H_2O$/ $Ce(NO_3)_3{\cdot}6H_2O$	500 °C for 3 h	CuO/CeO_2 nanofibers	–	[68]
PVA/$Zn(CH_3COO)_2{\cdot}2H_2O/AlCl_3{\cdot}6H_2O$ + PVA/$Zn(CH_3COO)_2{\cdot}2H_2O$/ $Co(CH_3COO)_2{\cdot}4H_2O$	450, 500, 550, 600 °C for 2 h	Co- and Al-doped ZnO nanofibers	Nanogenerator	[69]
PVA/$SnCl_2{\cdot}2H_2O$	1,000 °C for 2 h	SnO_2 nanofibers	–	[70]
PVA/$Zn(CH_3COO)_2{\cdot}2H_2O$	550 °C	ZnO nanofibers	Nanogenerator	[71]
PVA/$Mn(CH_3COO)_2{\cdot}4H_2O$	1,000 °C for 1 h	Mn_3O_4 nanofibers	–	[72]
PVA/$Al(NO_3)_3{\cdot}9H_2O$/ $Al[OCH(CH_3)_2]_3$	1,200 °C	Al_2O_3 nanofibers	–	[73]

Table 2.3: Mild solvents and solvent mixtures that belong to Class 3 used in green electrospinning.

Mild solvents	
Water	Ethanol
Acetic acid	Ethanol/water
Acetic acid/formic acid	Ethanol/acetic acid
Acetic acid/water	2-Methoxy ethanol
Acetic acid/2-methoxy ethanol	Ethyl acetate
Acetone	Formic acid
Dimethyl sulfoxide	Limonene

Mild Solvents

Many types of polymers, both synthetic and natural, have been used for the fabrication of "green fibers" using mild solvents. Table 2.3 summarizes the solvents used for green electrospinning that belongs to Class 3 (solvents with low toxic potential) according to the classification in ICH (International Conference on Harmonisation of Technical Requirements for Registration of Pharmaceuticals for Human Use) [74].

Literature documents the ability to electrospin synthetic polymers by using mild solvents instead of commonly used toxic ones. The investigated synthetic polymer/mild solvent systems are PVP/ethanol [75, 76], PVP/2-methoxy ethanol and acetic acid [77], PVA/ethanol/water [78], poly[*N*-isopropylacrylamide-*co*-(maleic acid)]/ethanol [79], PCL/glacial acetic acid [80], PCL/acetone [81], and nylon-6/formic acid [82].

Although the fibers from synthetic polymers are often a more economic and versatile alternative, biofibers are especially interesting because of their biocompatibility or biodegradability. Awal et al. [83] fabricated cellulose-based nanocomposite fibers electrospun from wood pulp and nylon 6,6 solution in formic acid. The same research group also obtained green composite fibers prepared by electrospinning PEO/wood pulp solution with a mixed solvent of ethanol and water [84]. Another wood component, lignin, is being used as an eco-friendly precursor [85]. The fabrication of lignin fibers through green electrospinning is problematic because of its low viscoelastic properties and water insolubility. The former case can be solved by using additional synthetic polymers to improve the spinnability. The water-soluble derivatives of lignin such as lignosulfonates can be an alternative for the latter case. Misra et al. [86] prepared sulfur-free anionic sodium carbonate lignin, chitosan, and PEO blend solution in aqueous acetic acid. The inherent negative charge of the lignin was neutralized by cationic chitosan. The free-sodium ions and PEO were removed from the fibers through a water-soaking process to obtain pure carbon fiber.

Chitosan, which is derived from chitin (the second most abundant natural polymer after cellulose), has also been used as a reducing and stabilizing agent in nanoparticle synthesis. For instance, Zhuang et al. [87] demonstrated the use of chitosan as a reducing agent and stabilizer for the fabrication of chitosan/gelatin nanofibers containing silver nanoparticles (AgNPs). A mixed solution of AgNPs–chitosan composite with gelatin containing acetic acid was used to produce nanofibers.

Green synthesis of metal nanoparticles on the electrospun nanofibers is also based on the addition of metallic salt(s) into the polymer solutions and further treating the nanofibers by using nontoxic and safe chemicals from plant extracts as reducing agents, annealing, and irradiation techniques (i.e., UV or plasma). As an example, Arvand et al. [88] described the synthesis process for antibacterial AgNP-modified chitosan-based fibers using Eucalyptus extract. Aqueous acetic acid solution of chitosan, PEO, and silver nitrate was electrospun, and the composite fibers were immersed in ethanolic extract of Eucalyptus as a reducing agent instead of using $NaBH_4$ to form AgNPs on the nanofibers. Similarly, use of *Falcaria vulgaris* extract as a green reductant was reported by Kohsari et al. [89]. Using a different approach, ex situ prepared AgNPs were added into the electrospinning solution of chitosan and PEO in aqueous acetic acid.

In addition to functionalization with metal nanoparticles, by combining chitosan with natural biocomponents leads to potential usage in biomedical applications. A new nanofibrous wound dressing containing green tea extract was proposed by Sadri and collaborators [90]. They prepared electrospinning solutions that dissolved chitosan/PEO/green tea extract in aqueous acetic acid solution. Chitosan and PEO were used as matrix, and green tea extract was used to improve the wound healing performance of the chitosan-based nanofibers. Li et al. [91] reported fabrication of electrospun zein fibers as carriers to stabilize green tea polyphenol, (−)-epigallocatechin gallate (EGCG). The fiber-forming solution was prepared by dissolving EGCG in an aqueous ethanol solution of zein. Encapsulating EGCG in zein fibers resulted in the enhanced stability of the bioactive compound.

Collagen scaffolds are preferred for use in tissue engineering, artificial skin, and drug delivery. Zhang et al. [92] prepared PVA/collagen micro-nanofibers from water/acetic acid mixtures. Zhou et al. [93] proposed a greener synthesis of electrospun collagen/hydroxyapatite (HAP) composite fibers from phosphate buffer saline/ethanol/water system. The desalination of collagen solution and the use of HAP sol led to a good dispersion of HAP particles in the electrospinning solution and the alignment of HAP particles in the collagen/HAP composites in the meantime improved the mechanical properties of the resulting composites. The use of acetic acid and water as a green solvent for the fabrication of a collagen/HAP composite fibers was also highlighted by Castilla-Casadiego et al. [94].

Nieuwland et al. [95] showed for the first time electrospinning of globular proteins under food-grade conditions. In the case of zein-based fibers, ethanol was used; while in gelatin-based fibers, warm water was used as a solvent. Uniform protein nanofibers were prepared by electrospinning of gelatin for air-filtration purposes. Instead of commonly used toxic solvents (e.g., TFA, TFE, or HFIP), a nontoxic solvent mixture of acetic acid/water mixture was used to fabricate uniform gelatin nanofiber mats as air-filtering material. In addition, the diameter of gelatin nanofibers was also reduced to be around 70 nm, which results in high surface area and high filtration efficiencies [96].

Another application is to use natural fibers as a skeleton material in gel polymer electrolyte (GPE) of lithium-ion batteries. Zhu et al. [97] fabricated soy protein isolate/PVA nanofibrous mats using acetic acid/water mixture as solvent. The biodegradable composite mat was then activated in liquid electrolytes providing high-performance and green skeleton material in GPEs.

Natural solvents can also be used as an alternative for electrospinning. For instance, limonene, which is the main component of lemon and orange oils, is a cyclic monoterpene and started to be used to obtain polystyrene [98–100], poly(vinyl alcohol) [101, 102], asolectin phospholipid [103], and CD nanofibers [104].

Ionic Liquids

ILs consist of ions influenced by the presence of Coulombic interactions and have different physicochemical properties compared to traditional VOCs. Many ILs are nonvolatile, more viscous than VOCs, highly conductive and they can provide dissolution of biopolymers not soluble in VOCs. Therefore, electrospinning of biopolymers, such as cellulose [105–110], heparin [106, 111], agarose [112], silk [113, 114], and chitin [115, 116], has attracted much attention. Unlike traditional solution process, using ILs requires dissolution of a polymer in the ILs as solvent followed by electrospinning into the coagulation bath which can be water, ethanol, or water/ethanol mixture as antisolvents. The most often used ILs are 1-butyl-3-methylimidazolium chloride ([C_4mim]Cl), bromide ([C_4mim]Br), and acetate ([C_4mim][OAc]), 1-ethyl-3-methylimidazolium benzoate ([C_2mim][BA]), acetate ([C_2mim][OAc]), and lactate ([C_2mim][OLac]). ILs containing basic acetate [OAc^-] or chloride [Cl^-] anions and alkyl imidazolium cations have been identified as the appropriate solvent for biopolymer dissolution [116, 117].

2.1.4 Colloid electrospinning

An improved method for the combination of complex nanostructures with fibers is referred to as colloid electrospinning, which leads to the development of physical and biological properties that encompass both materials [118]. This technique allows for the electrospinning of dispersions of polymer and/or inorganic nanoparticles/nanocapsules with a solution of an electrospinnable polymer. Colloid electrospinning can be classified into two main groups: as suspensions and as emulsions [118, 119]. The examples of organic/inorganic nanostructures used in suspension and emulsion electrospinning are summarized in Table 2.4. Inorganic nanostructures constitute the largest fraction of the suspension electrospinning category. They are advantageous when compared to organic colloids because inorganic nanostructures have a higher density than the polymer templates, and their location in the fiber is usually easily identified by electron microscopy [118, 120]. Metals, metalloids, and their oxides, fluorescent markers, and other minerals fall into the inorganic nanostructure groups used in suspension electrospinning. Besides inorganic nanostructures, hydrophilic/hydrophobic polymer particles and natural systems such as viruses and bacteria may be used in suspension electrospinning. In the case of emulsion electrospinning, water-in-oil, oil-in-water, and polymer/polymer (nonaqueous) emulsions can be electrospun to obtain core–sheath fibers, which are generally fabricated by coaxial electrospinning. Although this method allows for control of the nanofibers structure, it is relatively difficult to scale up. Conversely, emulsion electrospinning is advantageous in recent years as it allows the production of two-component nanofibers of controlled morphology. Furthermore, this approach is environmentally friendly, reducing the amount of organic solvent used during the process [119].

Polymers that are commonly used in colloid electrospinning are polystyrene [121–123], poly(ε-caprolactone) [124–126], poly(L-lactide) [127], poly(vinyl alcohol) [120, 124, 128–137], poly(vinyl acetate) [124, 138], poly(vinyl alcohol)/poly(acrylic acid) [139], poly(*N*-vinylpyrrolidone) [140–145], poly(vinyl formamide) [130], silk fibroin [146], poly(ethylene oxide) [147–150], *N*-carboxyethylchitosan [151], polyacrylonitrile [152, 153], polymethyl methacrylate [154–156], polyamide 6 [157–159], chitosan/gelatin [160], polyurethane [124, 161, 162], poly(ethylene terephthalate) [163], dextran [164], polyacrylamide [165–167], poly(L-lactide-*co*-glycolide) [168], methylcellulose [169], collagen [170], polyacrylic acid [171], polyvinylidene fluoride [172, 173], which act as a structural network for the organic/inorganic nanostructures. Since colloid electrospinning allows the use of almost any polymers and nanostructures, the resulting hierarchically structured or compartmented nanofibers pave the way for various applications such as catalysis [138], adsorption [143], membrane distillation [172, 173], optics [174–176], energy conversion and production [123, 177], drug delivery [131, 146], tissue engineering [113, 127, 178], and wound dressing [51, 134].

2.2 Green fibers

2.2.1 Natural material-based electrospinning

The use of eco-friendly products to reduce the consumption of synthetic plastics that appear in groundwater and soil is an obvious potential area for green electrospinning. Natural materials suggest that the material comes from a source in nature and therefore can be acquired from something that grows, for example, chitin, lignin, and silk. Many natural materials have the added benefit that they are more easily biodegradable than synthetic materials [251].

Some natural materials could be used as a solvent or a plasticizer to induce flow within the polymer fibers during the electrospinning process. For instance, D-limonene, which is a natural solvent made of monoterpene hydrocarbon and is the main component of orange or lemon peel oil, has been used as an electrospinning solvent [99, 252]. In addition, other natural molecules such as hyaluronic acid can be used in the formation of nanofibers without the difficulty of dissolution of much higher molecular weight polymers [253]. Natural polymers can roughly be made up of two classes of biomacromolecules; proteins like silk, collagen, and gelatin or polysaccharides like cellulose, chitosan, and dextran. [86, 254, 255]. Renewable resource-based molecules such as lignin have the potential for green electrospinning as they are abundantly available and cheap [256].

Again we are brought to the question of Chapter 1, "how green is green?," as many natural polymers do not necessarily dissolve with ecofriendly solvents, and other natural materials might only be a fraction of the final materials found in the nanofibers and do not necessarily biodegrade any faster a than synthetic polymers.

Table 2.4: Classification of the colloid electrospinning and the nanoparticles employed in the literature.

Colloid Electrospinning	
Electrospinning of suspensions	
Electrospinning of inorganic nanostructures	
Metals and metalloids	Ag [133,159,179], Au [121,180], Si [152], Ti, Zn [124], Co, Cu [181], Pt [144]
Metals and metalloid oxides	SiO_2 [122,134,135,139,143,147,154,165,172,173,182–188], TiO_2 [189–192], TiO_2/grapheme [123], Fe_3O_4 [193–198], ZnO [199], $BaTiO_3$ [200], ITO (In_2O_3/SnO_2) [148], MgO/ ZrO_2/Al_2O_3 [201], $Ni_{0.5}Zn_{0.5}Fe_2O_4$ [202], Boehmite (Al(O)-OH) [125, 158], $LiCoO_2/CeO_2$ [188], SnO_2/CeO_2 [135], WO_3 [145], graphene oxide [203, 204]
Fluorescent markers	Tris(8-hydroxyquinoline) aluminum(III) (Alq3) [164], CdS/CdSe/PdS [155, 205], ZnS/PdS [138], CoS [206]
Other minerals	Clay minerals (mica, zeolite, montmorillonite, bentonite) [207–210], boron nitride[124, 211], $CaCO_3$ [212, 213], calcium phosphate [214, 215], hydroxyapatite [124,155,161,168,170,216–219], hydroxyapatite + CNT [220, 221], curcumin [146], silicon carbide [222]
Electrospinning of polymer particles	Polystyrene [223–225], PS/PBA/P(S-*co*-BA)[129], P(S-*co*-BA), P(S-*co*-BA-*co*-AMA), P(S-*co*-BA-*co*-DAAM) [130], poly (styrenemethyl methacrylate-acrylic acid) (poly(St-MMA-AA)) [226, 227], PMMA [228], polyurethane [229, 230], PLGA [231], P(HA-*b*-EO) [232], PVDF [233], PVA [234], P(MAA-*co*-DVB)/P (MAA-*co*-TRIM) [163], P(NIPAM-*co*-AA-*co*-MBAAm) [235], poly (*N*-isopropylacrylamide-*co*-tert-butyl acrylate) [167], P(DMA-*co*-EDMA-*co*-4VP)/ P(DMA-*co*-EDMA-*co*-AMSA) [142]
Electrospinning of natural systems	M13 viruses [236], adipose-derived stem cells [237], *Lysinibacillus* sp. [238]
Electrospinning of emulsions	
Water-in-oil emulsion	PEG/PLA [239, 240], PDLLA [169,241,242], PLGD [243], cellulose nanocrystals/PLA [244]
Oil-in-water emulsion	PEO [245], PVA [246, 247], PVP [248, 249]
Polymer–polymer emulsion	PAN [156, 250]

Some solutions to this problem may come from taking advantage of the blend electrospinning technique discussed in the next section. Some natural polymers that may be difficult to electrospin in aqueous solutions such as chitosan and silk fibroin may be mixed with electrospinnable polymers like PVA, PEO, or PAA to increase their processability while including properties of the natural polymer. Through a careful polymer selection, nanofibers can have additional properties including antibacterial or antifouling properties, for example, chitosan nanofibers intrinsically exhibit antibacterial activity but require blending to make more easily spun [257].

2.2.2 Blend electrospinning

Blend electrospun fibers are produced from polymer solutions with or without other additives. Factors that influence the fiber formation are related to the properties of the polymer solution and electrospinning environment. In coaxial electrospinning, two solutions are electrospun simultaneously through different capillary channels into a nozzle. This generates composite nanofibers with a core–shell structure [258]. The term blend electrospinning can refer to a number of scenarios. Typically, blends are a mixture of two or more components, which are mixed together in order to gain properties from both materials; PEO/PVA for instance is a common blended fiber [259]. The term "emulsion electrospinning" has also been used to describe a type of blend electrospinning [118].

The main contribution blend electrospinning adds to the green electrospinning process is in the use of natural additives to reduce the synthetic materials. There are natural polymers that have relied heavily on addition of bioactive reagents, including essential oils, polyphenols, herbal extracts, honey, and other natural components [260, 261]. Cinnamaldehyde, for example, is a volatile essential oil that has been shown to eradicate pathogens nonspecifically and has been added to electrospun fibers [262].

"Green/Safe/Clean Electrospinning," which can be considered as an approach to toxicology, safety, and environmental issues, has been the focus of many research publications in recent years. Although the optimization of process parameters with benign solvents and biopolymers is more difficult, promising results have been obtained for the generation of green nanofibers particularly in biomedical and healthcare applications.

Abbreviations

AMSA	Acrylamidoethylsulfonic acid
BSA	Bovine Serum Albumin
CNT	Carbon nanotube
CMC	Carboxymethyl cellulose

CD	Cyclodextrin
DMAc	Dimethyl acetamide
DMSO	Dimethyl sulfoxide
DVB	Divinylbenzene
ECH	Epichlorohydrin
EGCG	Epigallocatechin-3-gallate
EDMA	Ethylene dimethacrylate
GPE	Gel polymer electrolyte
T_g	Glass transition temperature
HFIP	Hexafluoroisopropanol
HAP	Hydroxyapatite
HPβCD	Hydroxypropyl-β-cyclodextrin
HPγCD	Hydroxypropyl-γ-cyclodextrin
IC	Inclusion complex
ILs	Ionic liquids
T_m	Melting temperature
MAA	Methacrylic acid
MβCD	Methyl-β-cyclodextrin
NIPAM	*N*-isopropylacrylamide
DMF	*N,N*-dimethyl formamide
DMA	*N,N*-dimethylacrylamide
MBAAm	*N,N*-methylenebis(acrylamide)
ODA-MMT	Octadecyl amine-montromorillonite
O/W	Oil-in-water
PDLLA	Poly(DL-lactic acid)
PLA	Poly(lactic acid)
PLGA	Poly(lactic-*co*-glycolic acid)
PHA	Poly(hexamethylene adipate)
PBA	Poly(n-butyl acrylate)
P(S-*co*-BA-*co*-AMA)	Poly(styrene-*co*-*n*-butyl acrylate-*co*-allymethacrylate)
P(S-*co*-BA-*co*-DAAM)	Poly(styrene-*co*-*n*-butyl acrylate-*co*-diacetone acrylamide)
P(S-*co*-BA)	Poly(styrene-*co*-*n*-butyl acrylate)
Poly(St-MMA-AA)	Poly(styrenemethyl methacrylate-acrylic acid)
PCL-*b*-MPEG	Poly[(epsilon-caprolactone)-*block*-(methoxypolyethylene glycol)]
PHA-*b*-MPEG	Poly[(hexamethylene adipate)-*block*-(methoxypolyethylene glycol)]
PAA	Polyacrylic acid
PAN	Polyacrylonitrile
PEG	Polyethylene glycol
PEO	Polyethylene oxide
PEI	Polyethylenimine
PMMA	Polymethyl methacrylate
PU	Polyurethane
PVA	Polyvinyl alcohol
PVDF	Polyvinylidene fluoride
PVP	Polyvinylpyrrolidone
AgNPs	Silver nanoparticles
SCL	Sodium carbonate lignin
TTIP	Titanium(IV) isopropoxide
TFA	Trifluoroacetic acid

TFE	Trifluoroethanol
TRIM	Trimethylolpropane trimethacrylate
VOCs	Volatile Organic Compounds
W/O	Water-in-oil
WPU	Waterborne polyurethane
VP	4-vinylpyridine
EDC/NHS	1-ethyl-3-(3-dimethylaminopropyl)carbodiimide / *N*-hydroxysuccinimide
$[C_4mim]Cl$	1-butyl-3-methylimidazolium chloride
$[C_4mim]Br$	1-butyl-3-methylimidazolium bromide
$[C_4mim][OAc]$	1-butyl-3-methylimidazolium acetate
$[C_2mim][BA]$	1-ethyl-3-methylimidazolium benzoate
$[C_2mim][OAc])$	1-ethyl-3-methylimidazolium acetate
$[C_2mim][OLac]$	1-ethyl-3-methylimidazolium lactate

References

[1] McKee, MG., Layman, JM., Cashion, MP., & Long, TE. Phospholipid nonwoven electrospun membranes, Science, 2006, 311, 353–355.

[2] Hemp, ST., Hudson, AG., Allen, MH., Pole, SS., Moore, RB., & Long, TE. Solution properties and electrospinning of phosphonium gemini surfactants, Soft Matter, 2014, 10, 3970–3977.

[3] Yu, D-G., Zhu, L-M., White, K., & Branford-White, C. Electrospun nanofiber-based drug delivery systems, Health, 2009, 1, 67.

[4] Singh, G., Bittner, AM., Loscher, S., Malinowski, N., & Kern, K. Electrospinning of diphenylalanine nanotubes, Adv. Mater., 2008, 20, 2332–2336.

[5] Uyar, T., Havelund, R., Hacaloglu, J., Besenbacher, F., & Kingshott, P. Functional electrospun polystyrene nanofibers incorporating α-, β-, and γ-cyclodextrins: comparison of molecular filter performance, ACS Nano., 2010, 4, 5121–5130.

[6] Dror, Y., Ziv, T., Makarov, V., Wolf, H., Admon, A., & Zussman, E. Nanofibers made of globular proteins, Biomacromolecules, 2008, 9, 2749–2754.

[7] Celebioglu, A., & Uyar, T. Electrospinning of nanofibers from non-polymeric systems: Electrospun nanofibers from native cyclodextrins, J. Colloid Interface Sci., 2013, 404, 1–7.

[8] Ahn, Y., Kang, Y., Ku, M., Yang, Y-H., Jung, S., & Kim, H. Preparation of β-cyclodextrin fiber using electrospinning, RSC Adv., 2013, 3, 14983–14987.

[9] Celebioglu, A., & Uyar, T. Electrospun gamma-cyclodextrin (γ-CD) nanofibers for the entrapment of volatile organic compounds, RSC Adv., 2013, 3, 22891–22895.

[10] Kida, T., Sato, S-i., Yoshida, H., Teragaki, A., & Akashi, M. 1, 1, 1, 3, 3, 3-Hexafluoro-2-propanol (HFIP) as a novel and effective solvent to facilely prepare cyclodextrin-assembled materials, Chem. Commun., 2014, 50, 14245–14248.

[11] Celebioglu, A., & Uyar, T. Electrospinning of nanofibers from non-polymeric systems: polymer-free nanofibers from cyclodextrin derivatives, Nanoscale, 2012, 4, 621–631.

[12] Zhang, W., Chen, M., Zha, B., & Diao, G. Correlation of polymer-like solution behaviors with electrospun fiber formation of hydroxypropyl-β-cyclodextrin and the adsorption study on the fiber, Phys. Chem. Chem. Phys., 2012, 14, 9729–9737.

[13] Celebioglu, A., Sen, HS., Durgun, E., & Uyar, T. Molecular entrapment of volatile organic compounds (VOCs) by electrospun cyclodextrin nanofibers, Chemosphere, 2016, 144, 736–744.

[14] Chen, M., Nielsen, SR., Uyar, T. et al. Electrospun UV-responsive supramolecular nanofibers from a cyclodextrin–azobenzene inclusion complex, J. Mater. Chem. C, 2013, 1, 850–855.

[15] Vigh, T., Horváthová, T., Balogh, A. et al. Polymer-free and polyvinylpirrolidone-based electrospun solid dosage forms for drug dissolution enhancement, Eur. J. Phar. Sci., 2013, 49, 595–602.

[16] Celebioglu, A., Umu, OC., Tekinay, T., & Uyar, T. Antibacterial electrospun nanofibers from triclosan/cyclodextrin inclusion complexes, Coll. Surf. B: Biointerfaces, 2014, 116, 612–619.

[17] Balogh, A., Horváthová, T., Fülöp, Z. et al. Electroblowing and electrospinning of fibrous diclofenac sodium-cyclodextrin complex-based reconstitution injection, J. Drug Deliv. Sci. Technol., 2015, 26, 28–34.

[18] Aytac, Z., Yildiz, ZI., Kayaci-Senirmak, F., San Keskin, NO., Tekinay, T., & Uyar, T. Electrospinning of polymer-free cyclodextrin/geraniol–inclusion complex nanofibers: enhanced shelf-life of geraniol with antibacterial and antioxidant properties, RSC Adv., 2016, 6, 46089–46099.

[19] Celebioglu, A., Kayaci-Senirmak, F., İpek, S., Durgun, E., & Uyar, T. Polymer-free nanofibers from vanillin/cyclodextrin inclusion complexes: high thermal stability, enhanced solubility and antioxidant property, Food & Funct., 2016, 7, 3141–3153.

[20] Yildiz, ZI., Celebioglu, A., & Uyar, T. Polymer-free electrospun nanofibers from sulfobutyl ether7-beta-cyclodextrin (SBE7-β-CD) inclusion complex with sulfisoxazole: Fast-dissolving and enhanced water-solubility of sulfisoxazole, Int. J. Pharm., 2017, 531, 550–558.

[21] Celebioglu, A., & Uyar, T. Antioxidant vitamin E/cyclodextrin inclusion complex electrospun nanofibers: enhanced water solubility, prolonged shelf life, and photostability of vitamin E, J. Agric. Food Chem., 2017, 65, 5404–5412.

[22] Celebioglu, A., Aytac, Z., Kilic, ME., Durgun, E., & Uyar, T. Encapsulation of camphor in cyclodextrin inclusion complex nanofibers via polymer-free electrospinning: enhanced water solubility, high temperature stability, and slow release of camphor, J. Mater. Sci., 2018, 53, 5436–5449.

[23] Aytac, Z., Yildiz, ZI., Kayaci-Senirmak, F., Tekinay, T., & Uyar, T. Electrospinning of cyclodextrin/linalool-inclusion complex nanofibers: Fast-dissolving nanofibrous web with prolonged release and antibacterial activity, Food Chem., 2017, 231, 192–201.

[24] Celebioglu, A., & Uyar, T. Green and one-step synthesis of gold nanoparticles incorporated into electrospun cyclodextrin nanofibers, RSC Adv., 2013, 3, 10197–10201.

[25] Allais, M., Mailley, D., Hébraud, P. et al. Polymer-free electrospinning of tannic acid and cross-linking in water for hybrid supramolecular nanofibres, Nanoscale, 2018, 10, 9164–9173.

[26] Reijnders, L. Health hazards and recycling and life cycle assessment ofnanomaterials and their composites, Polym. Comp. Nanocomp., *Wiley- VCH Verlag GmbH & Co. KGaA*, 2013, 279–294.

[27] Zhang, B., Yan, X., He, H-W., Yu, M., Ning, X., & Long, Y-Z. Solvent-free electrospinning: opportunities and challenges, Polym. Chem., 2017, 8, 333–352.

[28] Bennion, B., & Bachu, S. Drainage and imbibition relative permeability relationships for supercritical CO2/brine and H2S/brine systems in intergranular sandstone, carbonate, shale, and anhydrite rocks, SPE Reserv. Eval. & Eng., 2008, 11, 487–496.

[29] Levit, N., & Tepper, G. Supercritical CO2-assisted electrospinning, J. Supercrit. Fluids, 2004, 31, 329–333.
[30] Barton, AF. CRC handbook of solubility parameters and other cohesion parameters: Routledge; 2017.
[31] Nayak, R. Review of Literature: Melt Electrospinning. Polypropylene Nanofibers: Springer, Cham, 2017, 9–39.
[32] Breitenbach, J. Melt extrusion: from process to drug delivery technology, Eur. J. Pharm. Biopharm., 2002, 54, 107–117.
[33] Brown, TD., Dalton, PD., & Hutmacher, DW. Melt electrospinning today: An opportune time for an emerging polymer process, Prog. Polym. Sci., 2016, 56, 116–166.
[34] Sun, JY., Bubel, K., Chen, F., Kissel, T., Agarwal, S., & Greiner, A. Nanofibers by green electrospinning of aqueous suspensions of biodegradable block copolyesters for applications in medicine, pharmacy and agriculture, Macromol. Rapid Commun., 2010, 31, 2077–2083.
[35] Cho, D., Nnadi, O., Netravali, A., & Joo, YL. Electrospun hybrid soy protein/PVA fibers, Macromol. Mater. Eng., 2010, 295, 763–773.
[36] Tebyetekerwa, M., Wang, XP., Marriam, I., Dan, P., Yang, SY., & Zhu, MF. Green approach to fabricate polyindole composite nanofibers for energy and sensor applications, Mater. Lett., 2017, 209, 400–403.
[37] Yang, XX., Fan, LP., Ma, LL. et al. Green electrospun manuka honey/silk fibroin fibrous matrices as potential wound dressing, Mater. Des., 2017, 119, 76–84.
[38] Li, Y., Jiao, MF., & Yang, MJ. In-situ grown nanostructured ZnO via a green approach and gas sensing properties of polypyrrole/ZnO nanohybrids, Sens. Actuator B-Chem., 2017, 238, 596–604.
[39] Lin, S., Chen, MX., Jiang, HY. et al. Green electrospun grape seed extract-loaded silk fibroin nanofibrous mats with excellent cytocompatibility and antioxidant effect, Colloid Surf. B-Biointerfaces, 2016, 139, 156–163.
[40] Morganti, P., Del Ciotto, P., Stoller, M., & Chianese, A. Antibacterial and Anti-inflammatory Green Nanocomposites, A Chianese, L DiPalma, E Petrucci, & M Stoller, eds. International Conference on Nanotechnology Based Innovative Applications for the Environment, 2016, 61–66.
[41] Tseng, KH., Chang, CY., Chung, MY., & Tang, YL. Bacteriostatic Substrate by Conductivity Method and Electric Spark Discharge Method Combined with Electrospinning for Silver Dressing. Int J Polym Sci 2016, 9425358.
[42] Sheng, XY., Fan, LP., He, CL., Zhang, KH., Mo, XM., & Wang, HS. Vitamin E-loaded silk fibroin nanofibrous mats fabricated by green process for skin care application, Int. J. Biol. Macromol., 2013, 56, 49–56.
[43] Cho, D., Netravali, AN., & Joo, YL. Mechanical properties and biodegradability of electrospun soy protein Isolate/PVA hybrid nanofibers, Polym. Degrad. Stabil., 2012, 97, 747–754.
[44] Rzayev, ZMO., Bunyatova, U., & Simsek, M. Multifunctional colloidal nanofiber composites including dextran and folic acid as electro-active platforms, Carbohydr. Polym., 2017, 166, 83–92.
[45] Simsek, M., Rzayev, ZMO., & Bunyatova, U. Multifunctional electroactive electrospun nanofiber structures from water solution blends of PVA/ODA-MMT and poly(maleic acid-alt-acrylic acid): effects of Ag, organoclay, structural rearrangement and NaOH doping factors, Adv. Nat. Sci.-Nanosci. Nanotechnol., 2016, 7, 025009.
[46] Shi, JJ., & Yang, EL. Green electrospinning and crosslinking of polyvinyl alcohol/ citric acid, J. Nano. Res., 2015, 32, 32–40.

[47] Shen, W., & Hsieh, YL. Biocompatible sodium alginate fibers by aqueous processing and physical crosslinking, Carbohydr. Polym., 2014, 102, 893–900.
[48] Bubel, K., Zhang, Y., Assem, Y., Agarwal, S., & Greiner, A. Tenside-free biodegradable polymer nanofiber nonwovens by "Green Electrospinning", Macromolecules, 2013, 46, 7034–7042.
[49] Toskas, G., Hund, RD., Laourine, E., Cherif, C., Smyrniotopoulos, V., & Roussis, V. Nanofibers based on polysaccharides from the green seaweed *Ulva Rigida*, Carbohydr. Polym., 2011, 84, 1093–1102.
[50] Fan, LP., Cai, ZX., Zhang, KH. et al. Green electrospun pantothenic acid/silk fibroin composite nanofibers: Fabrication, characterization and biological activity, Colloid Surf. B-Biointerfaces, 2014, 117, 14–20.
[51] Sridhar, R., Sundarrajan, S., Vanangamudi, A., Singh, G., Matsuura, T., & Ramakrishna, S. Green processing mediated novel polyelectrolyte nanofibers and their antimicrobial evaluation, Macromol. Mater. Eng., 2014, 299, 283–289.
[52] Gokmen, FO., Rzayev, ZMO., Salimi, K. et al. Novel multifunctional colloidal carbohydrate nanofiber electrolytes with excellent conductivity and responses to bone cancer cells, Carbohydr. Polym., 2015, 133, 624–636.
[53] Fan, LP., Wang, HS., Zhang, KH., He, CL., Cai, ZX., & Mo, XM. Regenerated silk fibroin nanofibrous matrices treated with 75% ethanol vapor for tissue-engineering applications, J. Biomater. Sci-Polym. Ed., 2012, 23, 497–508.
[54] Martinez-Rodriguez, MA., Garza-Navarro, MA., Moreno-Cortez, IE., Lucio-Porto, R., & Gonzalez-Gonzalez, VA. Silver/polysaccharide-based nanofibrous materials synthesized from green chemistry approach, Carbohydr. Polym., 2016, 136, 46–53.
[55] Wang, RZ., Wang, Z., Lin, S. et al. Green fabrication of antibacterial polymer/silver nanoparticle nanohybrids by dual-spinneret electrospinning, RSC Adv., 2015, 5, 40141–40147.
[56] Celebioglu, A., Aytac, Z., Umu, OCO., Dana, A., Tekinay, T., & Uyar, T. One-step synthesis of size-tunable Ag nanoparticles incorporated in electrospun PVA/cyclodextrin nanofibers, Carbohydr. Polym., 2014, 99, 808–816.
[57] Shi, DD., Wang, FJ., Lan, T., Zhang, YH., & Shao, ZQ. Convenient fabrication of carboxymethyl cellulose electrospun nanofibers functionalized with silver nanoparticles, Cellulose, 2016, 23, 1899–1909.
[58] Manjumeena, R., Elakkiya, T., Duraibabu, D., Ahamed, AF., Kalaichelvan, PT., & Venkatesan, R. 'Green' biocompatible organic-inorganic hybrid electrospun nanofibers for potential biomedical applications, J. Biomater. Appl., 2015, 29, 1039–1055.
[59] Lin, S., Wang, RZ., Yi, Y. et al. Facile and green fabrication of electrospun poly(vinyl alcohol) nanofibrous mats doped with narrowly dispersed silver nanoparticles, Int. J. Nanomed., 2014, 9, 3937–3947.
[60] Zhu, MM., Hua, DW., Pan, H. et al. Green electrospun and crosslinked poly(vinyl alcohol)/poly(acrylic acid) composite membranes for antibacterial effective air filtration, J. Colloid Interface Sci., 2018, 511, 411–423.
[61] Ezzat, M., Ghanim, M., Nageh, H., Hassanin, AH., & Abdel-Moneim, A. Antimicrobial activity of O-carboxymethyl chitosan nanofibers containing silver nanoparticles synthesized by green method, J. Nano. Res., 2016, 40, 136–145.
[62] Wang, W., Wang, K., He, JJ. et al. A synergistic strategy for nanoparticle/nanofiber composites towards p-nitrophenol catalytic hydrogenation, Chem. Res. Chin. Univ., 2015, 31, 1012–1017.
[63] Abdelgawad, AM., Hudson, SM., & Rojas, OJ. Antimicrobial wound dressing nanofiber mats from multicomponent (chitosan/silver-NPs/polyvinyl alcohol) systems, Carbohydr. Polym., 2014, 100, 166–178.

[64] Zhu, H., Du, ML., Zhang, M. et al. Facile fabrication of AgNPs/(PVA/PEI) nanofibers: High electrochemical efficiency and durability for biosensors, Biosens. Bioelectron., 2013, 49, 210–215.
[65] Zhu, H., Du, ML., Zhang, M. et al. Facile and green fabrication of small, mono-disperse and size-controlled noble metal nanoparticles embedded in water-stable polyvinyl alcohol nanofibers: High sensitive, flexible and reliable materials for biosensors, Sens. Actuator B-Chem., 2013, 185, 608–619.
[66] Ezzat, M., Ghanim, M., Nageh, H., Hassanin, AH., & Abdel-Moneim, A. Antimicrobial activity of O-carboxymethyl chitosan nanofibers containing silver nanoparticles synthesized by green method, J. Nano. Res., 2016, 40, 136–145.
[67] Chakravarty, A., Maiti, S., Mahanty, S., & De, G. Green synthesis of electrospun porous carbon nanofibers from sucrose and doping of Ag nanoparticle with improved electrical and electrochemical properties, ChemistrySelect, 2017, 2, 2265–2276.
[68] Pal, DB., Singh, P., & Mishra, PK. Composite ceria nanofiber with different copper loading using electrospinning method, J. Alloy. Compd., 2017, 694, 10–16.
[69] Subagiyo, D., Thoyib, M., Suyitno, Hadi S., Jamaluddin, A., & Hidayat, R. Effect of Codoping Cobalt and Aluminum on Enhancing the Piezoelectricity Properties of Fiber-Based Zinc Oxide. B Kristiawan, M Anwar, A Wijayanta, et al. eds. International Conference on Engineering, Science and Nanotechnology, 2017.
[70] Chandraiah, M., Sahoo, B., & Panda, PK. Preparation and characterization of SnO2 nanofibers by electrospinning, Trans. Indian Ceram. Soc., 2014, 73, 266–269.
[71] Suyitno, Huda S., Arifin, Z., Hadi, S., & Lambang, RL. Repeatability and Reproducibility of Fibre-Based Nanogenerator Synthesized by Electrospinning Machine. FL Gaol, J Webb, & J Ding, eds. 2014, International Conference on Manufacturing, Optimization, Industrial and Material Engineering 2014, 58, 012013.
[72] Sahoo, B., & Panda, PK. Synthesis and characterization of manganese tetroxide (Mn3O4) nanofibers by electrospinning technique, J. Adv. Ceram., 2013, 2, 26–30.
[73] Abdullah, R., Sreekantan, S., & Mohamad, H. Effects of The Polyvinyl Alcohol (PVA) on The Synthesis of Alumina Fibers Through Electrospinning Technique. ZA Ahmad, MA Yarmo, FHA Aziz, et al. eds. X-Ray and Related Techniques, Trans Tech Publications, Switzerland, 2011, 150–154.
[74] Guideline, IHT. Impurities: Guideline for residual solvents Q3C (R5), Current. Step., 2005, 4, 1–25.
[75] Li, L., Kang, WM., & Cheng, BW. Research on the process of a green method for preparing the alumina nanofibre. YH Kim, P Yarlagadda, XD Zhang, & ZJ Ai, eds. Advanced Materials and Structures, 335–336, 446–450.
[76] Rozent, O., Beilin, VV., Shter, GE., & Grader, GS. Deformation control during thermal treatment of electrospun PbZr0.52Ti0.48O3 nanofiber mats, J. Am. Ceram. Soc., 2016, 99, 1550–1556.
[77] Khajelakzay, M., & Taheri-Nassaj, E. Synthesis and characterization of PB(ZR(0.52), TI0.48)O-3 nanofibers by electrospinning, and dielectric properties of PZT-Resin composite, Mater. Lett., 2012, 75, 61–64.
[78] Shmueli, Y., Shter, GE., Assad, O. et al. Structural and electrical properties of single Ga/ZnO nanofibers synthesized by electrospinning, J. Mater. Res., 2012, 27, 1672–1679.
[79] Saithongdee, A., Varanusupakul, P., & Imyim, A. Preparation of thermally sensitive poly *N*-isopropylacrylamide-*co*-(maleic acid) hydrogel membrane by electrospinning using a green solvent, Green. Chem. Lett. Rev., 2014, 7, 220–227.
[80] Lawson, C., Stanishevsky, A., Sivan, M., Pokorny, P., & Lukas, D. Rapid fabrication of poly (epsilon-caprolactone) nanofibers using needleless alternating current electrospinning, J. Appl. Polym. Sci., 2016, 133, 43232.

[81] Bosworth, LA., & Downes, S. Acetone, a sustainable solvent for electrospinning poly(epsilon-Caprolactone) fibres: Effect of varying parameters and solution concentrations on fibre diameter, J. Polym. Environ., 2012, 20, 879–886.
[82] Park, M., Rabbani, MM., Shin, HK., Park, SJ., & Kim, HY. Dyeing of electrospun nylon 6 nanofibers with reactive dyes using electron beam irradiation, J. Ind. Eng. Chem., 2016, 39, 16–20.
[83] Awal, A., Sain, M., & Chowdhury, M. Preparation of cellulose-based nano-composite fibers by electrospinning and understanding the effect of processing parameters, Compos. Pt. B-Eng., 2011, 42, 1220–1225.
[84] Awal, A., & Sain, M. Cellulose-polymer based green composite fibers by electrospinning, J. Polym. Environ., 2012, 20, 690–697.
[85] Fang, W., Yang, S., Wang, X-L., Yuan, T-Q., & Sun, R-C. Manufacture and application of lignin-based carbon fibers (LCFs) and lignin-based carbon nanofibers (LCNFs), Green Chem., 2017, 19, 1794–1827.
[86] Schreiber, M., Vivekanandhan, S., Cooke, P., Mohanty, AK., & Misra, M. Electrospun green fibres from lignin and chitosan: a novel polycomplexation process for the production of lignin-based fibres, J. Mater. Sci., 2014, 49, 7949–7958.
[87] Zhuang, XP., Cheng, BW., Kang, WM., & Xu, XL. Electrospun chitosan/gelatin nanofibers containing silver nanoparticles, Carbohydr. Polym., 2010, 82, 524–527.
[88] Arvand, M., Mirzaei, E., Derakhshan, MA. et al. Fabrication of antibacterial silver nanoparticle-modified chitosan fibers using eucalyptus extract as a reducing agent, J. Appl. Polym. Sci., 2015, 132, 42133.
[89] Kohsari, I., Shariatinia, Z., & Pourmortazavi, SM. Antibacterial electrospun chitosan-polyethylene oxide nanocomposite mats containing bioactive silver nanoparticles, Carbohydr. Polym., 2016, 140, 287–298.
[90] Sadri, M., Arab-Sorkhi, S., Vatani, H., & Bagheri-Pebdeni, A. New wound dressing polymeric nanofiber containing green tea extract prepared by electrospinning method, Fiber. Polym., 2015, 16, 1742–1750.
[91] Li, Y., Lim, LT., & Kakuda, Y. Electrospun zein fibers as carriers to stabilize (-)-epigallocatechin gallate, J. Food Sci., 2009, 74, C233-C240.
[92] Zhang, XL., Tang, KY., & Zheng, XJ. Electrospinning and rheological behavior of poly (vinyl alcohol)/collagen blended solutions, J. Wuhan. Univ. Technol.-Mat. Sci. Edit., 2015, 30, 840–846.
[93] Zhou, YY., Yao, HC., Wang, JS., Wang, DL., Liu, Q., & Li, ZJ. Greener synthesis of electrospun collagen/hydroxyapatite composite fibers with an excellent microstructure for bone tissue engineering, Int. J. Nanomed., 2015, 10, 3203–3215.
[94] Castilla-Casadiego, DA., Maldonado, M., Sundaram, P., & Almodovar, J. "Green" electrospinning of a collagen/hydroxyapatite composite nanofibrous scaffold, MRS Commun., 2016, 6, 402–407.
[95] Nieuwland, M., Geerdink, P., Brier, P. et al. Food-grade electrospinning of proteins, Innov. Food. Sci. Emerg. Technol., 2013, 20, 269–275.
[96] Souzandeh, H., Wang, Y., & Zhong, WH. "Green" nano-filters: fine nanofibers of natural protein for high efficiency filtration of particulate pollutants and toxic gases, RSC Adv., 2016, 6, 105948–105956.
[97] Zhu, M., Tan, CY., Fang, Q., Gao, L., Sui, G., & Yang, XP. High performance and biodegradable skeleton material based on soy protein isolate for gel polymer electrolyte, ACS Sustain Chem. Eng., 2016, 4, 4498–4505.
[98] Shin, C., & Chase, GG. Nanofibers from recycle waste expanded polystyrene using natural solvent, Polym. Bullet., 2005, 55, 209–215.

[99] Khirandish, M., Borhani, S., Mallakpour, S., & Youssefi, M. Properties of PS/TiO 2 electrospun fibres using limonene as a solvent. Indian J. Fibre. Text., 2016, 41, 373–379.
[100] Wang, X., Yuan, Y., Huang, X., & Yue, T. Controlled release of protein from core–shell nanofibers prepared by emulsion electrospinning based on green chemical, J. Appl. Polym. Sci., 2015, 132, 41811.
[101] Camerlo, A., Vebert-Nardin, C., Rossi, RM., & Popa, A-M. Fragrance encapsulation in polymeric matrices by emulsion electrospinning, Eur. Polym. J., 2013, 49, 3806–3813.
[102] Camerlo, A., Bühlmann-Popa, A-M., Vebert-Nardin, C., Rossi, RM., & Fortunato, G. Environmentally controlled emulsion electrospinning for the encapsulation of temperature-sensitive compounds, J. Mater. Sci., 2014, 49, 8154–8162.
[103] Jørgensen, L., Qvortrup, K., & Chronakis, IS. Phospholipid electrospun nanofibers: Effect of solvents and co-axial processing on morphology and fiber diameter, RSC Adv., 2015, 5, 53644–53652.
[104] Aytac, Z., Yildiz, ZI., Kayaci-Senirmak, F. et al. Fast-dissolving, prolonged release, and antibacterial cyclodextrin/limonene-inclusion complex nanofibrous webs via polymer-free electrospinning, J. Agric. Food Chem., 2016, 64, 7325–7334.
[105] Zheng, Y., Miao, J., Maeda, N., Frey, D., Linhardt, RJ., & Simmons, TJ. Uniform nanoparticle coating of cellulose fibers during wet electrospinning, J. Mater. Chem. A, 2014, 2, 15029–15034.
[106] Hou, L., Udangawa, WRN., Pochiraju, A. et al. Synthesis of heparin-immobilized, magnetically addressable cellulose nanofibers for biomedical applications, ACS Biomater. Sci. Eng., 2016, 2, 1905–1913.
[107] Liu, Y., Nguyen, A., Allen, A., Zoldan, J., Huang, Y., & Chen, JY. Regenerated cellulose micro-nano fiber matrices for transdermal drug release, Mater. Sci. Eng: C, 2017, 74, 485–492.
[108] Xu, S., Zhang, J., He, A., Li, J., Zhang, H., & Han, CC. Electrospinning of native cellulose from nonvolatile solvent system, Polymer, 2008, 49, 2911–2917.
[109] Park, T-J., Jung, YJ., Park, H. et al. Photoluminescent synthetic wood fibers from an ionic liquid via electrospinning, Macromol. Res., 2011, 19, 317–321.
[110] Freire, MG., Teles, ARR., Ferreira, RA., Carlos, LD., Lopes-da-Silva, JA., & Coutinho, JA. Electrospun nanosized cellulose fibers using ionic liquids at room temperature, Green Chem., 2011, 13, 3173–3180.
[111] Viswanathan, G., Murugesan, S., Pushparaj, V., Nalamasu, O., Ajayan, PM., & Linhardt, RJ. Preparation of biopolymer fibers by electrospinning from room temperature ionic liquids, Biomacromolecules, 2006, 7, 415–418.
[112] Forget, A., Arya, N., Randriantsilefisoa, R. et al. Nonwoven carboxylated agarose-based fiber meshes with antimicrobial properties, Biomacromolecules, 2016, 17, 4021–4026.
[113] Srivastava, CM., & Purwar, R. Fabrication of robust antheraea assama fibroin nanofibrous mat using ionic liquid for skin tissue engineering, Mater. Sci. Eng: C, 2016, 68, 276–290.
[114] Fox, D., Fylstra, P., Hanley, M. et al. The preparation and characterization of bombyx mori silk nanocomposites using ionic liquids, ECS Trans., 2007, 3, 11–20.
[115] Shamshina, JL., Zavgorodnya, O., Bonner, JR., Gurau, G., Di Nardo, T., & Rogers, RD. "Practical" electrospinning of biopolymers in ionic liquids, Chem.Sus.Chem., 2017, 10, 106–111.
[116] Zavgorodnya, O., Shamshina, JL., Bonner, JR., & Rogers, RD. Electrospinning biopolymers from ionic liquids requires control of different solution properties than volatile organic solvents, ACS Sustainable Chem. Eng., 2017, 5, 5512–5519.
[117] Tan, SSY., & MacFarlane, DR. Ionic liquids in biomass processing. Ionic liquids, Springer, 2009, 311–339.

[118] Crespy, D., Friedemann, K., & Popa, AM. Colloid-electrospinning: fabrication of multicompartment nanofibers by the electrospinning of organic or/and inorganic dispersions and emulsions, Macromol. Rapid Commun., 2012, 33, 1978–1995.
[119] Agarwal, S., & Greiner, A. On the way to clean and safe electrospinning – green electrospinning: emulsion and suspension electrospinning, Polym. Adv. Technol., 2011, 22, 372–378.
[120] Jiang, S., He, W., Landfester, K., Crespy, D., & Mylon, SE. The structure of fibers produced by colloid-electrospinning depends on the aggregation state of particles in the electrospinning feed, Polymer, 2017, 127, 101–105.
[121] Kim, JK., & Ahn, H. Fabrication and characterization of polystyrene/gold nanoparticle composite nanofibers, Macromol. Res., 2008, 16, 163–168.
[122] Ding, B., Lin, J., Wang, X., Yu, J., Yang, J., & Cai, Y. Investigation of silica nanoparticle distribution in nanoporous polystyrene fibers, Soft. Matt., 2011, 7, 8376–8383.
[123] Asmatulu, R., Ceylan, M., & Nuraje, N. Study of superhydrophobic electrospun nanocomposite fibers for energy systems, Langmuir, 2010, 27, 504–507.
[124] Barakat, NA., Abadir, M., Sheikh, FA., Kanjwal, MA., Park, SJ., & Kim, HY. Polymeric nanofibers containing solid nanoparticles prepared by electrospinning and their applications, Chem. Eng. J., 2010, 156, 487–495.
[125] Hota, G., Kumar, BR., Ng, W., & Ramakrishna, S. Fabrication and characterization of a boehmite nanoparticle impregnated electrospun fiber membrane for removal of metal ions, J. Mater. Sci., 2008, 43, 212–217.
[126] Park, C., Hwang, YK., Hyun, DC., & Jeong, U. In situ localization of molecules in crosslinked particles during electrohydrodynamic process: Simple route to produce microcapsules and fibers with controlled release, Macromol. Rapid Commun., 2010, 31, 1713–1718.
[127] McKeon-Fischer, K., & Freeman, J. Characterization of electrospun poly (L-lactide) and gold nanoparticle composite scaffolds for skeletal muscle tissue engineering, J. Tissue Eng. Regen. Med., 2011, 5, 560–568.
[128] Stoiljkovic, A., Ishaque, M., Justus, U. et al. Preparation of water-stable submicron fibers from aqueous latex dispersion of water-insoluble polymers by electrospinning, Polymer, 2007, 48, 3974–3981.
[129] Stoiljkovic, A., Venkatesh, R., Klimov, E., Raman, V., Wendorff, J., & Greiner, A. Poly (styrene-co-n-butyl acrylate) nanofibers with excellent stability against water by electrospinning from aqueous colloidal dispersions, Macromolecules, 2009, 42, 6147–6151.
[130] Klimov, E., Raman, V., Venkatesh, R., Heckmann, W., & Stark, R. Designing nanofibers via electrospinning from aqueous colloidal dispersions: effect of cross-linking and template polymer, Macromolecules, 2010, 43, 6152–6155.
[131] Bijlard, A-C., Kaltbeitzel, A., Avlasevich, Y. et al. Dual-compartment nanofibres: separation of two highly reactive components in close vicinity, RSC Adv., 2015, 5, 97477–97484.
[132] Jiang, S., Lv, L., Li, Q., Wang, J., Landfester, K., & Crespy, D. Tailoring nanoarchitectonics to control the release profile of payloads, Nanoscale, 2016, 8, 11511–11517.
[133] Tseng, K-H., Chang, C-Y., Chung, M-Y., & Tang, Y-L. Bacteriostatic substrate by conductivity method and electric spark discharge method combined with electrospinning for silver dressing, Int. J. Polym. Sci., 2016, 9425358.
[134] Jiang, S., Ma, BC., Reinholz, J. et al. Efficient nanofibrous membranes for antibacterial wound dressing and UV protection, ACS Appl. Mater. Int., 2016, 8, 29915–29922.
[135] Jiang, S., Lieberwirth, I., Landfester, K., Muñoz-Espí, R., & Crespy, D. Nanofibrous photocatalysts from electrospun nanocapsules, Nanotechnology, 2017, 28, 405601.
[136] Jiang, S., Ma, BC., Huang, W. et al. Visible light active nanofibrous membrane for antibacterial wound dressing, Nanoscale Horiz., 2018, 3, 439–446.

[137] Jiang, S., Mable, CJ., Armes, SP., & Crespy, D. Directed assembly of soft anisotropic nanoparticles by colloid electrospinning, Macromol. Rapid Commun., 2016, 37, 1598–1602.
[138] Panthi, G., Barakat, NA., Park, M., Kim, H-Y., & Park, S-J. Fabrication of PdS/ZnS NPs doped PVAc hybrid electrospun nanofibers: Effective and reusable catalyst for dye photodegradation, J. Ind. Eng. Chem., 2015, 21, 298–302.
[139] Jiang, S., Lv, L-P., Landfester, K., & Crespy, D. Dual-responsive multicompartment nanofibers for controlled release of payloads, RSC Adv., 2016, 6, 43767–43770.
[140] Wang, Y., Li, Y., Sun, G. et al. Fabrication of Au/PVP nanofiber composites by electrospinning, J. Appl. Polym. Sci., 2007, 105, 3618–3622.
[141] Deniz, AE., Vural, HA., Ortaç, B., & Uyar, T. Gold nanoparticle/polymer nanofibrous composites by laser ablation and electrospinning, Mater. Lett., 2011, 65, 2941–2943.
[142] Piperno, S., Gheber, LA., Canton, P., Pich, A., Dvorakova, G., & Biffis, A. Microgel electrospinning: A novel tool for the fabrication of nanocomposite fibers, Polymer, 2009, 50, 6193–6197.
[143] Wu, C., Yuan, W., Al-Deyab, SS., & Zhang, K-Q. Tuning porous silica nanofibers by colloid electrospinning for dye adsorption, Appl. Surf. Sci., 2014, 313, 389–395.
[144] Jang, JS., Choi, SJ., Kim, SJ., Hakim, M., & Kim, ID. Rational design of highly porous SnO2 nanotubes functionalized with biomimetic nanocatalysts for direct observation of simulated diabetes, Adv. Funct. Mater., 2016, 26, 4740–4748.
[145] Kim, M-H., Jang, J-S., Koo, W-T. et al. Bi-modally porous WO3 microbelts functionalized with Pt catalysts for selective H2S sensors, ACS Appl. Mater. Int., 2018, 10, 20643–20651.
[146] Li, H., Zhu, J., Chen, S., Jia, L., & Ma, Y. Fabrication of aqueous-based dual drug loaded silk fibroin electrospun nanofibers embedded with curcumin-loaded RSF nanospheres for drugs controlled release, RSC Adv., 2017, 7, 56550–56558.
[147] Sharma, N., McKeown, SJ., Ma, X., Pochan, DJ., & Cloutier, SG. Structure– property correlations in hybrid polymer– nanoparticle electrospun fibers and plasmonic control over their dichroic behavior, ACS Nano., 2010, 4, 5551–5558.
[148] Munir, MM., Iskandar, F., Yun, KM., Okuyama, K., & Abdullah, M. Optical and electrical properties of indium tin oxide nanofibers prepared by electrospinning, Nanotechnology, 2008, 19, 145603.
[149] Maity, S., Downen, LN., Bochinski, JR., & Clarke, LI. Embedded metal nanoparticles as localized heat sources: An alternative processing approach for complex polymeric materials, Polymer, 2011, 52, 1674–1685.
[150] Aslanzadeh, S., Zhu, Z., Luo, Q., Ahvazi, B., Boluk, Y., & Ayranci, C. Electrospinning of colloidal lignin in poly (ethylene oxide) N, N-Dimethylformamide solutions, Macromol. Mater. Eng., 2016, 301, 401–413.
[151] Penchev, H., Paneva, D., Manolova, N., & Rashkov, I. Hybrid nanofibrous yarns based on N-carboxyethylchitosan and silver nanoparticles with antibacterial activity prepared by self-bundling electrospinning, Carbohydr. Res., 2010, 345, 2374–2380.
[152] Ji, L., Jung, K-H., Medford, AJ., & Zhang, X. Electrospun polyacrylonitrile fibers with dispersed Si nanoparticles and their electrochemical behaviors after carbonization, J. Mater. Chem., 2009, 19, 4992–4997.
[153] Lee, S., Lee, K., Moon, GD. et al. Preparation of macroporous carbon nanofibers with macroscopic openings in the surfaces and their applications, Nanotechnology, 2009, 20, 445702.
[154] Hsu, C-Y., & Liu, Y-L. Rhodamine B-anchored silica nanoparticles displaying white-light photoluminescence and their uses in preparations of photoluminescent polymeric films and nanofibers, J. Colloid Int. Sci., 2010, 350, 75–82.

[155] Mthethwa, T., Moloto, M., De Vries, A., & Matabola, K. Properties of electrospun CdS and CdSe filled poly (methyl methacrylate)(PMMA) nanofibres, Mater. Res. Bull., 2011, 46, 569–575.
[156] Bazilevsky, AV., Yarin, AL., & Megaridis, CM. Co-electrospinning of core– shell fibers using a single-nozzle technique, Langmuir, 2007, 23, 2311–2314.
[157] Shi, Q., Vitchuli, N., Nowak, J. et al. One-step synthesis of silver nanoparticle-filled nylon 6 nanofibers and their antibacterial properties, J. Mater. Chem., 2011, 21, 10330–10335.
[158] Yeom, BY., Shim, E., & Pourdeyhimi, B. Boehmite nanoparticles incorporated electrospun nylon-6 nanofiber web for new electret filter media, Macromol. Res., 2010, 18, 884–890.
[159] Zhao, H., Kang, W., Ma, X., Deng, N., Li, Z., & Cheng, B. Fabrication and catalytic behavior of hierarchically-structured nylon 6 nanofiber membrane decorated with silver nanoparticles, Chinese J. Catal., 2017, 38, 73–82.
[160] Zhuang, X., Cheng, B., Kang, W., & Xu, X. Electrospun chitosan/gelatin nanofibers containing silver nanoparticles, Carbohydr. Polym., 2010, 82, 524–527.
[161] Sheikh, F., Kanjwal, MA., Macossay, J., Barakat, N., & Kim, H. A simple approach for synthesis, characterization and bioactivity of bovine bones to fabricate the polyurethane nanofiber containing hydroxyapatite nanoparticles, EXPRESS Polym. Lett., 2012, 6, 41–53.
[162] Kim, S., Park, SG., Kang, SW., & Lee, KJ. Nanofiber-based hydrocolloid from colloid electrospinning toward next generation wound dressing, Macromol. Mater. Eng., 2016, 301, 818–826.
[163] Yoshimatsu, K., Ye, L., Lindberg, J., & Chronakis, IS. Selective molecular adsorption using electrospun nanofiber affinity membranes, Biosens. Bioelectron., 2008, 23, 1208–1215.
[164] Collins, AM., Olof, SN., Mitchels, JM., & Mann, S. Facile preparation and processing of aqueous dispersions of tris (8-hydroxyquinoline) aluminium (iii) photoluminescent nanoparticles, J. Mater. Chem., 2009, 19, 3950–3954.
[165] Lim, J-M., Yi, G-R., Moon, JH., Heo, C-J., & Yang, S-M. Superhydrophobic films of electrospun fibers with multiple-scale surface morphology, Langmuir, 2007, 23, 7981–7989.
[166] Lim, J-M., Moon, JH., Yi, G-R., Heo, C-J., & Yang, S-M. Fabrication of one-dimensional colloidal assemblies from electrospun nanofibers, Langmuir, 2006, 22, 3445–3449.
[167] Mu, Q., Zhang, Q., Gao, L. et al. Structural evolution and formation mechanism of the soft colloidal arrays in the core of PAAm nanofibers by electrospun packing, Langmuir, 2017, 33, 10291–10301.
[168] Nie, H., & Wang, C-H. Fabrication and characterization of PLGA/HAp composite scaffolds for delivery of BMP-2 plasmid DNA, J. Control. Rel., 2007, 120, 111–121.
[169] Yang, Y., Li, X., Cui, W., Zhou, S., Tan, R., & Wang, C. Structural stability and release profiles of proteins from core-shell poly (DL-lactide) ultrafine fibers prepared by emulsion electrospinning, J. Biomed. Mater. Res. Part A: An Off. J. Soc. Biomater., Japanese Soc. Biomater., Aust. Soc. Biomater. Korean Soc. Biomater., 2008, 86, 374–385.
[170] Stanishevsky, A., Chowdhury, S., Chinoda, P., & Thomas, V. Hydroxyapatite nanoparticle loaded collagen fiber composites: microarchitecture and nanoindentation study, J. Biomed. Mater. Res. Part A: An Off. J. Soc. Biomater., Japanese Soc. Biomater., Aust. Soc. Biomater. Korean Soc. Biomater., 2008, 86, 873–882.
[171] Wang, M., Li, X., Hua, W., Shen, L., Yu, X., & Wang, X. Electrospun poly (acrylic acid)/silica hydrogel nanofibers scaffold for highly efficient adsorption of lanthanide ions and its photoluminescence performance, ACS Appl. Mater. Int., 2016, 8, 23995–24007.
[172] Li, X., Yu, X., Cheng, C., Deng, L., Wang, M., & Wang, X. Electrospun superhydrophobic organic/inorganic composite nanofibrous membranes for membrane distillation, ACS Appl. Mater. Int., 2015, 7, 21919–21930.

[173] Su, C., Li, Y., Dai, Y., Gao, F., Tang, K., & Cao, H. Fabrication of three-dimensional superhydrophobic membranes with high porosity via simultaneous electrospraying and electrospinning, Mater. Lett., 2016, 170, 67–71.
[174] Chen, J., Yang, P., Wang, C. et al. Ag nanoparticles/PPV composite nanofibers with high and sensitive opto-electronic response, Nano. Res. Lett., 2011, 6, 121.
[175] Zhang, K., Yuan, W., Zhou, N., & Wu, C. Multicomponent nanofibers via electrospinning of polymers and colloidal dispersions for environmental and optical applications. Electrospun Nanofibers for Energy and Environmental Applications, Springer, China, 2014, 403–431.
[176] Yu, J., & Kan, C-W. Review on fabrication of structurally colored fibers by electrospinning, Fibers, 2018, 6, 70.
[177] Tebyetekerwa, M., Wang, X., Marriam, I., Dan, P., Yang, S., & Zhu, M. Green approach to fabricate polyindole composite nanofibers for energy and sensor applications, Mater. Lett., 2017, 209, 400–403.
[178] Fan, L., Wang, H., Zhang, K., He, C., Cai, Z., & Mo, X. Regenerated silk fibroin nanofibrous matrices treated with 75% ethanol vapor for tissue-engineering applications, J. Biomater. Sci., Polym. Ed., 2012, 23, 497–508.
[179] Abayazid, M., Op Den Buijs, J., de Korte, CL., & Misra, S. Effect of skin thickness on target motion during needle insertion into soft-tissue phantoms. Biomedical robotics and biomechatronics (BioRob), 2012 4th IEEE RAS & EMBS International Conference on; 2012: IEEE. p. 755–760.
[180] Saquing, CD., Manasco, JL., & Khan, SA. Electrospun nanoparticle–nanofiber composites via a one-step synthesis, Small, 2009, 5, 944–951.
[181] Li, Z., Huang, H., & Wang, C. Electrostatic forces induce poly (vinyl alcohol)-protected copper nanoparticles to form copper/poly (vinyl alcohol) nanocables via electrospinning, Macromol. Rapid Commun., 2006, 27, 152–155.
[182] Kanehata, M., Ding, B., & Shiratori, S. Nanoporous ultra-high specific surface inorganic fibres, Nanotechnology, 2007, 18, 315602.
[183] Zhao F, Wang X, Ding B, et al. Nanoparticle decorated fibrous silica membranes exhibiting biomimetic superhydrophobicity and highly flexible properties, RSC Adv., 2011, 1, 1482–1488.
[184] Friedemann, K., Corrales, T., Kappl, M., Landfester, K., & Crespy, D. Facile and large-scale fabrication of anisometric particles from fibers synthesized by colloid-electrospinning, Small, 2012, 8, 144–153.
[185] Jin, Y., Yang, D., Kang, D., & Jiang, X. Fabrication of necklace-like structures via electrospinning, Langmuir, 2009, 26, 1186–1190.
[186] Shi, Q., Vitchuli, N., Ji, L. et al. A facile approach to fabricate porous nylon 6 nanofibers using silica nanotemplate, J. Appl. Polym. Sci., 2011, 120, 425–433.
[187] Di, J., Chen, H., Wang, X. et al. Fabrication of zeolite hollow fibers by coaxial electrospinning, Chem. Mater., 2008, 20, 3543–3545.
[188] Horzum, N., Muñoz-Espí, R, Glasser, G., Demir, MM., Landfester, K., & Crespy, D. Hierarchically structured metal oxide/silica nanofibers by colloid electrospinning, ACS Appl. Mater. Int., 2012, 4, 6338–6345.
[189] Drew, C., Wang, X., Samuelson, LA., & Kumar, J. The effect of viscosity and filler on electrospun fiber morphology, J. Macromol. Sci., Part A, 2003, 40, 1415–1422.
[190] Choi, SK., Kim, S., Lim, SK., & Park, H. Photocatalytic comparison of TiO_2 nanoparticles and electrospun TiO_2 nanofibers: effects of mesoporosity and interparticle charge transfer, J. Phys. Chem. C, 2010, 114, 16475–16480.
[191] Song, M., Pan, C., Li, J., Zhang, R., Wang, X., & Gu, Z. Blends of TiO_2 nanoparticles and poly (*N*-isopropylacrylamide)-*co*-polystyrene nanofibers as a means to promote the biorecognition of an anticancer drug, Talanta, 2008, 75, 1035–1040.

[192] Roso, M., Sundarrajan, S., Pliszka, D., Ramakrishna, S., & Modesti, M. Multifunctional membranes based on spinning technologies: the synergy of nanofibers and nanoparticles, Nanotechnology, 2008, 19, 285707.
[193] Yang, D., Lu, B., Zhao, Y., & Jiang, X. Fabrication of aligned fibrous arrays by magnetic electrospinning, Adv. Mater., 2007, 19, 3702–3706.
[194] Wang, S., Wang, C., Zhang, B. et al. Preparation of Fe3O4/PVA nanofibers via combining in-situ composite with electrospinning, Mater. Lett., 2010, 64, 9–11.
[195] Sharma, N., Jaffari, GH., Shah, SI., & Pochan, DJ. Orientation-dependent magnetic behavior in aligned nanoparticle arrays constructed by coaxial electrospinning, Nanotechnology, 2010, 21, 085707.
[196] Xin, Y., Huang, Z., Peng, L., & Wang, D. Photoelectric performance of poly (p-phenylene vinylene)/Fe_3O_4 nanofiber array. J. Appl. Phys., 2009, 105, 086106.
[197] Zhang, C., Li, X., Yang, Y., & Wang, C. Polymethylmethacrylate/Fe_3O_4 composite nanofiber membranes with ultra-low dielectric permittivity, Appl. Phys. A, 2009, 97, 281–285.
[198] Kalra, V., Lee, J., Lee, JH. et al. Controlling nanoparticle location via confined assembly in electrospun block copolymer nanofibers, Small, 2008, 4, 2067–2073.
[199] Lee, S. Multifunctionality of layered fabric systems based on electrospun polyurethane/zinc oxide nanocomposite fibers, J. Appl. Polym. Sci., 2009, 114, 3652–3658.
[200] Mimura, K-i., Moriya, M., Sakamoto, W., & Yogo, T. Synthesis of $BaTiO_3$ nanoparticle/poly (2-hydroxyethyl methacrylate) hybrid nanofibers via electrospinning, Comp. Sci. Technol., 2010, 70, 492–497.
[201] Czech, T., Ramakrishna, S., & Sundarrajan, S. Electrospinning and electrospraying techniques for nanocomposite non-woven fabric production, Fibres Text. East Eur., 2009, 17, 77–81.
[202] Andrew, J., & Clarke, D. Enhanced ferroelectric phase content of polyvinylidene difluoride fibers with the addition of magnetic nanoparticles, Langmuir, 2008, 24, 8435–8438.
[203] Park, HS., Choi, BG., Hong, WH., & Jang, S-Y. Controlled assembly of graphene oxide nanosheets within one-dimensional polymer nanostructure, J. Colloid Int. Sci., 2013, 406, 24–29.
[204] Wahab, IF., Razak, SIA., Azmi, NS., Dahli, FN., Yusof, AHM., & Nayan, NHM. Electrospun graphene oxide-based nanofibres, Advances in Carbon Nanostructures, InTech, London, 2016, 101–120.
[205] Unnithan, AR., Barakat, NA., Nirmala, R., Al-Deyab, SS., & Kim, HY. Novel electrospun nanofiber mats as effective catalysts for water photosplitting, Ceram. Int., 2012, 38, 5175–5180.
[206] Panthi, G., Barakat, NA., Khalil, KA., Yousef, A., Jeon, K-S., & Kim, HY. Encapsulation of CoS nanoparticles in PAN electrospun nanofibers: effective and reusable catalyst for ammonia borane hydrolysis and dyes photodegradation, Ceram. Int., 2013, 39, 1469–1476.
[207] Davarpanah Jazi, R., Rafienia, M., Salehi Rozve, H., Karamian, E., & Sattary, M. Fabrication and characterization of electrospun polylactic-co-glycolic acid/zeolite nanocomposite scaffolds using bone tissue engineering, J. Bioact. Compat. Polym., 2018, 33, 63–78.
[208] Torres-Giner, S., & Lagaron, JM. Zein-based ultrathin fibers containing ceramic nanofillers obtained by electrospinning, I. Morphology and thermal properties. J. Appl. Polym. Sci., 2010, 118, 778–789.
[209] Ferrández-Rives, M., Beltrán-Osuna, ÁA., Gómez-Tejedor, JA., & Gómez Ribelles, JL. Electrospun PVA/bentonite nanocomposites mats for drug delivery, Materials, 2017, 10, 1448.
[210] Júnior, LP., Silva, DBdS., de Aguiar, MF., de Melo, CP., & Alves, KG. Preparation and characterization of polypyrrole/organophilic montmorillonite nanofibers obtained by electrospinning, J. Mol. Liq., 2019, 275, 452–462.

[211] Nagarajan, S., Belaid, H., Pochat-Bohatier, C. et al. Design of boron nitride/gelatin electrospun nanofibers for bone tissue engineering, ACS Appl. Mater. Int., 2017, 9, 33695–33706.
[212] Fujihara, K., Kotaki, M., & Ramakrishna, S. Guided bone regeneration membrane made of polycaprolactone/calcium carbonate composite nano-fibers, Biomaterials, 2005, 26, 4139–4147.
[213] Cao, Z., Wang, D., Lyu, L., Gong, Y., & Li, Y. Fabrication and characterization of PCL/$CaCO_3$ electrospun composite membrane for bone repair, RSC Adv., 2016, 6, 10641–10649.
[214] Erisken, C., Kalyon, DM., & Wang, H. A hybrid twin screw extrusion/electrospinning method to process nanoparticle-incorporated electrospun nanofibres, Nanotechnology, 2008, 19, 165302.
[215] Fu, Q-W., Zi, Y-P., Xu, W. et al. Electrospinning of calcium phosphate-poly (D, L-lactic acid) nanofibers for sustained release of water-soluble drug and fast mineralization, Int.J. Nanomed., 2016, 11, 5087.
[216] Teng, S-H., Lee, E-J., Wang, P., & Kim, H-E. Collagen/hydroxyapatite composite nanofibers by electrospinning, Mater. Lett., 2008, 62, 3055–3058.
[217] Zhang, Y., Venugopal, JR., El-Turki, A., Ramakrishna, S., Su, B., & Lim, CT. Electrospun biomimetic nanocomposite nanofibers of hydroxyapatite/chitosan for bone tissue engineering, Biomaterials, 2008, 29, 4314–4322.
[218] Puppi, D., Piras, AM., Chiellini, F. et al. Optimized electro-and wet-spinning techniques for the production of polymeric fibrous scaffolds loaded with bisphosphonate and hydroxyapatite, J. Tissue Eng. Regen. Med., 2011, 5, 253–263.
[219] Ramier, J., Bouderlique, T., Stoilova, O. et al. Biocomposite scaffolds based on electrospun poly (3-hydroxybutyrate) nanofibers and electrosprayed hydroxyapatite nanoparticles for bone tissue engineering applications, Mater. Sci. Eng: C, 2014, 38, 161–169.
[220] Mei, F., Zhong, J., Yang, X. et al. Improved biological characteristics of poly (L-lactic acid) electrospun membrane by incorporation of multiwalled carbon nanotubes/hydroxyapatite nanoparticles, Biomacromolecules, 2007, 8, 3729–3735.
[221] Zhao, X., Chen, X., Zhang, L. et al. Preparation of nano-hydroxyapatite coated carbon nanotube reinforced hydroxyapatite composites, Coatings, 2018, 8, 357.
[222] Fuad, A., Fatriani, N., Yogihati, C., Taufiq, A., & Latifah, E. Electrospinning β-SiC fibers from SiC nanoparticles dispersed in various polymer solutions as the electrospinning agents, J. Phys: Conf. Ser., 2018, IOP Publishing. 012013.
[223] Friedemann, K., Turshatov, A., Landfester, K., & Crespy, D. Characterization via two-color STED microscopy of nanostructured materials synthesized by colloid electrospinning, Langmuir, 2011, 27, 7132–7139.
[224] Herrmann, C., Turshatov, A., & Crespy, D. Fabrication of polymer ellipsoids by the electrospinning of swollen nanoparticles, ACS Macro Lett., 2012, 1, 907–909.
[225] Yuan, W., & Zhang, K-Q. Structural evolution of electrospun composite fibers from the blend of polyvinyl alcohol and polymer nanoparticles, Langmuir, 2012, 28, 15418–15424.
[226] Wang, J., Wen, Y., Ge, H. et al. Simple fabrication of full color colloidal crystal films with tough mechanical strength, Macromol. Chem. Phys., 2006, 207, 596–604.
[227] Yuan, W., Zhou, N., Shi, L., & Zhang, K-Q. Structural coloration of colloidal fiber by photonic band gap and resonant mie scattering, ACS Appl. Mater. Int, 2015, 7, 14064–14071.
[228] Jo, E., Lee, S., Kim, KT. et al. Core-sheath nanofibers containing colloidal arrays in the core for programmable multi-agent delivery, Adv. Mater., 2009, 21, 968–972.
[229] Helgeson, ME., Grammatikos, KN., Deitzel, JM., & Wagner, NJ. Theory and kinematic measurements of the mechanics of stable electrospun polymer jets, Polymer, 2008, 49, 2924–2936.

[230] Buruaga, L., Sardon, H., Irusta, L., González, A., Fernández-Berridi, MJ., & Iruin, JJ. Electrospinning of waterborne polyurethanes, J. Appl. Polym. Sci., 2010, 115, 1176–1179.
[231] Beck-Broichsitter, M., Thieme, M., Nguyen, J. et al. Novel 'Nano in Nano'composites for sustained drug delivery: Biodegradable nanoparticles encapsulated into nanofiber nonwovens, Macromol. Biosci., 2010, 10, 1527–1535.
[232] Sun, J., Bubel, K., Chen, F., Kissel, T., Agarwal, S., & Greiner, A. Nanofibers by green electrospinning of aqueous suspensions of biodegradable block copolyesters for applications in medicine, pharmacy and agriculture, Macromol. Rapid Commun., 2010, 31, 2077–2083.
[233] Rietveld, IB., Kobayashi, K., Yamada, H., & Matsushige, K. Morphology control of poly (vinylidene fluoride) thin film made with electrospray, J. Colloid Int. Sci., 2006, 298, 639–651.
[234] Dong, B., Smith, ME., & Wnek, GE. Encapsulation of multiple biological compounds within a single electrospun fiber, Small, 2009, 5, 1508–1512.
[235] Díaz, JE., Barrero, A., Márquez, M., Fernández-Nieves, A., & Loscertales, IG. Absorption properties of microgel-PVP composite nanofibers made by electrospinning, Macromol. Rapid Commun., 2010, 31, 183–189.
[236] Zussman, E. Encapsulation of cells within electrospun fibers, Polym. Adv. Technol., 2011, 22, 366–371.
[237] Chen, H., Liu, Y., & Hu, Q. A novel bioactive membrane by cell electrospinning, Exp. Cell Res., 2015, 338, 261–266.
[238] San Keskin, NO., Celebioglu, A., Sarioglu, OF., Uyar, T., & Tekinay, T. Encapsulation of living bacteria in electrospun cyclodextrin ultrathin fibers for bioremediation of heavy metals and reactive dye from wastewater, Colloids Surf B: Biointerfaces, 2018, 161, 169–176.
[239] Xu, X., Yang, L., Xu, X. et al. Ultrafine medicated fibers electrospun from W/O emulsions, J. Control. Rel., 2005, 108, 33–42.
[240] Li, X., Zhang, H., Li, H., & Yuan, X. Encapsulation of proteinase K in PELA ultrafine fibers by emulsion electrospinning: preparation and in vitro evaluation, Colloid Polym. Sci., 2010, 288, 1113–1119.
[241] Yang, Y., Li, X., Qi, M., Zhou, S., & Weng, J. Release pattern and structural integrity of lysozyme encapsulated in core–sheath structured poly (DL-lactide) ultrafine fibers prepared by emulsion electrospinning, Eur. J. Pharm. Biopharm., 2008, 69, 106–116.
[242] Dai, Y., Niu, J., Liu, J., Yin, L., & Xu, J. In situ encapsulation of laccase in microfibers by emulsion electrospinning: Preparation, characterization, and application, Biores. Technol, 2010, 101, 8942–8947.
[243] Wei, K., Li, Y., Lei, X. et al. Emulsion electrospinning of a collagen-like protein/PLGA fibrous scaffold: Empirical modeling and preliminary release assessment of encapsulated protein, Macromol. Biosci., 2011, 11, 1526–1536.
[244] Li, Y., Ko, FK., & Hamad, WY. Effects of emulsion droplet size on the structure of electrospun ultrafine biocomposite fibers with cellulose nanocrystals, Biomacromolecules, 2013, 14, 3801–3807.
[245] Angeles, M., Cheng, HL., & Velankar, SS. Emulsion electrospinning: composite fibers from drop breakup during electrospinning, Polym. Adv. Technol., 2008, 19, 728–733.
[246] Arecchi, A., Mannino, S., & Weiss, J. Electrospinning of poly (vinyl alcohol) nanofibers loaded with hexadecane nanodroplets, J. Food Sci., 2010, 75, N80–N88.
[247] Pal, J., Skrifvars, M., Nandan, B., & Srivastava, RK. Electrospun composite matrices from tenside-free poly(ε-caprolactone)-grafted acrylic acid/hydroxyapatite oil-in-water emulsions, J. Mater. Sci., 2017, 52, 2254–2262.

[248] Chen, H., Di, J., Wang, N. et al. Fabrication of hierarchically porous inorganic nanofibers by a general microemulsion electrospinning approach, Small, 2011, 7, 1779–1783.
[249] Choi, S-H., Youn, D-Y., Jo, SM., Oh, S-G., & Kim, I-D. Micelle-mediated synthesis of single-crystalline β (3C)-SiC fibers via emulsion electrospinning, ACS Appl. Mater. Int., 2011, 3, 1385–1389.
[250] Zander, NE., Strawhecker, KE., Orlicki, JA., Rawlett, AM., & Beebe, Jr TP. Coaxial electrospun poly (methyl methacrylate)–polyacrylonitrile nanofibers: Atomic force microscopy and compositional characterization, J. Phys. Chem. B, 2011, 115, 12441–12447.
[251] Du, Y., Wu, T., Yan, N., Kortschot, MT., & Farnood, R. Fabrication and characterization of fully biodegradable natural fiber-reinforced poly(lactic acid) composites, Comp. Part B: Eng., 2014, 56, 717–723.
[252] Balakrishnan, P., Geethamma, V., Sreekala, MS., & Thomas, S. Polymeric biomaterials: State-of-the-art and new challenges. Fundamental Biomaterials: Polymers: Woodhead Publishing, India, 2018, 1–20.
[253] Um, IC., Fang, D., Hsiao, BS., Okamoto, A., & Chu, B. Electro-spinning and electro-blowing of hyaluronic acid, Biomacromolecules, 2004, 5, 1428–1436.
[254] Sheikh, FA., Ju, HW., Lee, JM. et al. 3D electrospun silk fibroin nanofibers for fabrication of artificial skin, Nanomed.: Nanotechnol. Biol. Med., 2015, 11, 681–691.
[255] Horzum, N., Boyacı, E., Eroglu, AE., Shahwan, T., & Demir, MM. Sorption efficiency of chitosan nanofibers toward metal ions at low concentrations, Biomacromolecules, 2010, 11, 3301–3308.
[256] Kai, D., Jiang, S., Low, ZW., & Loh, XJ. Engineering highly stretchable lignin-based electrospun nanofibers for potential biomedical applications, J. Mater. Chem. B, 2015, 3, 6194–6204.
[257] Grimmelsmann, N., Homburg, SV., & Ehrmann, A. Needleless electrospinning of pure and blended chitosan. IOP Conference Series: Materials Science and Engineering; 2017: IOP Publishing. p. 012098.
[258] Sun, Z., Zussman, E., Yarin, AL., Wendorff, JH., & Greiner, A. Compound core–shell polymer nanofibers by co-electrospinning, Adv. Mater., 2003, 15, 1929–1932.
[259] Arik, N., Inan, A., Ibis, F. et al. Modification of electrospun PVA/PAA scaffolds by cold atmospheric plasma: alignment, antibacterial activity, and biocompatibility, Polym. Bull., 2019, 76, 797–812.
[260] Sarhan, WA., & Azzazy, HM. High concentration honey chitosan electrospun nanofibers: biocompatibility and antibacterial effects, Carbohydr. Polym., 2015, 122, 135–143.
[261] Pham, QP., Sharma, U., & Mikos, AG. Electrospinning of polymeric nanofibers for tissue engineering applications: a review, Tissue Eng., 2006, 12, 1197–1211.
[262] Rieger, KA., & Schiffman, JD. Electrospinning an essential oil: Cinnamaldehyde enhances the antimicrobial efficacy of chitosan/poly (ethylene oxide) nanofibers, Carbohydr. Polym., 2014, 113, 561–568.

Shani L. Levit, Rebecca C. Walker,
Amanda L. Pham, Christina Tang

3 Polymer-free electrospinning

Abstract: Traditionally, electrospinning has been limited to high-molecular-weight polymers. Generally, polymer networks are thought to provide molecular entanglement and elasticity to achieve fiber formation. Polymer-free electrospinning explores fabricating fibers from biomolecules, supramolecular assemblies, and colloidal nanoparticles for new applications. There have been two main approaches to electrospinning these systems: (1) leveraging intermolecular interactions to achieve "pseudoentangled" systems or (2) inducing aggregation to facilitate fiber formation. These exciting advances are a result of interdisciplinary approaches that leverage principles from supramolecular chemistry and polymer processing. Understanding the underlying mechanisms that facilitate fiber formation from polymer-free systems may lead to innovation in functional materials and sustainable processing of electrospun nanofibers.

Keywords: aggregation, amphiphiles, cyclodextrin, entanglement, flocculation, melt electrospinning, peptide, protein, supramolecular assembly, trisamides

3.1 Introduction

Electrospinning, a process that uses electrostatic interactions to uniaxially draw one-dimensional nanostructures, is a simple method for creating nonwoven nanofibers from a variety of materials for numerous applications [1–9]. A solid fiber is produced as the electrospinning solution or melt is continuously stretched due to the application of an electric field and the solvent evaporates or fiber cools, respectively [10–15]. To electrospin nanofibers, preventing breakup of the jet into droplets associated with electrospraying is required. Since droplet formation is often observed with low viscosity solutions, initial reports of electrospinning used solutions of high-molecular-weight polymer with high viscosity and viscoelasticity. It is commonly observed that with increasing concentration, there is an initial transition from beads to beaded fibers and a second transition to uniform fibers (Figure 3.1). The ability to form uniform fibers has been frequently attributed to polymer entanglement at higher polymer concentrations [16–18].

Shani L. Levit, Rebecca C. Walker, Amanda L. Pham, Christina Tang, Chemical and Life Science Engineering, Virginia Commonwealth University, Richmond, VA

https://doi.org/10.1515/9783110581393-003

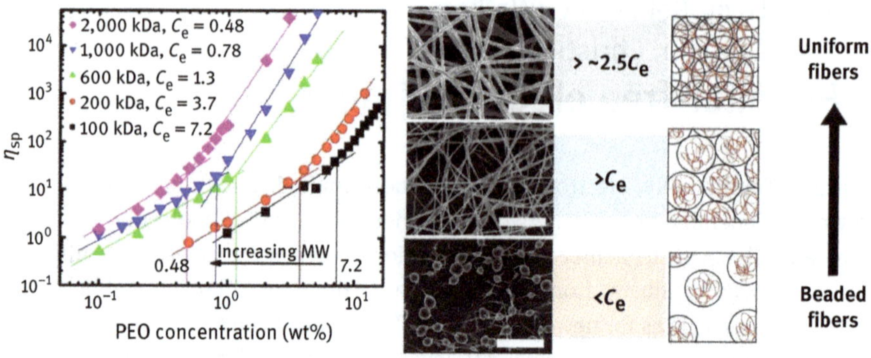

Figure 3.1: Specific viscosity as a function of polymer concentration to determine entanglement concentration for polyethylene oxide (PEO) of various molecular weights. SEM (PEO 600 kDa) showing the transition from beaded fibers to uniform fibers as the polymer concentration increases above the entanglement concentration (5 µm scale bar). For neutral polymers in a good solvent, for example, aqueous PEO, concentrations above ~2.5× the entanglement concentration form uniform fibers. Reprinted with permission. Copyright 2008 American Chemical Society.

To predict electrospinnability, semiempirical relationships have been used to quantify the degree of polymer entanglement to achieve uniform fibers [19–21]. The entanglement concentration (C_e) can be estimated by measuring the viscosity (zero shear) as a function of polymer concentration ([C]) and examining the scaling relationship between the specific viscosity (eq. (3.1)) and concentration. Note that the specific viscosity (η_{sp}) accounts for the viscosity of the solvent:

$$\eta_{sp} = \frac{\eta_0 - \eta_s}{\eta_s} \tag{3.1}$$

where η_0 is the zero-shear viscosity of the polymer solution and η_s is the viscosity of the solvent. Upon the onset of polymer entanglement, there is a sharp increase in the scaling relationship. Based on the theory for semidilute solutions, below the entanglement concentration $\eta_{sp} \propto [C]^{1.25}$ and above the entanglement concentration $\eta_{sp} \propto [C]^{4.8}$. For linear, neutral polymers in a good solvent, for example, aqueous solutions of polyethylene oxide, the transition from beaded fibers to uniform fibers generally occurs at a polymer concentration of approximately 2–2.5× the entanglement concentration (Figure 3.1). This approach has worked well for a number of polymer systems, for example, polyvinyl alcohol (PVA, aq.), polyethylene oxide (aq.), polystyrene in tetrahydrofuran, poly(D-lactic acid) in *N,N′*-dimethylformamide (DMF), and poly(L-lactic acid) in dichloromethane [19, 20]. Analogous approaches have been developed for polyelectrolyte solutions. Polyelectrolytes transition from droplets to fibers at approximately 8× C_e. The viscosity scaling relationship for polyelectrolytes to determine the entanglement concentration is an increase in the scaling relationship from $\eta_{sp} \propto [C]^{0.5}$ to $\eta_{sp} \propto [C]^{1.5}$ [21].

However, the presence of entanglement is a sufficient but not necessary condition for the electrospinning solution to demonstrate strong elastic properties, and polymer solutions with concentrations below the entanglement concentration have been electrospun [22]. Rutledge and coworkers attribute the ability to form uniform fibers to elasticity of the polymer solution. To prevent beaded fibers, the Rayleigh breakup instability must be suppressed. The instability can be slowed or suppressed by the viscoelasticity of the polymer solution. Therefore, electrospinning of polymer solutions with concentrations below the entanglement transition is possible when the relaxation time is longer than the time of extensional deformation. The timescales for instability and the viscoelasticity can be quantified with the Deborah number (*De*), that is, the ratio of the fluid relaxation time and instability growth time. At high Deborah number ($\gg$1), the elastic response delays jet breakup. Deborah numbers above 6 result in uniform fibers because the instability is completely suppressed by elastic forces or arrested at a very early stage of instability growth [22].

To increase the versatility of electrospinning for new applications, polymer-free electrospinning has also been explored. Polymer-free electrospinning has successfully formed continuous fibers from proteins, peptides, small molecules, surfactants, and colloidal dispersions which offer potential for new applications. Since topics such as protein electrospinning and colloid electrospinning have been reviewed elsewhere [23, 24], we provide an overview of the key results to date. We focus on advances in fundamental understanding of molecular interactions that facilitate fiber formation. There have been two main approaches to electrospinning polymer-free fibers: (1) leveraging intermolecular interactions to achieve "pseudoentangled" systems or (2) inducing aggregation to facilitate fiber formation.

3.2 "Pseudoentangled" systems

Polymer-free electrospinning has been achieved by leveraging intermolecular interactions. Biological materials such as proteins and peptides have been successfully electrospun into nanofibers. Nanofibers from self-assembled supramolecular structures have also been reported.

3.2.1 Protein electrospinning

For biomedical engineering applications such as wound healing, tissue regeneration, biosensors, and drug delivery, protein nanofibers are of particular interest because they are biodegradable and biocompatible [23, 25–27]. Typically, to produce biofunctional fibers, biomolecules are either introduced into the polymer electrospinning solution prior to electrospinning or grafted onto fibers after electrospinning [23]. Direct electrospinning of proteins and peptides has also been considered. Unlike entangled polymer solutions, proteins and peptides have complex secondary and tertiary structures and lack viscoelastic properties, which leads to challenges when electrospinning [28]. Recent advances in electrospinning have facilitated direct electrospinning of biomolecules (Figure 3.2) leading to improved understanding of the mechanisms that induce molecular entanglement and fiber formation [19, 29]. An overview of proteins that have been successfully electrospun is provided in Table 3.1.

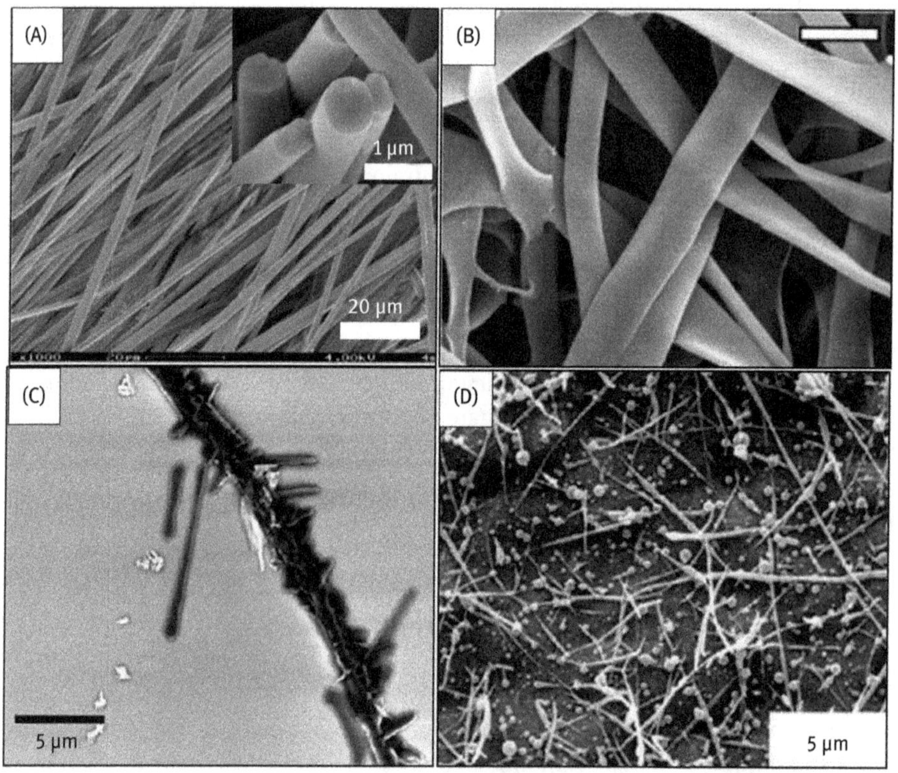

Figure 3.2: Fiber morphologies from electrospun proteins and peptides. (A) Cylindrical BSA fibers (scale bar = 20 μm) and their cross section (scale bar = 1 μm). Reprinted with permission [30]. Copyright 2008 American Chemical Society. (B) Ribbonlike fibers from silk-elastin-like protein (SELP) (scale bar = 1 μm). Reprinted with permission from Ref. [25]. (C) Fmoc–Phe–Gly peptide needlelike fibers (scale bar = 5 μm). Reproduced from Ref. [28] with permission from the Royal Society of Chemistry. (D) Beads and beaded fibers from coassembling peptides (CAP) (scale bar = 5 μm). Reprinted with permission from source 52.

Table 3.1: Overview of protein electrospinning.

Protein	Concentration	Solvent	Morphology	Diameter	Reference
Bovine serum albumin	10–12 wt%	TFE	Cylindrical with pores	–	[30]
Albumin from egg white	7 wt%	HFIP	Irregular with branching	Several micrometers	
		TFA	Cylindrical with branching	300–800 nm	[28]
Collagen (type I)	8 w/v%	HFIP	Cylindrical with pores	460 nm (100–1,200 nm)	[31]
Collagen (type I)	50 and 180 mg mL^{-1}	HFIP	Cylindrical	Nanometer	[32]
	180 mg mL^{-1}	TFE	Cylindrical	Nanometer	
Gelatin	14–29 w/v%	Acetic acid	Smooth and beaded fibers	113–145 µm	[33]
	20–23 w/v%	Formic acid		114–137 µm	
Elastin-like protein (ELP)	20–35 wt%	Aqueous solution	Ribbonlike	1.2–1.8 µm (thickness 190 ± 60 nm)	[34]
Silk-elastin-like protein (SELP)	6–18 wt%	DI water	Ribbonlike	25–1,800 nm	[25]
Dragline spider silk protein	15 wt%	HFIP	Cylindrical	~300 nm	[35]
Silk fibroin	11–17 wt%	DI water	Cylindrical to ribbonlike	0.21–0.49 µm	[36, 37]
Wheat gluten	10 w/v%	HFIP	Cylindrical and ribbonlike	100–5,000 nm	[38]
Zein	20–30%	Ethanol	Ribbonlike	0.6–8.0 mm	[39]
	27–30%	Isopropyl alcohol	Ribbonlike	1.4–6.4 mm	
	20–30%	Acetic acid	Cylindrical	0.4–10.1 mm	
	35%	DMF	Beaded fibers	0.3–0.8 mm	
Zein and soy protein isolate (SPI)	23.75% Zein and 0.25% SPI	Acetic acid	Cylindrical and ribbonlike	0.32–0.42 µm	[40]
Zein	300 mg mL^{-1}	Acetic acid	Cylindrical	186 ± 70 nm	[41]

Table 3.1 (continued)

Protein	Concentration	Solvent	Morphology	Diameter	Reference
Hordein	150 mg mL^{-1}	Acetic acid	Cylindrical	184 ± 70 nm	
Gliadin	200 mg mL^{-1}	Acetic acid	Cylindrical	192 ± 61 nm	
Amaranth protein isolate (API)	5 wt%	HFIP	Beaded fibers	1,961 ± 997 nm	[42]
	8–10 wt%	HFIP	Fibers	390 ± 115 nm	
		Formic acid	Beaded fibers	1,388 ± 357 nm	
81 kDa elastin mimetic peptide	5 wt%	DI water	Beaded fibers	–	
	10 wt%	DI water	Thin filaments	300–400 nm	[43]
	20 wt%	DI water	Broad ribbonlike	~3μm (thickness 10–20 nm)	

Globular proteins

Using bovine serum albumin (BSA) as a model, globular proteins have been successfully electrospun. Technically, electrospinning globular proteins such as BSA is challenging because their solutions are of low viscosity. To electrospin, BSA was partially denatured in a mixture of trifluoroethanol (TFE) and water and then treated with β-mercaptoethanol (β-ME) to reduce the intramolecular disulfide bonds. Fiber formation was attributed to the spontaneously formed inter- and intramolecular bonds (Figure 3.3(A)). These spontaneously reformed disulfide bonds also significantly impacted the mechanical properties of the resulting fibers. Due to the nature of proteins, pH of the electrospinning solution was found to affect fiber formation. Interestingly, extreme pH improved the fiber formation, whereas pH near the isoelectric point protein aggregation disrupted electrospinning. At acidic pH, BSA undergoes conformational transition taking an expanded form. Basic conditions (pH ~ 9) promote reduction of disulfide bonds thereby creating more sulfhydryl sites that undergo spontaneous reformation [30].

Electrospinning of globular proteins has also been reported via microliter electrospinning. Microliter electrospinning is a variation of electrospinning in which a droplet is placed on a platinum wire to which voltage is applied [28]. Dry air can also be applied (~3,500 cm^3 min^{-1}) to this setup above the wire to encourage jet formation [27, 28]. When the albumin was electrospun using hexafluoroisopropanol

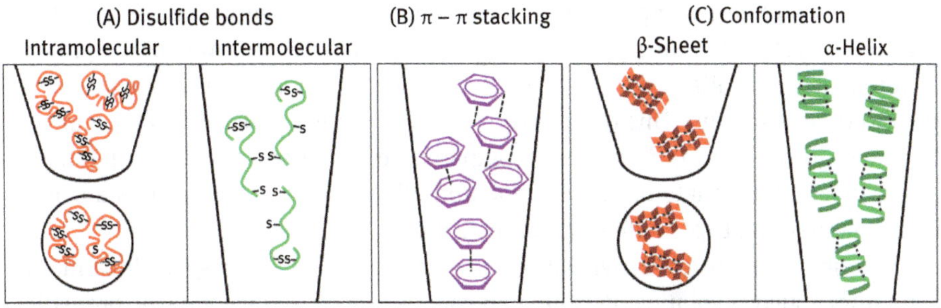

Figure 3.3: Schematic of factors that promote fiber formation during electrospinning of proteins and peptides. (A) Intermolecular disulfide bonds between protein enable molecular entanglement. (B) π–π stacking of aromatic residues between peptides facilitates molecular entanglement. (C) α-Helices and coil conformation compared to β-sheets, possibly due to flexibility of helices compared to sheets.

(HFIP) as the solvent (7 wt%), irregular micron-scale fibers were obtained. Electrospinning albumin using trifluoroacetic acid (TFA) as the solvent (7 wt%), 300–800 nm diameter fibers were reported. This result demonstrates the impact of solvent and protein interactions. Circular dichroism spectroscopy of albumin in HFIP indicates that HFIP induces minimal denaturation. The ability to form uniform fibers from TFA was attributed to molecular flexibility analogous to polymer systems. TFA increases the protonation of protein resulting in denaturation, which is thought to promote fiber formation [28]. Circular dichroism spectroscopy indicates a predominately disordered structure due to denaturation. Notably, denaturation is not observed when analyzing the dry fiber which suggests the protein partially refolds after electrospinning [28, 30].

Fibrous proteins

Collagen and gelatin

Electrospun collagen fibers are of interest for medical treatments such as wound healing because collagen is a major component of the extracellular matrix [31, 44]. The challenge has been determining an appropriate solvent to achieve sufficient protein concentrations to produce uniform fibers. Both type I and III collagen have been electrospun from HFIP solutions and produced fibers with different properties, such as variability in fiber diameter [44]. Further studies on electrospinning type 1 collagen (8% collagen in HFIP) produced fibers with diameters between 100 and 1,200 nm and average of 460 nm. Post-electrospinning, the collagen fibers are soluble in water, limiting their potential use. Therefore, fibers were cross-linked with glutaraldehyde vapor. Cross-linking decreased the pore volume (89–71%) and increased the fiber density (0.096–0.452 mg mL^{-1}). The cross-linked fibers exhibited mechanical strength comparable to commercial products with tensile strength of ~12 MPa [31]. However, using HFIP as solvent results in denaturation of the proteins in the fibers. Zeugolis et al. [32] attribute the denaturation to the solvent and not the electrical stimulus. This denaturation results in a lower denaturation temperature and lower crystallinity when compared to unprocessed collagen [31, 32]. Alternatively, phosphate-buffered saline and ethanol have been used as a solvent while maintaining triple-helical structure [23].

Gelatin, a denatured derivative of collagen, has been electrospun. It cannot be spun from aqueous solution, despite high concentrations and viscoelastic properties. This behavior was attributed to its many ionizable groups leading to polyelectrolyte characteristics and strong hydrogen bonding that limits the mobility of the polymer molecules in solution. Analogous to the synthetic polymers, fiber formation has been solvent dependent. Organic solvents such as TFE, formic acid, and acetic acid have been considered [45–47]. Formic acid or acetic acid and water mixtures have also been used. For example, fibers ranging from 100 to 760 nm were obtained using acetic acid (aq.) or formic acid (aq.). Analogous to polymer solutions, high concentrations (10–29 w/v%) of gelatin are needed to form uniform fibers, whereas beads or beaded fibers were obtained at lower gelatin concentrations. Increasing gelatin concentration led to increased viscosity and increased conductivity. Decreasing pH also resulted in increased shear viscosity, but decreased conductivity as well as expanded the coil structure of gelatin. Solvent selection is important to obtain suitable solution properties (conductivity, surface tension, and viscosity) to form fibers [33]. Interestingly, electrospinning collagen using HFIP as the solvent results in similar properties to electrospun gelatin [32].

Recombinant elastin

Elastin is a key structural protein of the extracellular matrices of many types of tissues. Thus, electrospun elastin fibers are of significant interest for tissue engineering scaffolds. Fibers electrospun from natural sources of elastin have been reported using HFIP as a solvent. Recombinant proteins have been successfully electrospun into fiber with elastin-like properties [34, 43]. A 81 kDa recombinant protein with 39 repeating elastomeric peptides ((Val-Pro-Gly-Val-Gly)$_4$(Val-Pro-Lys-Gly)) rich in glycine was electrospun from aqueous solution. Fibers were produced with diameters ranging from 200 to 3,000 nm and three morphologies depending on the solution concentration: beaded and triangle morphology (5 wt%), thin fibers (10 wt%), and ribbonlike fibers with widths of 10–20 nm (20 wt%). The produced fibers had ultimate tensile strength of fabrics with 35 MPa and elastic modulus of 1.8 GPa [43]. Another recombinant elastin-like protein (37.6 kDa) with repeating units of Val-Pro-Gly-Lys-Gly has been electrospun in aqueous solution into fibers with ribbonlike morphology. Lysine was incorporated into the protein to enable specific cross-linking post-electrospinning using a vapor phase (glutaraldehyde) followed by aqueous completion (NaCl buffer) cross-linking, which prevents disassembly and instability of fibers in aqueous solution and mechanical properties comparable to tissue [34].

Silk

Silk-based biomaterials are biocompatible with high strength and toughness compared to alternative biopolymer-based materials [48, 49]. Due to its unique mechanical properties, electrospun silk fibers are of interest for tissue scaffolds. In particular, spider silk is one of the strongest proteins that has been attributed to its conformation of highly crystalline β-sheets surrounded by glycine-rich elastic region [36, 37]. Electrospinning recombinant spider silk in HFIP solvent was found to alter the conformation of β-sheets into α-helixes due to application of the electric field [35]. However, the α-helix transition is reversible. Therefore, fibers can be treated with methanol after electrospinning to induce β-sheets to increase the mechanical strength of fibers [25, 35]. Recombinant silk-elastin-like protein (SELP) made up of silk and elastin subunits has been electrospun from aqueous solution. The silk units provide strength while the elastic blocks provide elastic behavior and increase solubility in water. Electrospinning SELP between 6 and 18 wt% in aqueous solution provided uniform fibers with a ribbonlike morphology with widths between 25 and 1,800 nm when electrospinning parameters varied. The SELP fibers were found to have ultimate tensile strength of 30.8 MPa and initial modulus of 0.88 GPa. Similar to the spider-silk proteins, electrospinning was found to induce conformational changes of β-sheets into α-helixes due to the slow rate of β-sheets crystallization compared to solvent evaporation. Treating fibers with methanol increased β-sheets in electrospun SELP fibers [25]. Silk nanofibers were also produced from electrospinning 8–21 wt% silk fibroin in aqueous solutions. The diameter of fibers was found to increase with increased concentration from 200 to 500 nm and decrease with increased electrical conductivity from 500 to 120 nm. Treatment of fibers with ethanol improved β-sheet formation after electrospinning similar to the SELP fiber [37].

Plant-based proteins

Using plant proteins for electrospinning fiber scaffolds is of interest due to their low cost and availability. Wheat gluten protein has been electrospun into fibers with cylindrical and flat morphologies having diameters between 100 nm and 5 µm with HFIP as the solvent. The fibers were found to have a skin–core structure due to rapid evaporation of the solvent at the surface forming a skinlike barrier with the core contained 30 nm pores [50]. The fiber formation is attributed to chain entanglement of the protein molecules and reversible disulfide bonds that act similar to physical cross-links rather than covalent bonds [38].

Zein, a by-product of bioethanol production from corn, has also been successfully electrospun. Zein dissolved in alcohols produced ribbonlike fibers. Using acetic acid as the solvent, cylindrical fibers on the millimeter scale were obtained. The difference in morphology is likely due to protein–solvent interaction; however, there was no difference in secondary and tertiary structure [39]. Wang et al. further investigated the electrospinning of zein as well as hordein and gliadin, which are major proteins in cereal gains. All three proteins were electrospun from acetic acid/water solutions and produced fibers with average diameters between 184 and 192 nm. The study found that electrospinnability and the mechanical properties of fibers were affected by the protein conformation in the electrospinning solution. Hordein is unfolded and flexible in acidic solution enabling it to form intermolecular hydrogen bonds that allowed fiber formation at relatively low concentrations, 150 mg mL^{-1} hordein compared to 200 mg mL^{-1} and 300 mg mL^{-1} for gliadin and zein, respectively, but produced brittle fibers. In contrast, gliadin and zein are compact in acidic solution and when electrospun form mechanically strong fibers. When protein aggregation occurred in solution, no fibers were electrospun [41]. Addition of zein to other proteins promoted fiber formation of nonelectrospinnable proteins such as soy protein isolate [40]. Another technique that is used to encourage fiber formation of proteins has been heat-induced denaturation. For example, denatured amaranth protein isolate prior to electrospinning along with sufficient viscosity, conductivity, and surface tension induced fiber formation in HFIP [42].

3.2.2 Peptides

Peptide (<50 amino acids) nanofibers provide biocompatible support with biorecognition sites due to the presence of embedded bioactive moieties [51]. Electrospinning would enable patterning of peptides [26]. However, they are more difficult to electrospin than proteins due to low viscosity, low solubility, and lack of molecular entanglement for fiber formation. An overview of peptides that have been successfully electrospun is provided in Table 3.2.

An approach to electrospinning peptides is to leverage peptide self-assembly to provide sufficient molecular interactions to electrospin the supramolecular structures. For example, amino acids with aromatic residue interactions have been found to aid in crystal and fiber formation as well as self-assembly into tubular structures due to π–π interactions [26]. Diphenylalanine (Phe–Phe) was electrospun into fibers between 400 and 900 nm using HFIP as the solvent [26]. The minimum concentration to form uniform fibers was 13 wt%. The high content of aromatic residues induce π–π stacking and hydrogen bond formation resulting in self-assembly into tubular structures with diameters on the order of 100 nm. The self-assembled

Table 3.2: Overview of peptide electrospinning.

Peptides	Concentration	Solvent	Morphology	Diameter	Reference
Diphenylalanine (Phe–Phe)	6 wt%	HFIP	Initial fiber formation	260–750 nm	[26]
	13.2 wt%	HFIP	Fibers and beaded fibers	400–800 nm	
	> 13.2 wt%	HFIP	Beadless fibers	350–975 nm	
Monoamino-tetraphenylporphyrine ($TPP\text{-}NH_2$)	9.1 wt%	HFIP	Short needlelike fibers	840 ± 570 nm	[27]
Phe-Phe-TPP	9.1 wt%	HFIP	Length > 100 um	400 ± 290	
Phe-Phe-TPP + Phe-Phe	9.1 wt% (1:9 molar ratio)	HFIP	Quasi-infinite long with some beads	450 ± 240 nm	
Self-assembling peptides (SAPs)	20 w/v%	HFIP	Fibers and beaded fibers	80–260 nm	[52]
Coassembling peptides (CAPs)	30 w/v%	TFA	Beaded fibers	236 ± 109 nm	
Fmoc-Phe-Gly (Fmoc-FG)	18 wt%	HFIP		~300 nm	[28]
	> 18 wt%	HFIP	Bandlike composed of small needles	300–400 nm	

supramolecular structures interact analogous to molecular entanglement (i.e., “pseudoentanglements”) to facilitate fiber formation [27] (Figure 3.3(B)). Analogous to polymer electrospinning, the study found that solution concentration, viscosity, and applied voltage affected fiber morphology and diameter [26]. Additional functionality has been incorporated into the electrospun fiber by conjugating tetraphenylporphyrin (TPP) compounds to diphenylalanine. Phe–Phe–TPP and Phe–Phe at a 1:9 molar ratio produced fibers similar to Phe–Phe–TPP with an average fiber diameter to 450 ± 240 nm with some bead defects. π–π stacking of TPP provided molecular interactions to facilitate fiber formation during electrospinning [27]; however, crystallinity prevented continuous jet formation.

Recently, the effect of secondary structure on ability to form fibers via electrospinning has been investigated using a series of coassembling peptides (CAP) and self-assembling peptides (SAP). Standard SAPs are comprised of a self-assembling backbone with repeat units of LDLK, LKLK, CDLK, or LDLD to provide a tendency to self-assemble with glycine spaces and various functional motifs. The authors explored various sequences (self-assembling backbone and functional motifs), number of repeating nuts, molecular weights, and net charge. Due to the short chain length and limited solubility, the peptides could not be electrospun from aqueous solutions. CAP and SAP fibers were fabricated with diameters between 100 and 400 nm from solutions of HFIP, TFA, and TFE. Solvents were selected to promote high solubility of the peptide, rapid fiber solidification (i.e., high volatility), and reduce charge density during electrospinning (i.e., low dielectric constant). Fiber morphology was affected by peptide length; longer peptides promoted fiber formation. Maleki et al. determined that high concentrations (>20 w/v%) of peptides that provide physical entanglement of the backbone were required to facilitate fiber formation. Further, secondary structure of the peptide in the electrospinning solution was an important consideration. Using cyclodextrin (CD) to probe secondary structure, the authors conclude that higher random coil/α-helical structure conformations in the solution promoted fiber formation, whereas β-sheet aggregation interfered with fiber formation. This finding was attributed to the high elasticity of the α-helix compared to the rigid nature of the β-sheet (Figure 3.3(C)). Notably, β-sheets in the fibers increase the mechanical strength of the fibers and provide means for further biofunctionalization. Biotin also promoted the peptide self-assembly into nanofibrils or flat, tubular nanofibers, and the resulting hierarchical structures were successfully electrospun into uniform fibers [52].

Overall, electrospinning is a simple technique for fabricating protein/peptide fibers with controllable diameters for various biomedical applications. Similar to traditional polymers, reaching critical concentration of proteins and peptides to facilitate molecular entanglement and increase viscosity is necessary to achieve uniform, electrospun fibers [26, 33, 37, 43]. To achieve sufficient concentrations, solvent selection is an important consideration. Solvents that have been commonly used for electrospinning proteins and peptides are HFIP, TFE, TFA, formic acid, and acetic acid. Leveraging intermolecular interactions has been key to successful protein or peptide formation. Partial denaturing of proteins or peptides and disrupting disulfide bonds allows for spontaneous reformation of disulfide bridges between globular protein molecules. Intermolecular disulfide and hydrogen interactions provide physical molecular entanglement (Figure 3.3(A)) [26, 30, 38]. Electrospinning peptides has generally involved physical interactions of supramolecular structures, that is, self-assembled peptide nanofibrils due to π–π stacking between aromatic residues [26, 27] (Figure 3.3(B)). Fiber formation of peptides also affect the secondary structure, α-helical conformation encourages fiber formation but decreases the mechanical strength of fibers, while β-sheets promote protein aggregation which decrease spinnability but increase mechanical strength of fibers (Figure 3.3(C)). Solvents such as HFIP have been found to promote α-helices conformation [25, 28, 30, 35, 52]. Fibers can be treated post-electrospinning with alcohol to change α-helices into β-sheets to increase mechanical strength or by cross-linking to decrease solubility and increase mechanical strength of the fibers [31, 33, 37]. Based on recent advances in electrospinning proteins and peptides, the approaches that have been developed can be expanded to electrospinning other biomolecules for various potential applications.

3.2.3 Supramolecular assemblies

An approach to electrospinning solutions of small molecules has been to use self-assembling systems such as phospholipids and surfactants. The ability to form fibers has been attributed to interactions of the supramolecular assemblies.

Phospholipids

Nanofibrous, phospholipid membranes are of interest as drug-delivery platforms, biosensors, and biocompatible scaffolds. Phospholipids are amphiphilic building blocks of cell membranes. They have a charged head group and hydrocarbon tails. Due to their amphiphilic molecular structure, they spontaneously self-assemble in aqueous solution. Long and coworkers reported electrospinning of lecithin, a mixture of phospholipids and neutral lipids, using a chloroform and DMF mixture as the solvent. Using dynamic light scattering and solution rheology to probe the lecithin self-assembly in the nonaqueous environment, they observed 9 nm reverse spherical micelles (with the polar headgroup oriented toward the hydrophilic core of the micelle) at low concentration. As the concentration increases, the spherical micelles undergo a one-dimensional growth to form cylinders. As the concentration increases further, entanglement of the wormlike micelles occurs. Upon electrospinning, there is a transition from droplets to beaded fibers with increasing lecithin concentration. A transition to electrospinning uniform fibers (~3 µm) occurred at 45 wt%, above the entanglement concentration of 35 wt%. Therefore, successful electrospinning was attributed to intermolecular associations between lecithin micelles. This transition is similar to associating polymers [53, 54]. Asolectin (a mixture of phospholipid less purified than lecithin) has been electrospun from solutions of chloroform and DMF mixtures, cyclohexane, and iso-octane. A coaxial electrospinning setup with phospholipid as the core solution, and DMF as the sheath solution facilitated formation of smaller fibers [55]. Atomic force microscopy analysis of the asolectin fibers indicated that the fibers are primarily hydrophobic with elastic modulus comparable to other phospholipid-based structures [56].

Gemini surfactants

Electrospinning surfactants is of interest for achieving charged hydrophilic surfaces for controlled release, scaffolds, and coatings [57, 58]. Low-molecular-weight, self-assembling compounds such as surfactants have also been electrospun. Specifically, Gemini surfactants are two monomeric surfactants covalently bound at their head groups through a spacer. The head groups are hydrophilic; common head groups include ammoniums, imidazoliums, sulfonates, and carboxylates. Spacers may be alkyl (flexible) or aromatic (rigid). Gemini surfactants demonstrate lower critical micelle concentrations and increased solubilization compared to conventional solvents. Short spacer lengths constrain the charged head groups at closer proximity than possible with the electrostatic repulsive forces in monomeric surfactant systems. This proximity leads to higher packing parameters and lower spontaneous curvature resulting in efficient packing into cylindrical geometries. Ammonium gemini surfactant (*N,N'*-dido-decyl-*N,N,N'*, *N'*-tetramethyl-*N,N'*-ethanediammonium dibromide (12-2-12)) was electrospun from mixtures of water and methanol. In polar solvents, the hydrophilic headgroups of the ammonium gemini surfactant orient toward the solvent environment. At low concentrations, spherical micelles are observed. The spherical micelles transition into threadlike micelles with increasing concentration. With further increases in concentration, the supramolecular threadlike micelles entangle. Therefore, there are three concentration regimes: (1) dilute in which micelle size increases with concentration, (2) semidilute in which micelles grow rapidly, and (3) concentrated in which the aggregation number depends on the net charge of the endgroups. The overlap concentration occurs at the transition from the dilute to semidilute regime, which can be identified using solution rheological scaling behavior. In water, the overlap concentration was 1.5 wt% in water and methanol. This concentration corresponded with a transition from linear, entangled to branched threadlike micelles. However, aqueous systems did not produce electrospun fibers at any concentration. In methanol–water mixtures, the overlap concentration was 11 wt% and there was a transition from globular micelles to overlapped micelles. Threadlike micelles were observed at double the overlap concentration (22 wt%) that corresponded with a transition from electrospraying droplets to beaded fibers (Figure 3.4).

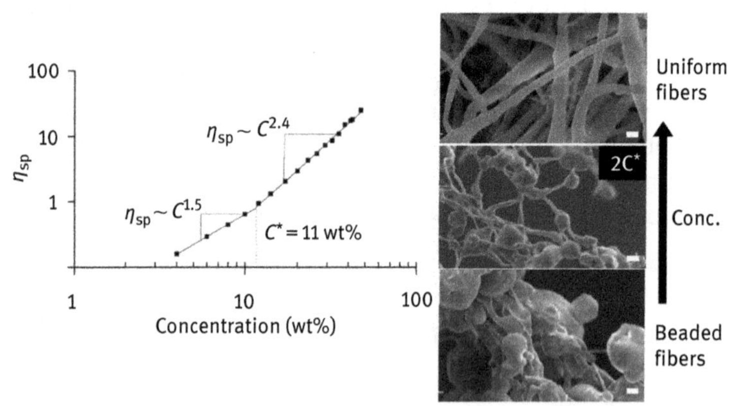

Figure 3.4: Specific viscosity as a function of Gemini surfactant concentration to overlap concentration (C*). SEM showing the transition from beaded fibers to uniform fibers as the surfactant concentration increases above the overlap concentration analogous to entangled polymer solutions (scale bar = 2 μm). Reprinted with permission from Ref. [58]. Copyright 2010 American Chemical Society.

At 30 wt% the fibers (1–6 µm) were irregular, but continuous without bead defects. Ability to form fibers was attributed to entanglement of the threadlike micelles [58]. Presence of methanol as a cosolvent was advantageous for fiber formation due to the lower surface tension, higher volatility, and higher surfactant concentration leading to higher solution conductivity [58]. Electrospinning of phosphonium gemini surfactants has also been explored. Fibers were achieved at 52 wt% of the phosphonium gemini surfactants in chloroform as well as 42–44 wt% in water–methanol mixtures. The molecular structure was important to forming fibers. Specifically, two to three methylene spacers were required to stabilize the electrospinning jet. With longer spacers, it was presumed that there were insufficient supramolecular entanglements to successfully form fibers [58, 59].

Cyclodextrin (native)

CDs are naturally occurring cyclic oligosaccharides produced from the enzymatic conversion of starch. The form of CDs depends on the number of glucose units with α-, β-, and γ-CDs having 6, 7, and 8 glucose units in each compound, respectively. These oligosaccharides have a toroid-shaped molecular structure with a subnanometer-sized cavity. Due to this cavity, CDs can form noncovalent, host–guest inclusion complexes with various molecules. Their ability to form inclusion complexes makes CD useful for a number of applications, for example, drug delivery and filtration/separation/purification [60–62].

CDs can self-assemble and aggregate due to hydrogen bonding. Fabrication of materials, for example, hydrogels and microcubes from supramolecular assemblies of CD has been well studied with solvent quality being an important consideration for materials processing. Electrospinning native CD fibers has been a challenge due to their low solubility [62]. Their solubility in water is limited due to intramolecular hydrogen bonding that is preferable to CD–water interactions. To successfully electrospin native CD fibers, solvent selection has been critical. α- and β-CD have been electrospun from highly concentrated (~150 w/v%) solutions using 10 w/v% NaOH aqueous solution. Electrospinning was comparable to polymer systems with transitions from droplets/beads to beaded fibers to uniform fibers with increasing CD concentration. Ability to form fibers was attributed to hydrogen bonding as fibers were not produced in the presence of urea which disrupts hydrogen bonding. Rheological characterization indicates that these concentrated CD solutions have significant viscoelasticity (storage modulus greater than the loss modulus for the frequency range tested) facilitating fiber formation [62].

Native β-CD has also been electrospun from ionic liquid (1-ethyl-3-methylimidazolium acetate) DMF mixtures. DMF was necessary to increase the solubility of the CD. β-CD concentrations of at least 60 wt% were required to form fibers. The resulting fibers were ~1–3 µm in diameter [60]. γ-CD was electrospun from dimethyl sulfoxide–water mixtures [63]. HFIP has also been used as a solvent to electrospin α-, β-, and γ-CDs. The CD concentration required to form uniform fibers and resulting fiber size decreased with increasing molecular size, that is, γ-CD < β-CD < α-CD. Electrospinning lower CD concentrations from HFIP facilitated the formation of smaller fibers ~750 nm. However, the differences between electrospinning α-, β-, and γ-CDs are not fully understood. The type of CD is expected to affect its self-assembly and size of the supramolecular assemblies [61]. Similarly, electrospinning hydroxypropyl-β-cyclodextrin (HPβCD) from DMF was attributed to aggregation of wormlike micelles due to hydrogen bonding [64].

3.2.4 Supramolecular polymers

Supramolecular polymers are self-organized structures of low-molecular-weight monomers assembled by reversible noncovalent interactions such as hydrogen bonding. Such materials are of interest in the development of smart functional materials whose physical properties and function could be modulated by external stimuli. An example of an electrospinnable supramolecular structure is benzo-21-crown-7 with dialkylammonium salts that form threaded linear supramolecular polymers via guest–host interactions. Strong guest–host associations led to high viscosity solutions that contributed to fiber formation. Additionally, salts with long, flexible alkyl chains favored entanglement of the linear supramolecular polymer that promoted fiber formation [65].

Z-isomers of stilbene (1,1-biindane with two ureidopyrimidinone moieties), a photoresponsive supramolecular polymer, have also been electrospun into fluorescent fibers. The supramolecular polymers of the Z-isomer formed by ring-chain polymerization have been attributed to quadruple hydrogen bonded polymers. High concentrations of the Z-isomer were electrospun, indicating formation of supramolecular polymers with high molecular weights to facilitate fiber formation. This system provides an approach to smart nanofibers with switchable properties.

3.2.5 Melt electrospinning

To avoid the use of hazardous solvents, melt electrospinning in which fibers are spun from molten polymer rather than a polymer solution has been considered. The mass throughput of melt electrospinning is ~5–10-fold compared to solution processing. Typically, melt electrospinning is performed with high-molecular-weight polymers. However, it has been demonstrated that supramolecular assemblies can be melt electrospun to form uniform, homogenous fibers [66, 67].

Trisamides are considered one of the simplest and most versatile motifs in supramolecular chemistry. Variation on the core structure and lateral substituents dictates the self-assembly and resulting phase behavior. The crystalline phases depend significantly on the molecular structure. Columnar nematic liquid crystal phases have been observed with cyclohexane-based trisamides [68]. These liquid crystal phases have facilitated fiber formation. For example, a derivative of cyclohexane-cis-cis-1,3,5-tricarboxylic acid (Figure 3.5) forms a columnar rectangular phase and a columnar nematic phase. The phases were found to affect fiber formation. Interestingly, the columnar rectangular phase could not be electrospun due to its high viscosity. The columnar nematic phase formed fibers with a large distribution of diameters. Increasing the temperature to achieve an optical isotropic melt improved fiber uniformity [66, 67]. The ability to form fibers from the optical isotropic melt was attributed to the presence of short columnar aggregate macrodipoles that interact with the applied electric field.

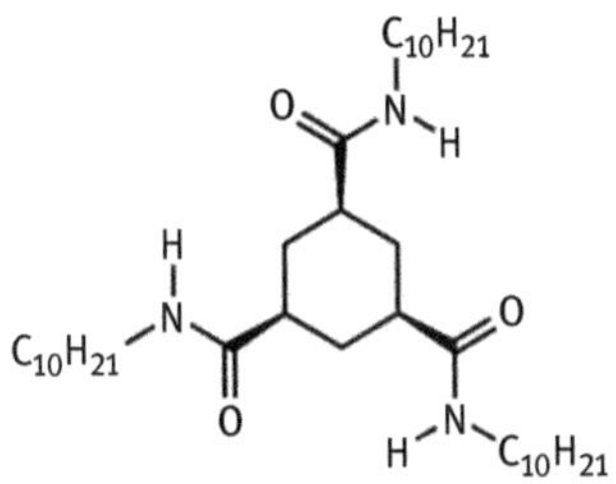

1

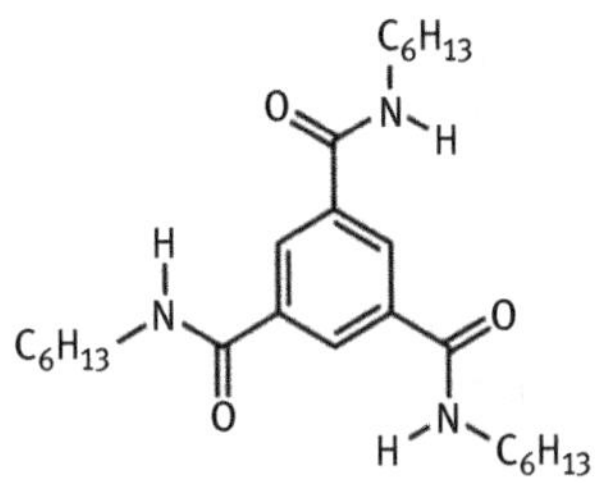

2

Figure 3.5: Structures of representative trisamides that have been melt electrospun into uniform fibers. Adapted from Ref. [67].

Building on these results, Singer et al. explored the effect of melt electrospinning process parameters using a trisamide based on trimesic acid with *n*-hexyl substituents (compound 2, Figure 3.5). They determined that with increasing temperature, fiber formation decreased and sphere formation increased due to decreasing hydrogen bond interactions. The strength and the direction of the applied electric field are also important factors in successful fiber formation. Applying a negative charge on the collector plate results in a positively charged jet. These conditions improved the homogeneity of the electrospinning process and thinner, more uniform fibers were obtained. Fibers could be formed with an applied positive voltage at lower electric field strengths [66, 67].

Further, Singer et al. explored structure–property relationships between molecular structure and electrospinnability. They systematically investigated the size and flexibility of the substituents, amide connectivity to the core, alicyclic or aromatic rings as the core, and the amount of hydrogen bonds in the molecule [66, 67]. Using benzenetrisamides as a model system, they determined that branched substituents formed spheres or beaded fibers, whereas linear substituents formed uniform, thin fibers. Comparable results were obtained with cyclohexanetrisamides. Linear substituents allowed for more packing of the melt, leading to a more stable electrospinning jet, whereas branched substituents sterically disrupted columnar packing during fiber formation [66, 67]. Molecular symmetry was also an important consideration. Symmetric C- and N-centered trisamides were electrospun into uniform fibers. However, compounds with mixed amide connectivities resulted in jet breakup. This result was attributed to reduced molecular packing in the case of asymmetric compounds [66, 67]. The stereochemistry was also a factor in forming fibers. Bisamides based on cyclohexane cores were electrospun, with only the *cis*-isomer forming uniform fibers and the *trans*-isomer forming beads. However, this result is not fully understood [66, 67].

Monoamides, bisamides with aromatic cores, and sorbitol derivatives did not form fibers because the intermolecular interactions were too weak to prevent jet breakup. Notably, molecules with extended π-conjugations, such as perylene bisimide derivatives and glass-forming tertiary benzenetrisamides, were also electrospun into fibers. These results demonstrate that amide hydrogen bonds and π–π interactions provide sufficient intermolecular interactions to facilitate fiber formation [66, 67].

Overall, symmetric benzenetrisamides and cyclohexanetrisamides with linear substituents generally formed fibers due to strong intermolecular interactions, whereas bisamides and sorbitol derivatives lacked sufficient intermolecular interactions to prevent jet breakup. For trisamides, linear substituents promoted the formation of uniform fibers compared to branched substituents. Hydrogen bonding and π-conjugated structures provide sufficient intermolecular interactions to facilitate fiber formation [66, 67].

3.3 Aggregated systems

3.3.1 Modified cyclodextrins

As mentioned earlier, CDs, a family of cyclic oligosaccharides, are rings of sugar units with hydrophobic cores and hydrophilic exteriors. Because of their unique shape, they possess the ability to entrap many hydrophobic/nonpolar substances within their cavities forming inclusion complexes [69]. These complexes can be useful for a variety of applications including ones with CD nanofibers such as antimicrobial fibers with inclusion of limonene, an antibacterial essential oil or as water filtration systems [69–72].

Electrospinning CDs and CD inclusion complexes into large nanofibers with high specific surface area is of interest. Although electrospinning native CDs is possible, it has generally required toxic solvents. Another approach has been to work with modified CDs. For example, modified CDs such as HPβCD and methyl-β-cyclodextrin have been electrospun. Unlike native CD solutions, aqueous solutions of modified CDs showed no significant viscoelasticity. The viscosity of HPβCD solutions (10–80 wt%) shows a nearly exponential rise with increasing HPβCD concentration (Figure 3.6), which may be due to the formation of aggregates. Therefore, fiber formation has been attributed due to hydrogen-bonding-induced aggregation behavior indicated by the increase in solution viscosity [70, 73, 74]

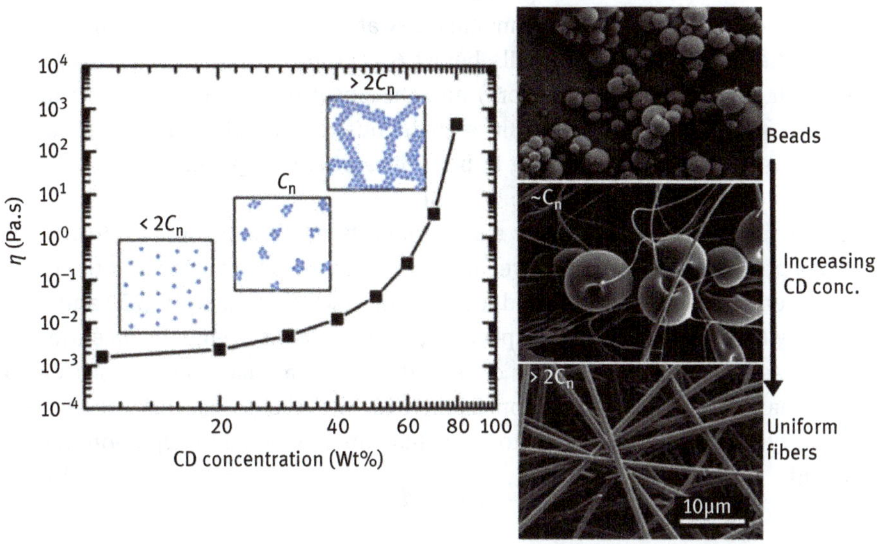

Figure 3.6: Zero-shear viscosity for HPβCD solutions. The cartoons (not drawn to scale) demonstrate how the networks may grow larger with increasing HPβCD to the point where they have enough cohesion above 65 wt% to electrospin into fibers. Reproduced from Ref. [75], with permission from the Royal Society of Chemistry.

Differential scanning calorimetry (DSC) was used to quantify free and bound water. Above 60 wt% HPβCD, a reduction to almost 0% free water was obtained, indicating a dramatic increase in bound water via hydrogen bonding between the CD molecules. Based on the amount of bound water and the size of the aggregates determined by dynamic light scattering, the aggregates contain on the order of 30 CD molecules (MW 104 g mol^{-1}). Thus, these aggregates are considerably smaller than polymer chains typically electrospun [75].

This mechanism of aggregation was thought to be analogous to depletion flocculation. As the HPβCD concentration increases, the interactions between the HPβCD molecules increase as the molecules participate in intermolecular hydrogen bonding with each other instead of with water molecules. Therefore, the intermolecular hydrogen bonding between HPβCD and solvent provided sufficient molecular cohesion to prevent jet breakup during electrospinning, thus facilitating formation of uniform fibers. The electrospinnability is disrupted by the addition of urea which disrupts hydrogen bonding. This result indicates that the CD–water network is a result of hydrogen bonding [75].

The observed increase in bound water corresponds with the increase in viscosity, and the exponential increase of viscosity as a function of weight fraction (approximate volume fraction) in Figure 3.6 is comparable to trends observed in concentrated dispersions of weakly attracted colloidal particles [52]. Further, the onset of fiber formation could be predicted based on the hypothesis that the exclusion of water leads to an electrospinnable CD–water network. According to the DSC analysis, the hydration number falls below 6 at approximately 30 wt% (C_n, the onset of an electrospinnable network) and the onset of fiber formation occurred at ~2–2.5 × C_n at 70 wt%. This result is analogous to entanglement concentration (C_e for conventional electrospinning of polymers, where the polymer concentration required for electrospinning is 2–2.5 × C_e [75].

Interestingly, adding relatively small amounts of PVA (0.1 w/w%) to HPβCD 60 w/w% produces uniform bead-free fibers despite the observation that the minimum concentration required to produce uniform fibers from neat HPβCD and neat PVA is 70 wt% and 5 wt% PVA, respectively, and the entanglement concentration for PVA was ~2.5 wt%. The ability of the addition of a small amount of PVA to facilitate nanofiber formation is consistent with the hypothesis that the electrospinnability of neat HPβCD is due to a mechanism analogous to depletion flocculation [76].

3.3.2 Colloid electrospinning

Colloid electrospinning is another variation of electrospinning that has recently been developed to create multifunctional fibers/nonwoven fabrics (Figure 3.7). In colloid electrospinning, colloidal nanoparticles (organic or inorganic) are blended with a small amount (~few wt%) of polymer solution. The nanoparticles are embedded within the fibers during electrospinning which avoids the need for post-treatment. The resulting nanofibers have multiple functionalities depending on the properties of the colloidal nanoparticles. Examples include superhydrophobic nonwoven fibers, magnetic nanofibers, and nanofibers for ultrafiltration [24, 77–81].

Fundamentally, it is important to understand the effect of processing on the hybrid nanofiber structure. Notably, the location of particles can be affected by the solvent. For example, silica particles were transferred from the core to the shell of the nanofibers as the vapor pressure of the solvent decreased [24]. Using, polystyrene nanoparticles dispersed in PVA as a model system, Yuan and Zhang [82] systematically investigated structures resulting from colloid electrospinning. Fiber formation is dictated by polymer concentration. At low PVA concentrations, aggregates of nanoparticles connected by PVA nanofibers are formed. Increasing the

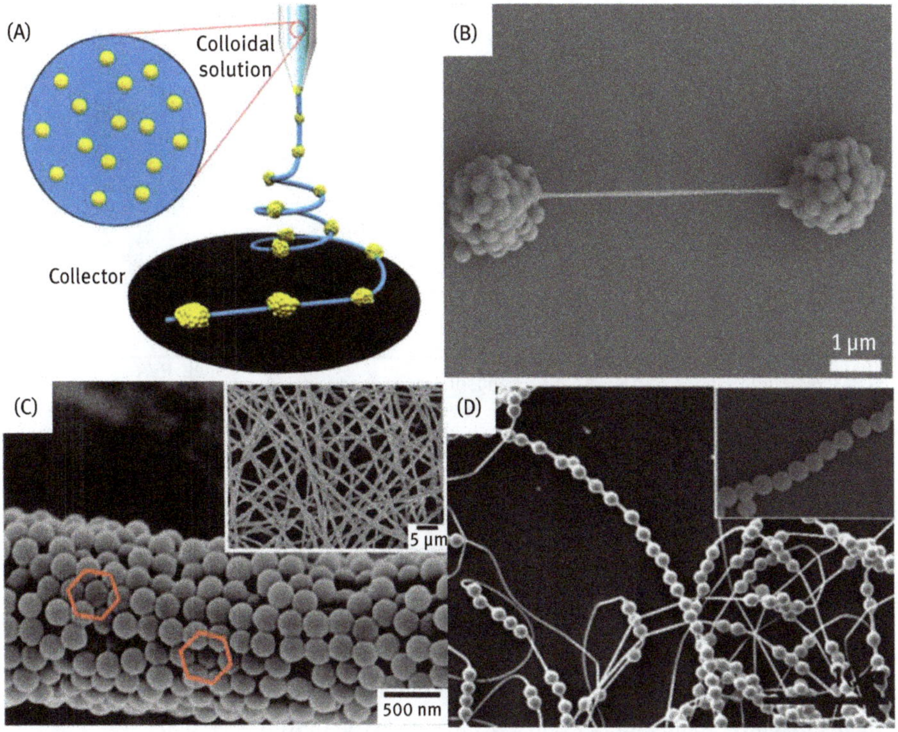

Figure 3.7: (A) Overview of colloid electrospinning. Overview of the structure of hybrid nanofibers obtained from colloid electrospinning (B) nanodumbbells (1 µm scale bar), (C) hexagonally close packed (500 nm scale bar), and (D) necklacelike (scale bar = 10 µm). (A) and (B) reproduced from Ref. [78] with permission from the Royal Society of Chemistry. (C) reprinted with permission from Ref. [82]. Copyright 2012 American Chemical Society. (D) reprinted with permission from Ref. [77]. Copyright 2006 American Chemical Society.

concentration of PVA, polystyrene nanoparticles were embedded in the PVA fibers. PVA is thought to act as a binder for the nanoparticles. At high nanoparticle loading (1:4 wt PVA:wt nanoparticles) with the addition of a nonionic surfactant (Tx-100), hexagonal packing of the nanoparticles on the surface of the fibers is observed. Due to the timescale of particle arrangement (milliseconds), the resulting structure is less ordered than the classical hexagonal packing observed in films. The close packing achieved has been used to achieve structurally colored fibers [83]. Further increasing the nanoparticle loading disrupted fiber formation. The PVA concentration and polystyrene nanoparticle loading were the key parameters that dictated hybrid nanofiber structure, for example, necklace-like, hexagonally packed (Figure 3.7) [82]. The particle packing in the fiber has been attributed to adhesion between the particles due to the presence of the polymer (Figure 3.8). Since PVA is a well-established flocculant [84], the mechanism of aggregation may be bridging flocculation.

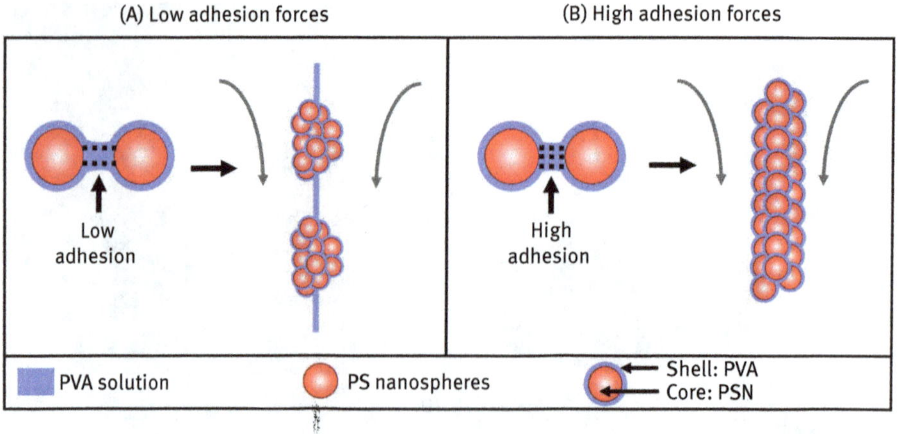

Figure 3.8: Schematic of colloid electrospinning. Aggregation state of the colloidal particle dictates structure of the hybrid nanofiber. Reprinted with permission from Ref. [82]. Copyright 2012 American Chemical Society.

Hybrid nanofibers produced with silver nanoparticles have been used in antibacterial applications [24]. Similarly, cells such as bacteria, viruses, yeast, and mammalian cells have also been electrospun from suspension [24, 81, 85, 86]. Although mammalian cells dehydrate during fiber formation [86], yeast and viruses are more robust and show some biological activity after electrospinning [24, 85].

3.4 Outlook

Although initially limited to high-molecular-weight polymers, electrospinning has proven a versatile platform to achieving fibers from biomolecules, supramolecular assemblies, and colloids. Fibers from these materials have promising applications in scaffolds, functional coatings, separations, sensing, and so on. Overall, advances in polymer-free electrospinning means to create micro- and nanofibers with novel properties with significant potential for new applications. These exciting advances are a result of interdisciplinary approaches that combine principles from supramolecular chemistry and polymer processing. Understanding the underlying mechanisms that facilitate fiber formation from polymer-free systems may lead to innovation in both materials and sustainable processing of functional electrospun nanofibers.

Acknowledgments: This work was partially supported by startup funding at Virginia Commonwealth University, Virginia Commonwealth University Presidential Research Quest Fund, and NSF (award number 1651957).

Abbreviations

API	Amaranth protein isolate
β-ME	β-Mercaptoethanol
BSA	Bovine serum albumin
$[C]$	Polymer concentration
C_e	Entanglement concentration
C_n	Onset of an electrospinnable network
CAP	Coassembling peptides
CD	Cyclodextrin
De	Deborah number
DMF	*N,N′*-Dimethylformamide
DMSO	Dimethyl sulfoxide
ELP	Elastin-like protein
HFIP	Hexafluoroisopropanol
HPβCD	Hydroxypropyl-β-cyclodextrin
η_0	Zero-shear viscosity
η_s	Viscosity of the solvent
η_{sp}	Specific viscosity
PEO	Polyethylene glycol
Phe–Phe	Diphenylalanine
PVA	Polyvinyl alcohol
SAP	Self-assembling peptides
SEM	Scanning electron microscopy
SELP	Silk-elastin-like protein
SAP	Self-assembling peptides
TFA	Trifluoroacetic acid
TFE	Trifluoroethanol
TPP	Tetraphenylporphyrin

References

[1] Agarwal, S., Wendorff, J.H., & Greiner, A. Use of electrospinning technique for biomedical applications, Polymer Guildf, 2008, 49, 5603–5621. doi:10.1016/j.polymer.2008.09.014

[2] Agarwal, S., Wendorff, J.H., & Greiner, A. Progress in the field of electrospinning for tissue engineering applications, Adv. Mater., 2009, 21, 3343–3351. doi:10.1002/adma.200803092

[3] Ahmed, F.E., Lalia, B.S., & Hashaikeh, R. A review on electrospinning for membrane fabrication : Challenges and applications, Desalination, 2014, 356, 15–30. doi:10.1016/j.desal.2014.09.033

[4] Cavaliere, S., Subianto, S., Savych, I., Jones, D.J., & Roziere, J. Electrospinning: designed architectures for energy conversion and storage devices, Energy Environ. Sci., 2011, 4, 4761–4785. doi:10.1039/b000000x

[5] Chigome, S., & Torto, N. A review of opportunities for electrospun nanofibers in analytical chemistry, Anal. Chim. Acta., 2011, 706, 25–36. doi:10.1016/j.aca.2011.08.021

[6] Ma, H., Hsiao, B.S., & Chu, B. Functionalized electrospun nanofibrous microfiltration membranes for removal of bacteria and viruses, J. Memb. Sci., 2014, 452, 446–452. doi:10.1016/j.memsci.2013.10.047

[7] Mao, X.W., Hatton, T.A., & Rutledge, G.C. A review of electrospun carbon fibers as electrode materials for energy storage, Curr. Org. Chem., 2013, 17, 1390–1401. 10.2174/1385272811317130006
[8] Yoon, K., Hsiao, B.S., & Chu, B. Functional nanofibers for environmental applications, J. Mater. Chem., 2008, 18, 5326. 10.1039/b804128h
[9] Thavasi, V., Singh, G., & Ramakrishna, S. Electrospun nanofibers in energy and environmental applications, Energy Environ. Sci., 2008, 1, 205. 10.1039/b809074m
[10] Reneker, D.H., & Chun, I. Nanometre diameter fibres of polymer, produced by electrospinning, Nanotechnology, 1996, 7, 16–223. doi:10.1088/0957-4484/7/3/009
[11] Helgeson, M.E., Grammatikos, K.N., Deitzel, J.M., & Wagner, N.J. Theory and kinematic measurements of the mechanics of stable electrospun polymer jets, Polymer , Guildf, 2008, 49, 2924–2936. doi:10.1016/j.polymer.2008.04.025
[12] Rutledge, G.C., & Fridrikh, S. V. Formation of fibers by electrospinning, Adv. Drug Deliv. Rev., 2007, 59, 1384–1391. doi:10.1016/j.addr.2007.04.020
[13] Thompson, C.J., Chase, G.G., Yarin, A.L., & Reneker, D.H. Effects of parameters on nanofiber diameter determined from electrospinning model, Polymer (Guildf), 2007, 48, 6913–6922. doi:10.1016/j.polymer.2007.09.017
[14] Teo, W.E., & Ramakrishna, S. A review on electrospinning design and nanofibre assemblies., Nanotechnology, 2006, 17, R89–R106. doi:10.1088/0957-4484/17/14/R01
[15] Li, D., & Xia, Y. Electrospinning of nanofibers: Reinventing the wheel?, Adv Mater., 2004, 16, 1151–1170. doi:10.1002/adma.200400719
[16] Andrady, A.L. Science and Technology of Polymer Nanofibers, John Wiley & Sons, Hoboken, New Jersey, 2008.
[17] Li, Z., & Wang, C. One-dimensional Nanostructures, Springer-Verlag Berlin, Heidelberg, 2013, doi:10.1007/978-3-642-36427-3
[18] Ramakrishna, S. Electrospinning process, in: An introd. to electrospinning nanofibers, world scientific, 2005.
[19] Shenoy, S.L., Bates, W.D., Frisch, H.L., & Wnek, G.E. Role of chain entanglements on fiber formation during electrospinning of polymer solutions: Good solvent, non-specific polymer-polymer interaction limit, Polymer (Guildf), 2005, 46, 3372–3384. doi:10.1016/j.polymer.2005.03.011
[20] McKee, M.G., Wilkes, G.L., Colby, R.H., & Long, T.E. Correlations of solution rheology with electrospun fiber formation of linear and branched polyesters, Macromolecules, 2004, 37, 1760–1767. doi:10.1021/ma035689h
[21] McKee, M.G., Hunley, M.T., Layman, J.M., & Long, T.E. Solution rheological behavior and electrospinning of cationic polyelectrolytes, Macromolecules, 2006, 39, 575–583. doi:10.1021/ma051786u
[22] Yu, J.H., Fridrikh, S. V., & Rutledge, G.C. The role of elasticity in the formation of electrospun fibers, Polymer (Guildf), 2006, 47, 4789–4797. doi:10.1016/j.polymer.2006.04.050
[23] Khadka, D.B., & Haynie, D.T. Protein- and peptide-based electrospun nanofibers in medical biomaterials, nanomedicine nanotechnology, Biol. Med., 2012, 8, 1242–1262. doi:10.1016/j.nano.2012.02.013
[24] Crespy, D., Friedemann, K., & Popa, A.M. Colloid-electrospinning: Fabrication of multicompartment nanofibers by the electrospinning of organic or/and inorganic dispersions and emulsions, Macromol Rapid Commun., 2012, 33, 1978–1995. 10.1002/marc.201200549
[25] Ner, Y., Stuart, J.A., Whited, G., & Sotzing, G.A. Electrospinning nanoribbons of a bioengineered silk-elastin-like protein (SELP) from water, Polymer (Guildf), 2009, 50, 5828–5836. doi:10.1016/j.polymer.2009.09.017

[26] Singh, G., Bittner, A.M., Loscher, S., Malinowski, N., & Kern, K. Electrospinning of diphenylalanine nanotubes, Adv. Mater., 2008, 20, 2332–2336. doi:10.1002/adma.200702802
[27] Nuansing, W., Georgilis, E., de Oliveira, T.V.A.G., Charalambidis, G., Eleta, A., Coutsolelos, A.G., Mitraki, A., & Bittner, A.M. Electrospinning of tetraphenylporphyrin compounds into wires, Part. Part. Syst. Charact., 2014, 31, 88–93. doi:10.1002/ppsc.201300293
[28] Nuansing, W., Frauchiger, D., Huth, F., Rebollo, A., Hillenbrand, R., & Bittner, A. M. Electrospinning of peptide and protein fibres: approaching the molecular scale, Faraday Discuss., 2013, 166, 209–221. doi:10.1039/C3FD00069A
[29] Kong, L., & Ziegler, G.R. Role of molecular entanglements in starch fiber formation by electrospinning, Biomacromolecules, 2012, 13, 2247–2253. doi:10.1021/bm300396j
[30] Dror, Y., Ziv, T., Makarov, V., Wolf, H., Admon, A., & Zussman, E. Nanofibers made of globular proteins, Biomacromolecules, 2008, 9, 2749–2754. doi:10.1021/bm8005243
[31] Rho, K.S., Jeong, L., Lee, G., Seo, B.-M., Park, Y.J., Hong, S.-D., Roh, S., Cho, J.J., Park, W.H., & Min, B.-M. Electrospinning of collagen nanofibers: Effects on the behavior of normal human keratinocytes and early-stage wound healing, Biomaterials, 2006, 27, 1452–1461. doi:10.1016/j.biomaterials.2005.08.004
[32] Zeugolis, D.I., Khew, S.T., Yew, E.S.Y., Ekaputra, A.K., Tong, Y.W., Yung, L.-Y.L., Hutmacher, D.W., Sheppard, C., & Raghunath, M. Electro-spinning of pure collagen nano-fibres – Just an expensive way to make gelatin?, Biomaterials, 2008, 29, 2293–2305. doi:10.1016/j.biomaterials.2008.02.009
[33] Songchotikunpan, P., Tattiyakul, J., & Supaphol, P. Extraction and electrospinning of gelatin from fish skin, Int. J. Biol. Macromol., 2008, 42, 247–255. doi:10.1016/j.ijbiomac.2007.11.005
[34] Benitez, P.L., Sweet, J.A., Fink, H., Chennazhi, K.P., Nair, S. V., Enejder, A., & Heilshorn, S.C. Sequence-specific crosslinking of electrospun, elastin-like protein preserves bioactivity and native-like mechanics, Adv, Healthc. Mater., 2013, 2, 114–118. doi:10.1002/adhm.201200115
[35] Stephens, J.S., Fahnestock, S.R., Farmer, R.S., Kiick, K.L., Chase, D.B., & Rabolt, J.F. Effects of electrospinning and solution casting protocols on the secondary structure of a genetically engineered dragline spider silk analogue investigated via Fourier transform Raman spectroscopy, Biomacromolecules, 2005, 6, 1405–1413. doi:10.1021/bm049296h
[36] Altman, G.H., Horan, R.L., Lu, H.H., Moreau, J., Martin, I., Richmond, J.C., & Kaplan, D.L. Silk matrix for tissue engineered anterior cruciate ligaments, Biomaterials, 2002, 23, 4131–4141. doi:10.1016/S0142-9612(02)00156-4
[37] Cao, H., Chen, X., Huang, L., & Shao, Z. Electrospinning of reconstituted silk fiber from aqueous silk fibroin solution, Mater. Sci. Eng. C., 2009, 29, 2270–2274. doi:10.1016/j.msec.2009.05.012
[38] Woerdeman, D.L., Ye, P., Shenoy, S., Parnas, R.S., Wnek, G.E., & Trofimova, O. Electrospun fibers from wheat protein: Investigation of the interplay between molecular structure and the fluid dynamics of the electrospinning process, Biomacromolecules, 2005, 6, 707–712. doi:10.1021/bm0494545
[39] Selling, G.W., Biswas, A., Patel, A., Walls, D.J., Dunlap, C., & Wei, Y. Impact of solvent on electrospinning of zein and analysis of resulting fibers, Macromol. Chem. Phys., 2007, 208, 1002–1010. doi:10.1002/macp.200700056
[40] Phiriyawirut, M., Rodchanacheewa, N., Nensiri, N., & Supaphol, P. Morphology of electrospun mats of soy protein isolate and its blend, Adv. Mater. Res., 2008, 55–57, 733–736. doi:10.4028/www.scientific.net/AMR.55-57.733
[41] Wang, Y., & Chen, L. Electrospinning of prolamin proteins in acetic acid: The effects of protein conformation and aggregation in solution, Macromol. Mater. Eng., 2012, 297, 902–913. doi:10.1002/mame.201100410

[42] Aceituno-Medina, M., Lopez-Rubio, A., Mendoza, S., & Lagaron, J.M. Development of novel ultrathin structures based in amaranth (Amaranthus hypochondriacus) protein isolate through electrospinning, Food Hydrocoll, 2013, 31, 289–298. doi:10.1016/j.foodhyd.2012.11.009
[43] Huang, L., McMillan, R.A., Apkarian, R.P., Pourdeyhimi, B., Conticello, V.P., & Chaikof, E.L. Generation of synthetic elastin-mimetic small diameter fibers and fiber networks, Macromolecules, 2000, 33, 2989–2997. doi:10.1021/ma991858f
[44] Matthews, J.A., Wnek, G.E., Simpson, D.G., & Bowlin, G.L. Electrospinning of collagen nanofibers, Biomacromolecules, 2002, 3, 232–238. doi:10.1021/bm015533u
[45] Huang, Z.M., Zhang, Y.Z., Ramakrishna, S., & Lim, C.T. Electrospinning and mechanical characterization of gelatin nanofibers, Polymer (Guildf), 2004, 45, 5361–5368. doi:10.1016/j.polymer.2004.04.005
[46] Ki, C.S., Baek, D.H., Gang, K.D., Lee, K.H., Um, I.C., & Park, Y.H. Characterization of gelatin nanofiber prepared from gelatin-formic acid solution, Polymer (Guildf), 2005, 46, 5094–5102. doi:10.1016/j.polymer.2005.04.040
[47] Song, J.H., Kim, H.E., & Kim, H.W. Production of electrospun gelatin nanofiber by water-based co-solvent approach, J. Mater. Sci. Mater. Med., 2008, 19, 95–102. doi:10.1007/s10856-007-3169-4
[48] Zhang, X., Reagan, M.R., & Kaplan, D.L. Electrospun silk biomaterial scaffolds for regenerative medicine, Adv. Drug Deliv. Rev., 2009, 61, 988–1006. doi:10.1016/j.addr.2009.07.005
[49] Vendrely, C., & Scheibel, T. Biotechnological production of spider-silk proteins enables new applications, Macromol. Biosci., 2007, 7, 401–409. doi:10.1002/mabi.200600255
[50] Koombhongse, S., Liu, W., & Reneker, D.H. Flat polymer ribbons and other shapes by electrospinning, J. Polym. Sci. Part B Polym. Phys., 2001, 39, 2598–2606. doi:10.1002/polb.0000
[51] Nune, M., Kumaraswamy, P., Krishnan, U.M., & Sethuraman, S. Self-assembling peptide nanofibrous scaffolds for tissue engineering: novel approaches and strategies for effective functional regeneration, Curr Protein Pept Sci., 2013, 14, 70–84.
[52] Maleki, M., Natalello, A., Pugliese, R., & Gelain, F. Fabrication of nanofibrous electrospun scaffolds from a heterogeneous library of co- and self-assembling peptides, Acta Biomater., 2017, 51, 268–278. doi:10.1016/j.actbio.2017.01.038
[53] McKee, M.G., Layman, J.M., Cashion, M.P., & Long, T.E. Phospholipid nonwoven electrospun membranes, Science, 2006, 311, 353–355. doi:10.1126/science.1119790
[54] Hunley, M.T., McKee, M.G., & Long, T.E. Submicron functional fibrous scaffolds based on electrospun phospholipids, J. Mater. Chem., 2007, 17, 605–608. doi:10.1039/B613474B
[55] Jørgensen, L., Qvortrup, K., & Chronakis, I.S. Phospholipid electrospun nanofibers: effect of solvents and co-axial processing on morphology and fiber diameter, RSC Adv., 2015, 5, 53644–53652. doi:10.1039/C5RA10498J
[56] Mendes, A.C., Nikogeorgos, N., Lee, S., & Chronakis, I.S. Nanomechanics of electrospun phospholipid fiber, Appl. Phys. Lett., 2015, 106, doi:10.1063/1.4922283
[57] Hemp, S.T., Hudson, A.G., Allen, M.H., Pole, S.S., Moore, R.B., & Long, T.E. Solution properties and electrospinning of phosphonium gemini surfactants, 2014, 10, 3970–3977. doi:10.1039/C4SM00271G
[58] Cashion, M.P., Li, X., Geng, Y., Hunley, M.T., & Long, T.E. Gemini surfactant electrospun membranes, Langmuir ACS J Surfaces Colloids, 2010, 26, 678–683. doi:10.1021/la902287b
[59] Hemp, S.T., Hudson, A.G., Allen, M.H., Pole, S.S., Moore, R.B., & Long, T.E. Solution properties and electrospinning of phosphonium gemini surfactants, Soft Matter, 2014, 10, 3970–3977. doi:10.1039/C4SM00271G

[60] Ahn, Y., Kang, Y., Ku, M., Yang, Y.-H., Jung, S., & Kim, H. Preparation of β-cyclodextrin fiber using electrospinning, RSC Adv., 2013, 3, 14983. doi:10.1039/c3ra42339e

[61] Kida, T., Sato, S., Yoshida, H., Teragaki, A., & Akashi, M. 1,1,1,3,3,3-Hexafluoro-2-propanol (HFIP) as a novel and effective solvent to facilely prepare cyclodextrin-assembled materials, Chem. Commun., 2014, 50, 14245–14248. doi:10.1039/C4CC06690A.

[62] Celebioglu, A., & Uyar, T. Electrospinning of nanofibers from non-polymeric systems: Electrospun nanofibers from native cyclodextrins, ,J Colloid Interface Sci., 2013, 404, 1–7. doi:10.1016/j.jcis.2013.04.034

[63] Celebioglu, A., & Uyar, T. Electrospun gamma-cyclodextrin (γ-CD) nanofibers for the entrapment of volatile organic compounds, RSC Adv., 2013, 3, 22891. doi:10.1039/c3ra44870c

[64] Zhang, W., Chen, M., Zha, B., & Diao, G. Correlation of polymer-like solution behaviors with electrospun fiber formation of hydroxypropyl-β-cyclodextrin and the adsorption study on the fiber, Phys. Chem. Chem. Phys., 2012, 14, 9729. doi:10.1039/c2cp41092c

[65] Yan, X., Zhou, M., Chen, J., Chi, X., Dong, S., Zhang, M., Ding, X., Yu, Y., Shao, S., & Huang, F. Supramolecular polymer nanofibers via electrospinning of a heteroditopic monomer, Chem Commun., 2011, 47, 7086. doi:10.1039/c1cc11790d

[66] Singer, J.C., Giesa, R., & Schmidt, H.W. Shaping self-assembling small molecules into fibres by melt electrospinning, Soft Matter, 2012, 8, 9972–9976. doi:10.1039/C2sm26469b

[67] Singer, J.C., Ringk, A., Giesa, R., & Schmidt, H.W. Melt electrospinning of small molecules, Macromol. Mater. Eng., 2015, 300, 259–276. 10.1002/mame.201400296

[68] Timme, A., Kress, R., Albuquerque, R.Q., & Schmidt, H.W. Phase behavior and mesophase structures of 1,3,5-benzene- and 1,3,5-cyclohexanetricarboxamides: Towards an understanding of the losing order at the transition into the isotropic phase, Chem. - A Eur. J., 2012, 18, 8329–8339. doi:10.1002/chem.201103216

[69] Aytac, Z., Yildiz, Z.I., Kayaci-Senirmak, F., San Keskin, N.O., Kusku, S.I., Durgun, E., Tekinay, T., & Uyar, T. Fast-dissolving, prolonged release, and antibacterial cyclodextrin /limonene-inclusion complex nanofibrous webs via polymer-free electrospinning, J. Agric. Food Chem., 2016, 64, 7325–7334. doi:10.1021/acs.jafc.6b02632

[70] Celebioglu, A., & Uyar, T. Electrospinning of polymer-free nanofibers from cyclodextrin inclusion complexes, Langmuir, 2011, 27, 6218–6226. doi:10.1021/la1050223

[71] Celebioglu, A., Yildiz, Z.I., & Uyar, T. Electrospun crosslinked poly-cyclodextrin nanofibers: Highly efficient molecular filtration thru host-guest inclusion complexation, Sci. Rep., 2017, 7, 1–11. doi:10.1038/s41598-017-07547-4

[72] Celebioglu, A., & Uyar, T. Green and one-step synthesis of gold nanoparticles incorporated into electrospun cyclodextrin nanofibers, RSC Adv., 2013, 3, 10197. 10.1039/c3ra41175c

[73] Celebioglu, A., Umu, O.C.O., Tekinay, T., & Uyar, T. Antibacterial electrospun nanofibers from triclosan/cyclodextrin inclusion complexes, colloids surfaces, B Biointerfaces, 2014, 116, 612–619. doi:10.1016/j.colsurfb.2013.10.029

[74] Vigh, T., Horváthová, T., Balogh, A., Sóti, P.L., Drávavölgyi, G., Nagy, Z.K., & Marosi, G. Polymer-free and polyvinylpyrrolidone-based electrospun solid dosage forms for drug dissolution enhancement, Eur. J. Pharm. Sci. Off. J. Eur. Fed. Pharm. Sci., 2013, 49, 595–602. doi:10.1016/j.ejps.2013.04.034

[75] Manasco, J.L., Saquing, C.D., Tang, C., & Khan, S.A. Cyclodextrin fibers via polymer-free electrospinning, RSC Adv., 2012, 2, 3778–3784. doi:10.1039/C2RA00004K

[76] Manasco, J.L., Tang, C., Burns, N. a., Saquing, C.D., & Khan, S. A. Rapidly dissolving poly(vinyl alcohol)/cyclodextrin electrospun nanofibrous membranes, RSC Adv., 2014, 4, 13274. doi:10.1039/c3ra43836h

[77] Lim, J.M., Moon, J.H., Yi, G.R., Heo, C.J., & Yang, S.M. Fabrication of one-dimensional colloidal assemblies from electrospun nanofibers, Langmuir, 2006, 22, 3445–3449. doi:10.1021/la053057d

[78] Xue, M., Li, F., Wang, Y., Cai, X., Pan, F., & Chen, J. Ultralow-limit gas detection in nano-dumbbell polymer sensor via electrospinning, Nanoscale, 2013, 5, 1803. doi:10.1039/c3nr34090b

[79] Wu, C., Yuan, W., Al-Deyab, S.S., & Zhang, K.Q. Tuning porous silica nanofibers by colloid electrospinning for dye adsorption, Appl. Surf. Sci., 2014, 313, 389–395. doi:10.1016/j.apsusc.2014.06.002

[80] Friedemann, K., Turshatov, A., Landfester, K., & Crespy, D. Characterization via two-color STED microscopy of nanostructured materials synthesized by colloid electrospinning, Langmuir, 2011, 27, 7132–7139. doi:10.1021/la104817r

[81] Zussman, E. Encapsulation of cells within electrospun fibers, Polym. Adv. Technol., 2011, 22, 366–371. doi:10.1002/pat.1812

[82] Yuan, W., & Zhang, K.Q. Structural evolution of electrospun composite fibers from the blend of polyvinyl alcohol and polymer nanoparticles, Langmuir, 2012, 28, 15418–15424. doi:10.1021/la303312q

[83] Yuan, W., Zhou, N., Shi, L., & Zhang, K.-Q. Structural coloration of colloidal fiber by photonic band gap and resonant Mie scattering, ACS Appl. Mater. Interfaces, 2015, 7, 14064–14071. doi:10.1021/acsami.5b03289

[84] Gregory, J., & Barany, S. Adsorption and flocculation by polymers and polymer mixtures, Adv. Colloid Interface Sci., 2011, 169, 1–12. doi:10.1016/j.cis.2011.06.004

[85] Canbolat, M.F., Gera, N., Tang, C., Monian, B., Rao, B.M., Pourdeyhimi, B., & Khan, S. A. Preservation of cell viability and protein conformation on immobilization within nanofibers via electrospinning functionalized yeast, ACS Appl. Mater. Interfaces, 2013, 5, 9349–9354. doi:10.1021/am4022768

[86] Canbolat, M. Fatih., Tang, C., Bernacki, S.H., Pourdeyhimi, B., & Khan, S. Mammalian cell viability in electrospun composite nanofiber structures, Macromol. Biosci., 2011, 11, 1346–1356. doi:10.1002/mabi.201100108

Valeria Annibaldi

4 Melt electrospinning

Abstract: Amongst the solvent-free electrospinning techniques Melt Electrospinning (ME) has emerged as an alternative to solution electrospinning. In ME the electric potential difference is applied to a polymer melt rather than to a polymer solution. This more environmentally friendly polymer process technology offers noteworthy advantages in terms of cost reduction and post processing as solvent extraction is not required, there is no dealing with residual toxic solvents and the utilisation of the precursors is highly efficient. Moreover, ME allows to process materials which do not dissolve easily and polymers can be used as received. This technology also enables the deposition of structures with a 3D architecture. All these aspects suggest that ME is a better candidate for scale up and applications in fields such as tissue engineering and regenerative medicine. In addition to the biomedical engineering area, the landscape of potential applications is eclectic, including textiles, filtration and sensors amongst several others.

This chapter offers an overview on the ME principles, process parameters, instrumentation required and materials processed. Challenges and limitations are also presented and applications discussed.

Keywords: Melt Electrospinning, Environmentally Friendly, Polymers, Applications, Process Parameters, Melt Electospinning Writing, Solvent Free, 3D Scaffolds, Tissue Engineering

4.1 Introduction

Among the green electrospinning techniques, melt electrospinning (ME) has received growing attention in the last decade. In ME, materials are processed as received and heat is used to melt the polymers, without the need of using solvents. The modus operandi of ME has similarities with the sister technology solution electrospinning (SE), extensively explored since the early 1990s. SE is the most popular electrospinning technique because it offers high versatility with a relatively simple setup, to the extent numerous research groups build their own home-made electrospinning kit. However, this technique also presents numerous challenges, the most significant being recovery and disposal of organic solvents, potential environmental pollution due to solvent accumulation and residue of highly toxic organic solvents in the material fabricated. These drawbacks have a negative impact on both mass production and applications such as drug delivery and tissue engineering. Moreover, SE is incapable

Valeria Annibaldi, Avectas, Maynooth, Ireland

https://doi.org/10.1515/9783110581393-004

of processing nonsoluble polymers such as polypropylene (PP) and high-density polyethylene (PE). ME is a great alternative for overcoming the problems associated with SE. Furthermore, the absence of solvents can promote high throughput and make ME a desirable technique for large-scale production. Despite the increasing number of research groups working on ME and the variety of polymers that have been melt electrospun [1], it is still a niche technology, mostly due to the complicated apparatus required. This chapter provides an overview on this emerging technology and its potential applications. It also offers an insight into the principles governing the ME process, the various setups and components and the process parameters. Polymers processed, size and morphology of the fibers are also discussed. It is impossible to cover in a chapter the intricacies of this developing technology; hence, relevant literature will be recommended for supplementary readings and more in-depth analysis.

4.2 Principles of melt electrospinning

The principles governing the ME process are analogous to SE. In both cases, a fluid (liquid solution for SE or molten polymer for ME) is pushed through a small aperture under the effect of an electric field generated by a high-potential difference applied between such aperture (called spinneret, needle, emitter or nozzle) and a collector (flat plate, dish, mandrel or drum). Charge repulsion causes the droplet originating at the tip of the spinneret to deform generating the characteristic Taylor cone (Tc) [2]. When the charge repulsion of the electrified fluid overcomes the surface tension, an electrified jet is formed that then lands on the collector. Highly conductive fluids form unstable jets that break up while insulating materials do not sustain surface charge; hence, low electrostatic drawing forces are generated on the emerging jet. For conductivity of the fluid between 10^{-6} and 10^{-8} S m^{-1}, it is possible to form a stable Tc [1]. The main difference between the two techniques is that in SE fibers are formed due to solvent evaporation, while in ME they are formed as a result of cooling of the polymer melt while travelling to or onto the collector. Moreover, in SE the jet originating from the Tc is electrically charged. The surface charge density generates jet instabilities that result in a whipping motion. This spiraling motion allows for solvent evaporation and additional stretching of the jet that causes thinning of the deposited fibers [3]. The solvent is the main contributor to the surface charge because polymers are nonconductive. The absence of solvent in ME greatly reduces the surface charge density on polymer melts; hence, the perturbations generating bending instabilities are dampened [4]. Those instabilities are responsible for whipping and stretching of the jet; therefore, their reduction or suppression produce fibers in the submicron or micron scale rather than nanofibers. Moreover, the tendency of the polymer to solidify as it flows out of the spinneret also hinders the stretching of the jet, thus resulting in bigger fibers. This is valid for nonisothermal systems, which represent the largest portion of ME apparatus.

Conversely, some research groups have modeled isothermal conditions, which means the temperature is maintained uniformly in the ME zone and in this latter case thinning of the jet has been predicted [5]. In SE it is far less challenging to fabricate nanofibers. In ME this is also now possible, thanks to the dedicated work of various research groups (see Section 4.5). For more in-depth study of the principles governing ME, Yamane and Yamamoto published an extensive body of work on the jet kinematics, characterizing the jet from its genesis to fiber deposition. They analyzed the jet initiation, jet break-up and jet trajectory. In their investigation of the jet profile, they identified the parameters that lead to bending and/or whipping of the jet and the dominant stretching forces (Figure 4.1) in addition to studying solidification points, elongation rate, and jet diameter [6].

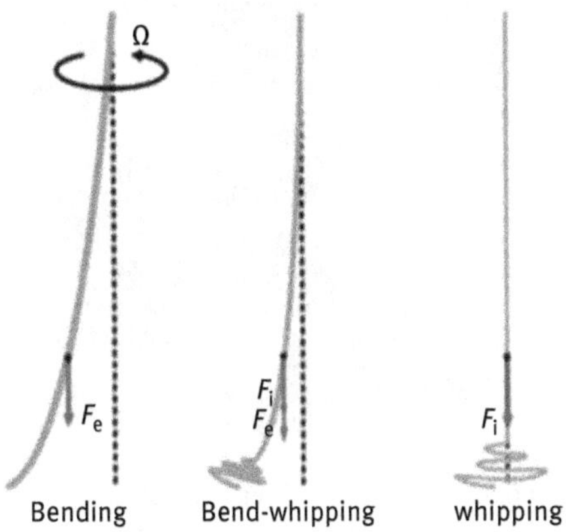

Figure 4.1: Schematics of jet profiles and the dominant stretching forces in the case of bending, bending-whipping and whipping (Fe electrical force, Fi inertial force). Image reprinted from Ref. [6] with permission from Elsevier.

ME can be thought of as an extrusion technique, with the addition of high voltage to further stretch the fibers, allowing fabrication of smaller size fibers in comparison to other printing techniques.

4.3 Instruments, components and setup

The majority of the literature available on ME is the result of work carried out with home-made systems. Various research groups utilized materials/piece of equipment they have access to and built customized instruments using off- the-shelf products as components. For instance, there is no single conformation nor a configuration that can be defined better than others available as each of them presents advantages and disadvantages. For example, using a plastic syringe as polymer reservoir is straight-forward and inexpensive, but plastic cannot withstand elevated temperatures, limiting the choice of polymers that can be melted. Stainless steel reservoirs, where polymer feed is pressure driven, can tolerate very high temperatures, but the flow rate is more difficult to measure. All types of ME configuration have five fundamental modules in common, which will be discussed in this section: heating device, polymer feeding system, high-voltage supply, spinneret and collector substrate. Another mutual characteristic and advantage over conventional SE is that ME instruments can be operated on the bench, as no toxic solvents are involved in the process. Examples of ME setups are illustrated in Figure 4.2.

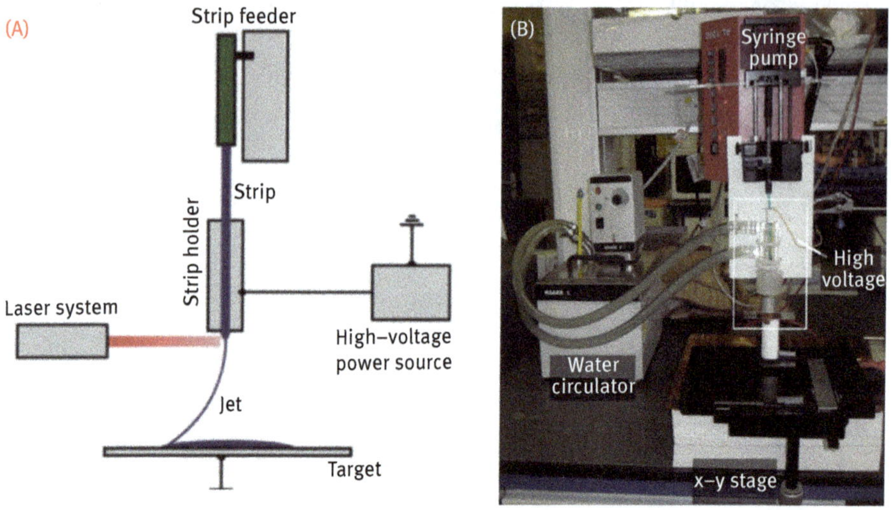

Figure 4.2: Examples of ME setup. (A) Schematic of configuration with laser heating, polymer rod feed. Image reprinted from Ref. [6] with permission from Elsevier. (B) Image of apparatus utilizing circulating water as heating source, polymer fed through syringe pump, x–y stage collector. © IOP Publishing. Reproduced with permission of Ref. [7]. All rights reserved.

4.3.1 Heating

The various strategies to generate polymer melt described in the literature are summarized in this section and advantages and disadvantages of each system are highlighted.

Electrical heating

Electrical heating is the most common heating method employed in ME [8–15]. The fact that it is easy to operate, heats up the sample quickly and provides uniform heating, makes electrical heating popular. The main disadvantage is the electrical interference with the high voltage that raises both operational and safety concerns. To solve these issues, various research groups have opted for connecting high voltage to the collector and grounding the spinneret so that the electrical heater is located away from the high voltage [13, 16, 17]. Melting of the polymer is achieved through one or more heating jackets surrounding the polymer reservoir. Ideally, the heating jackets are placed as close as possible to the spinneret to avoid a drop in temperature.

Circulating fluid

Water and oil have been used as heating fluids in ME. Water is suitable for polymers with melting intervals of 55–78 °C [18–23], while oil can be used as a circulating liquid for polymers with optimal working temperature ranging from 210 to 255 °C [24–26]. The use of circulating fluids offers a uniform temperature distribution of the polymer reservoir. Care must be paid to isolating the heating fluids from the reservoir containing the polymer for safety reasons and to avoid contamination of the polymer. The narrow range of applicable temperatures limits the application of this heating method in ME.

Hot air

Dalton et al. and Long et al. reported of hot air used to melt polymers [21, 27]. Heat guns are the most common devices to generate hot air [21, 27, 28]. The temperature of the air can reach up to 600 °C and the heat generated can be transferred to melt polymers even with relatively high melting intervals, for example, different type of PP with ME temperature of 270 and 320 °C [21]. Heat guns are inexpensive and simple to use, but do not offer accurate temperature control.

Laser heating

Ogata's team is one of the leading research groups using laser technology to generate polymer melts [6, 29–32]. This is an efficient method, especially for polymers exhibiting high melting points like poly(ethylene terephthalate) (PET), with a melting point of about 260 °C and polyalirate, which melts at about 330 °C [30]. Moreover, it does not interfere with the applied voltage. The laser light is directed to the tip of the spinneret where the polymer is mechanically driven. This method is also helpful for thermally sensitive polymers as the heating time is short. However, the application is limited by the excessive cost of the laser source and complexity of operation [29, 33].

Other

Other heating methods have been reported in various publications. They include fire heating and microwave heating [34].

4.3.2 Polymer feed

Unlike SE, in ME fluid displacement cannot occur in tubing. The polymer reservoir must be in line with, and as close as possible to, the spinneret to minimize the portion that needs to be heated and to avoid solidification of the polymer before travelling in the ME zone. Moreover, for consistent ME the polymer needs to be constantly delivered to the tip of the spinneret. Fluctuations in the rate of delivery will impact the thickness of the jet and result in variable fiber diameter. Syringe pumps are not ideal for low flow rates and require long stabilization time [7, 35]. However, they represent the easiest way to mechanically displace polymers. A syringe pump connected to a syringe or a piston within the heating chamber is typically used to control the extrusion of the polymer melt [7, 9, 15, 19, 21, 22, 25, 36]. Several groups use screw extruders to drive the polymer to the spinneret [16, 17, 37]. Air pressure has also been used [14]. Alternatively, various groups have used polymer rods [29, 30], monofilaments [38] or bundles [39] that are mechanically fed into a zone where the polymer is melted by means of a laser. In this way the time during which the polymer is heated is minimized.

It is worth to note that the flow rate units may vary depending on the polymer feed method used. Some authors express flow rate as screw turns measured in rpm [16], for syringe pump the units adopted are mL h^{-1} or μL h^{-1} [40], m min^{-1} [38] are used to measure linear feed rate for rods of polymers, while psi [14] pressure units were reported for pressure-driven systems.

4.3.3 Spinneret

Stainless steel or brass needles, with blunt ends, are the most frequently used spinneret (Figure 4.3(A)). The metal needle is connected to the polymer reservoir and is included in the heated zone. However, other geometries have been reported. For example, some authors used capillaries with triangular or crossed shape in an attempt to augment surface area to volume ratio, while maintaining mechanical properties of the polymer [41]. In another approach, PP fibers were generated from a rotating iron or aluminum disc immersed in polymer melt, with the collector placed above the disc (Figure 4.3(B)) [42]. This interesting method did not offer great control over fiber size. Needleless spinning is also possible with laser heating technology. The tip of a polymer rod or bundle is heated and is subjected to the electric field. The polymer jet is formed directly from the rod [29, 33, 43] or bundle [39]. Several publications described spray heads where multiple jets are generated from multiple Taylor cones. An example is the inner-cone nozzle described by Yang and colleagues (Figure 4.3(C)) [44] or the umbrella-like spray head described in other manuscripts [45–47].

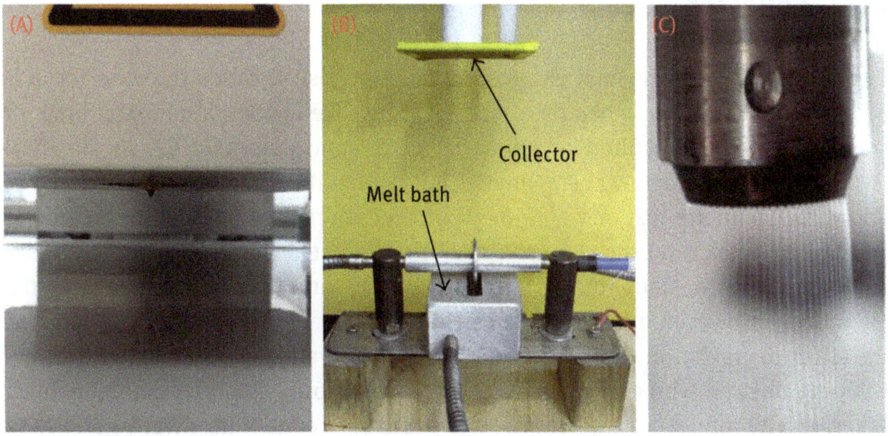

Figure 4.3: Various spinneret types: (A) brass nozzle i.d. 0.3 cm, (B) needleless disc, (C) multiple Tc needleless inner-cone nozzle. Image A is previously unpublished, image B is reproduced with permission from Ref. [42] and image C is reproduced from Ref. [44] with permission from Elsevier.

4.3.4 High voltage

In many ME instruments, the spinneret is connected to a positive source of high voltage and the collector is earthed [8, 10, 40]. Configurations whereby negative voltage is applied to the spinneret exist [21, 48] but are less common due to the risk of arcing. To prevent arcing from the spinneret to other metallic components of the instrument, the spinneret must be shielded. Some authors have used ceramic materials, which are electrically insulating but thermally conductive to enable the application of high voltage (up to 35 kV) and high temperature (up to 410 °C) minimizing risk of arcing [9–11]. Alternatively, for polymer with low melting temperatures, plastic syringes can be used as polymer reservoir to avoid arcing. The most common approach to protect the apparatus from electrical discharge is to ground the spinneret and connect the collector to the high-voltage supply [15–17, 35, 42]. This arrangement is prevalent in instruments with electrical heating modules [16, 17, 42]. The drawback of this configuration is that for a given voltage, the electrostatic drawing force acting on the polymer molten is diminished. Since the spinneret is earthed, the polymer is not charged directly, but indirectly through the electric field generated in the space between collector and spinneret. Moreover, the collector is usually a piece of metal with a much larger surface area in comparison to the spinneret; hence, the charge density acting on the polymer is lower. Therefore, if the high voltage is applied to the collector, a higher voltage is required to generate and maintain a stable Tc and polymer jet [16, 37]. To generate a strong enough electric filed, Hrynevich et al. applied a positive voltage of 7 kV to the spinneret and a smaller negative voltage of 1.5 kV to the collector [49].

4.3.5 Collector

The substrate where the melt electrospun fibers deposit is called the collector. It is a conductive material, generally a piece of metal like copper, aluminum, brass or stainless steel manufactured in various geometries and shapes [21, 37, 40, 50–52]. It is normally placed below the spinneret but configurations where the collector is above the spinneret [42] or where the spinneret is oriented parallel to the ground and the collector is on the side have also been reported [16, 21, 50]. Those studies suggest that electrostatic forces control the process and gravity is irrelevant to the ME process. However, a recent study by Hutmacher et al. [53] specifically addressed the effect of gravity on the ME process. Their study validated the possibility of developing devices with either top, side or upside down spinning configurations. When gravity influences the process, it is possible to control its effect by adjusting the system parameters accordingly, regardless of the ME orientation. Interestingly the authors found that gravity affects the Tc but not the fiber jet [53]. A limitation in the horizontal setup is represented by applications where the collector contains liquid, in that case the intrinsic nature of the work would be incompatible with a horizontal setup because liquid would spill [20, 35]. Huang et al. were among the first to collect fibers into a vessel filled with a conductive solution. They used this approach to reduce the diameter of poly(methyl methacrylate) (PMMA) fibers [35].

Other works avail of moving collectors to deposit linear fibers [37, 51]. The combination of ME with a moving stage represents an emerging field called melt electrospinning writing (MEW) and will be expanded in Section 4.6.

4.4 Process parameters and physical properties of the polymers

This section examines the key process parameters that impact the ME process and fiber dimensions. The relevant polymer properties are also considered. It is not always possible to distinguish the effect of a single variable, or to vary a parameter independently from the others as often there is a close interconnection among them. However, the objective is to provide the reader with a fundamental overview of the core parameters controlling the ME process.

4.4.1 Voltage and distance

The potential difference applied in ME is similar to the voltage applied in SE. Typical values range between 5 and 30 kV. In some configurations, the high voltage is applied to the spinneret, while in others, the emitter is earthed and the collector is connected to the high-voltage power supply. Some authors reported little effect of the voltage and distance on fiber size [54]. Others noticed an increment of fiber size at a lower voltage [55]. It is difficult to discriminate the effect of voltage and distance independently from each other as they both affect the electric field experienced by the polymer melt.

A striking difference between ME and SE is the distance from spinneret to collector. In SE the long travel between the emitter tip and the collection vessel enables solvent evaporation and stretching of the fibers. In ME the working distance is significantly reduced to avoid solidification of the fiber during flight, which would be detrimental to fiber quality and cause lack of adhesion to the collection vessel. On the other hand, depositing molten polymer is undesirable as fibers would deform, fuse with other fibers at the sites of intersection and shrink upon cooling. The optimal distance for ME is a balance between the two events and typically varies between 1 and 10 cm [34], depending on the polymer, experimental parameters and setup, although there are studies where distances up to 30 cm were assessed [34, 56]. A number of studies have evaluated distance in relation to fiber diameter from different perspectives. A comprehensive body of work by Brown et al. on polycaprolactone (PCL) [40] concluded that there is no clear trend on the effect of varying collector distance (Figure 4.4). Ogata

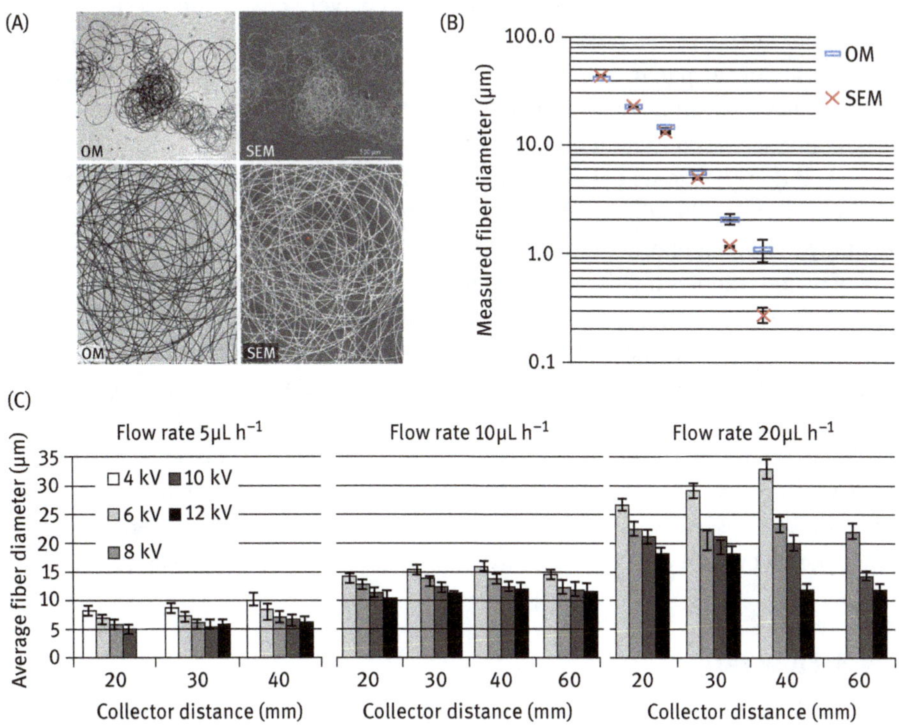

Figure 4.4: Effect of instrument parameters such as, flow rate, applied voltage and collector distance on fiber diameter, reproduced from Ref. [40] with permission from Elsevier.

et al. reported that the distance does not have a noteworthy effect on fiber diameter but in their study they varied the distance while keeping electric field constant, which means they adjusted the voltage applied [29]. Fang et al. noticed that with increasing distance the fiber diameter increases while maintaining the same applied voltage. This is probably a consequence of the reduced electric field intensity at larger distances [42]. Ko et al. [55] also reported an increment in fiber diameter at larger tip to collector distance. When the distance was increased from 5 to 10 cm at 20 kV and 80 °C, the fiber diameter changed from 49 ± 5 to 93 ± 12 µm. The authors also showed that by increasing the voltage applied, smaller fibers were deposited. This supports the concept that at larger spinneret to collector distance bigger fibers are formed because of the lower electric field intensity. As a matter of fact, even Brown et al. [40] reported increments in fiber diameter when the voltage decreases, at a given spinneret to collector distance.

4.4.2 Spinneret diameter, material and conformation

Spinneret diameters reported in the literature range from 0.15 to 2 mm. The viscosity of polymer melts is far greater than the viscosity of polymer solutions; hence, pumping fluid through a restricted capillary can be very challenging. The choice of the emitter size depends on the polymer and heating system and can be tuned to the viscosity of the polymer. In general, extrusion through tips smaller than 0.3 mm diameter is very challenging. If the diameter chosen is too small, the polymer melt electrospun is not consistent and can deposit as fragmented filaments. In some instances, it does not come out at all, causing a blockage of the spinneret. The impact of nozzle size on fiber diameter is more substantial than in SE, whereby other factors play a preponderant role on determining fiber size such as the solvent system, rate of evaporation and polymer concentration [3]. In ME increments in the emitter size will lead to bigger fiber diameter. For example, Ko et al. [55] carried out an extensive piece of work where they characterized PCL fibers generated with nozzle diameter varying from 150 to 1,700 µm and they reported fiber size progressively increasing from 12 ± 1 to 220 ± 9 µm.

Some scientists raised the question that the dependency of fiber size is not related to spinneret diameter per se but to the consequent lower mass flow rate caused by the use of a capillary with smaller inner diameter. However, other evidence points out at a direct dependency of fiber size on spinneret diameter [9, 55].

4.4.3 Flow rate

Scientists unanimously agree that out of all the process parameters, polymer flow rate has the greatest impact on fiber diameter. In general, fiber size increases with flow rate, as illustrated in Figure 4.4. In some cases, the two parameters are directly proportional as published by Hutmacher et al. for PCL blends [57] and Park and colleagues for PLA [58] and poly(lactic-*co*-glycolic acid) (PLGA) [54].

The SEM micrographs shown in Figure 4.5 represent poly(lactic acid)/poly(ethylene glycol) (PLA/PEG) samples collected at mass flow rate of 5.40, 2.54, 1.12, and 0.65 mL h^{-1}. The average diameter of the PLA fibers decreased significantly from 15 to approximately 7 μm with decreasing mass flow rate from 5.4 to 0.65 mL h^{-1}, while the other experimental parameters were unvaried [58].

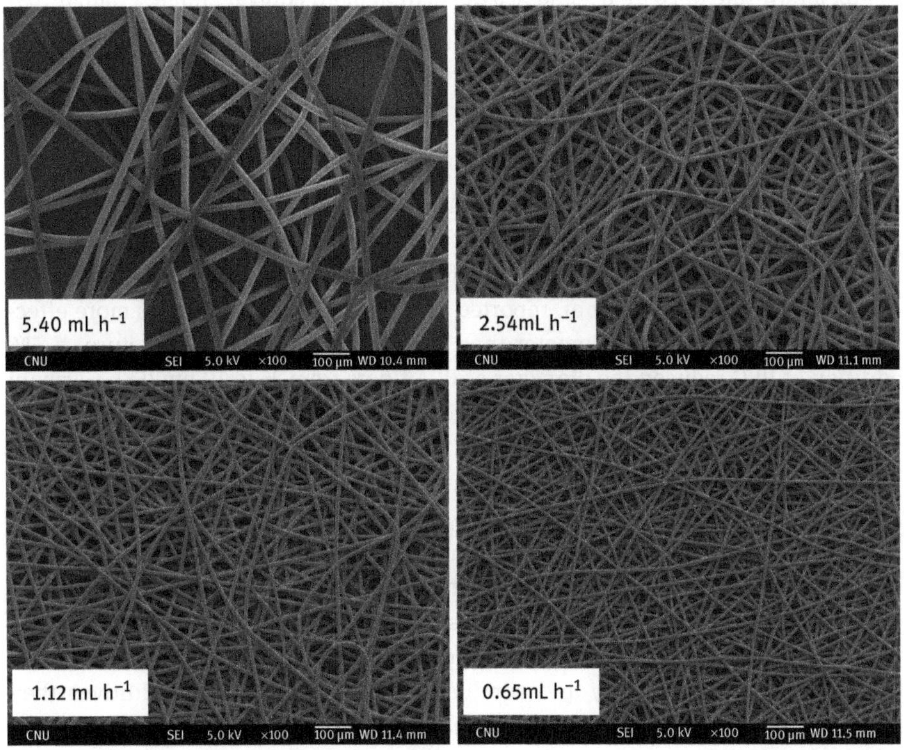

Figure 4.5: SEM images of melt electrospun PLA fiber containing 10% (w/w) PEG at different flow rates. Image reprinted with permission from Ref. [58].

Park and colleagues explained the substantial decrease in the fiber diameter, at the lower mass flow rate with the formation of a smaller Tc due to the decreased volume supplied from the polymer melt [54, 58]. Carroll and Joo explained that as the flow rate decreases, the electrical field strength increases; hence, the electrostatic drawing force thinning the fibers also increases [11]. On the other hand, it is well known from the literature that in SE fiber diameter is only moderately influenced by the flow rate [3].

The studies carried out by Hutmacher and Park also highlight the possibility to control fiber size by modulating only one experimental parameter while keeping all others constant. Nevertheless, it is important to keep in mind that in many instances with increasing flow rate, the voltage applied needs to increase to provide sufficient drawing force on the ejected melt [40].

It is also interesting to note that flow rates used in ME can be far lower in comparison to SE. For example, Hutmacher et al. [57] used flow rate of 5–20 μL h^{-1}. In ME all the material ejected is retrieved. In SE, the polymer only represents a percent of the solution and the yields are far lower after solvent evaporation. For instance, in ME, a flow rate of 20 μL h^{-1} enables deposition of an equivalent mass than a 10% solution at 200 μL h^{-1} in SE [4].

4.4.4 Additives

Dalton and colleagues demonstrated that the addition of viscosity reducing additives such as Irgatec CR 76 for PP significantly reduced fiber diameter below 1 μm [21]. PEG is also an effective plasticizer for reducing the diameter of electrospun PLA melt. In a study Yoon et al. reported that the addition of PEG was more effective in reducing the fiber diameter than increasing spinning temperature. A 5% addition of PEG reduced the diameter of PLA fibers from 25 to 10 μm [58]. Garmabi et al. also reduced the diameter of PLA fibers by blending the polymer with PEG, which acts as a viscosity-reducing agent. They also monitored thermal stability of the blends and reported that the smallest fibers were fabricated using a ratio of PLA/PEG 70:30 [52].

Two important material properties that affect the process are viscosity and conductivity. Viscosity mainly depends on the molecular weight and structure of the polymer, but it can be altered by the presence of additives. Likewise, the conductivity of the polymer melt can be improved with additives. Nayak et al. [15] investigated the effect of viscosity and conductivity on ME of PP. The additives they studied were PEG, poly(dimethyl siloxane) (PDMS) and sodium oleate (SO). Both PEG and PDMS were able to reduce the shear viscosity of the PP melt and finer fibers were fabricated. The fiber diameter was smaller and size distribution more uniform with specific amounts of SO, despite the fact that this latter additive increased the viscosity of the polymer melt. SO is a salt; therefore, it contributes to increasing the conductivity of the polymer melt, which in turn results in further stretching of the jet and smaller fibers are deposited [15]. As a matter of fact, the conductivity of PP increased from 10^{-10} to 10^{-7} S m^{-1} upon addition of 4% SO. If the concentration of the salt exceeds a certain limit, other side effects can occur, such as agglomeration of the salt molecules that have a detrimental effect on the quality of the fibers [15, 59].

4.4.5 Temperature and viscosity

The melting temperature of a polymer is closely interconnected to another physical property, the viscosity. It is important to clarify that, for a given polymer, the optimal temperature at which melt electrospinning occurs may not directly correspond to the melting temperature of the material. A polymer can be physically melted but fail at ME. It needs to reach the proper fluidity and hence viscosity, compatible with extrusion through the spinneret. It should also be capable of deforming when the high voltage is applied and form a stable Tc under the effect of the applied electric field. When working with a new polymer, if the melting interval is unknown because, for example, the material is newly synthesized or customized, it is recommended to take a small sample and study its thermal properties to gather useful information on the thermal behavior and identify the initial settings for ME. The answers to the following questions are very useful to optimize ME fabrication. How much can the temperature be increased before the polymer start decomposing? How long a certain processing temperature can be applied for before the polymer chains begin to breakdown? How long a certain mass loaded needs to be heated for to guarantee that all polymer loaded is melted? Finding the answers to these questions is important to identify the interval of confidence for manufacturing good quality and reproducible materials. Care must be taken to avoid overheating the polymer to the extent of degradation or decomposition. For example, PLA is prone to thermal degradation at temperature only 10° above its melting temperature (200 °C). Visual inspection of the polymer is a useful tool as many polymers change color, often turn yellowish upon degradation. Thermogravimetric analysis is another informative technique to study the thermal behavior of a polymer.

Once the polymer is sufficiently melted, further increase in heating temperature to decrease the viscosity does not have significant effect on the fiber diameter [29]. However, in a study on PCL by Ko et al. [55], increasing the melting temperature generated an increase in the fiber diameter. They explained that more polymer was drawn from the syringe as a result of the reduced viscosity at higher temperature [55]. However, it must be highlighted that with proper control of the mass flow rate, fiber diameter should decrease when viscosity is lowered.

Two different scenarios can occur in the formation of melt electrospun fibers, depending on the instrumentation used. Fibers can be formed under isothermal or non-isothermalconditions. With isothermal settings, the temperature is uniform in the ME zone. However, most instrumental configurations operate in nonisothermal settings characterized by a temperature drop away from the polymer tip or reservoir. Zhmayev and Jool have studied and modeled isothermal and nonisothermal conditions. The behavior they predicted in their models was supported by experimental work done with PLA [5]. In summary, at a given feed rate, under isothermal conditions an increase in the melting temperature corresponds to a decrease in the polymer viscosity that allows thinning of the polymer jet. Under nonisothermal conditions, the effect of temperature in thinning the jet is far less pronounced as the drop of temperature under the spinneret causes the airborne jet to solidify. Another elegant study by Joo and colleagues [9] is very valuable to appreciate the effect of applied temperature and viscosity. In a highly controlled manner, they created nonisothermal conditions and were able to evaluate effect of temperature on PLA fiber size. They varied the temperature from 180 to 255 °C and the related change in viscosity caused the fiber diameter to change from 4 to 2 µm. Considering an increment of 75 °C, it is a moderate variation.

Dalton et al. used a different approach to modulate viscosity of PP. They did not act on the working temperature; instead they altered the viscosity by adding a viscosity-reducing agent. With this alternative strategy, they managed to reduce viscosity from 75 to 33 Pa s, with a consequent fiber diameter reduction from 36 to 0.84 µm [21].

4.4.6 Molar mass

Scientists unanimously agree that the molar mass is the main property influencing the size of the deposited fibers. By analogy in SE fiber size is mainly determined by polymer concentration [3]. In general, it is a good practice, where possible, to characterize the molecular mass of the polymer intended for utilization, especially in the case of polymers that are lab synthesized. In the case of commercially available polymers, more than one author has experienced batch-to-batch variability or some degree of variations from the specifications reported by the manufacturers [56]. These alterations could be the result of postmanufacture degradation.

Most polymers suitable for ME are available in a wide range of molar masses. For example, PCL is commonly used from 14,000 to 90,000 g mol^{-1}. The choice of molar mass may be dictated by desired attributes of the end product such as load stress, modulus and tensile strength, degradability of the fabricated material or simply by ease of processing. Lower molar mass polymer melts result in reduction of the fiber diameter. The greater chain entanglement of a higher molecular mass polymer offers greater resistance to electrostatic stretching and pulling [16]. A manuscript by Lyons et al. [16] clearly shows the relationship between polymer mass and fiber size. They presented data for melt electrospun PP of various molar mass, 580 000 , 190,000 and 12,000 g mol^{-1} that gave fibers of 466 , 10.6 and 3.6 µm, respectively. In general, polymers with higher molar mass are more difficult to process as they are less conductive, have higher viscosity and require elevated temperatures to melt. Lower molar mass polymers may be added to the core ME material to facilitate the process. Hochleitner et al. [60] improved the ME of poly (lactide-block-ethylene glycol-block-lactide) (PLA-PEG-PLA) triblock copolymer by adding low molar mass PLA (18,000–28,000 g mol^{-1}). Using only the triblock copolymer, the fibers exhibited inhomogeneous diameter. With increasing PLA content, the melting temperature of the feed polymer decreased. Blend ratios of 5%, 10% and 20% (w/w) of PLA into the triblock copolymer were compared and 10 wt% PLA led to the most uniform fiber diameter. Other authors [9] suggested in situ thermal degradation of polymers with high molar mass as a way to facilitate the ME process and obtain smaller fibers. PLA mass was reported to change from 186,000 to 40,000 g mol^{-1} after 1 h at 200 °C. The thermal degradation occurring in the melt reservoir is mainly due to intramolecular transesterification. This can be coupled to some degree of mechanical scission that may take place when the melt is forced through the nozzle, and this effect is more prominent with spinneret of narrow inner diameter (0.16 mm) [9].

4.4.7 Tacticity

In a study published in 2004 Lyons and colleagues [16] realized that besides the molecular mass, the tacticity of the polymer has a high impact on the fiber size. Their study concluded that for a given molecular mass, the more oriented structure gave the smaller fibers. In their study they evaluated PP as this polymer is commercially available in a wide range of molecular weights and in different tacticities. They evaluated isotactic and atactic PP [16] and their results were in agreement with an independent study carried out in 2009 [17]. Kadomae and colleagues varied the tacticity of PP by blending atactic (a-PP) and isotactic PP (i-PP) in different ratios and showed that fiber size decreases with increasing ratio of isotactic PP. Although the higher tacticity is connected to a higher degree of crystallinity, the authors did not correlate fiber size to crystallinity. From measurements of the electric field, they appreciated that the electric field strength at the tip of cone of i-PP is stronger than that of a-PP, which in turn corresponds to higher elongation of the molten fiber. They believe that further research is required to investigate the reason why the higher tacticity leads to higher electric field at the tip of cone.

4.5 Polymers and fibers

As discussed in Section 4.2, a constraint of ME is that the fibers produced are typically in the micrometer diameter range, although there are various manuscripts reporting the successful fabrication of structures with a fiber diameter less than 1 μm. Submicron fibers have been attained with different configurations by various research groups [7, 9, 21, 33, 39]. The size of the smallest fibers fabricated was just below 300 nm [21, 22]. Certain laboratories also described a dual deposition, where samples were characterized by the presence of nano- and microfibers [7, 9, 21]. However, the majority of the literature published refers to micro-sized fibers for ME fabrication. Often polymers processed were designed for engineering use, thereby they are characterized by high molecular mass and it is well accepted that the polymer mass has an impact of the fiber produced (Section 4.4.6). Dalton and Hutmacher claim that the fibers of the highest quality are best made from polymers with molecular mass comprised between 40,000 and 80,000 g mol^{-1}. With the lower masses it is more plausible to aim for smaller fibers, even below the micron range. In most cases, the morphology of melt electrospun fibers consists in smooth cylindrical fibers. Some authors hypothesized that the smooth surface is due to the partial solidification that occurs as soon as the jet leaves the spinnerette and also to the fact that there is no solvent evaporation to cause unhomogeneity of the fiber surface [16, 61]. An exception was presented by Dalton et al. who showed large fibers of isotactic PP with a surface structure perpendicular to the direction of the fiber. They hypothesized that the nonsmooth surface was a consequence of extrusion instabilities, or shrinkage due to cooling [21].

Potentially, uniform fiber deposition is achievable. For example, Dalton et al. reported standard deviation ±6% for PEG/PCL fibers [7]. In addition, good size uniformity was achieved by Ogata and team with PLA [29] and Deng et al. with low density PE [62].

A number of polymers have been processed with ME over the last 15 years. A manuscript by Hutmacher et al. published in 2016 summarizes in great detail melt electrospun polymers, specifying experimental parameters and instrument setup [56]. With ME becoming a more established technology, this number is intended to grow in the future. PP is without doubt the most investigated polymer [12, 15–17, 21, 38, 42, 59]. i-PP and a-PP from different makers, with molecular mass from 5,000 to 190,000 g mol^{-1}, have been successfully melt electrospun [16]. Temperatures required to melt PP ranged from 200 °C to 410 °C, depending on the molecular mass and heating system. The smallest fibers reported were about 2 µm (tubular structures) without any additive [63] and 310 nm with additive (5% NaCl) [15]. In general, fibers below 1 µm are achievable with the aid of additives. The inclusion of Irgatec, a viscosity-reducing additive to a particular isotactic popypropylene, termed PP-15, actively reduced the polymer chain length and had a dramatic impact on the diameter of the PP-15 fibers that were reduced from 36 µm to 840 nm [21]. Common fibers size reported for PP varied between 5 and 60 µm depending on the polymer properties and process parameters. PP was the first thermoplastic polymer to be studied with ME because it does not readily dissolve in any solvent. It represents the qualities of an attractive plastic, especially for engineering applications due to its flexibility, toughness and inertness in most solvents.

Polyesters are another class of compounds rather investigated. Among them PCL is the most commonly studied. A number of aspects make PCL attractive: it is a biocompatible polymer, FDA approved [64] and it can be processed at a relatively low temperature, depending on its molecular mass. The first and second characteristics are appealing to scientists working in the biomedical field, and the third one is advantageous because PCL can be melt electrospun with unsophisticated heating apparatus. For example, medical grade PCL can be melted at 73 °C [49] or 76 °C, using a plastic syringe [53]. Another attractive polymer belonging to the class of polyesters is PLA [9, 45, 52]. This material is eco-friendly and can be biodegraded to produce carbon dioxide and water. A biodegradable material is again attractive for the biomedical industry as well as for the food and packaging industry. Zhou and coworkers accomplished the deposition of submicron fibers of 800 nm, despite describing some degree of variability [9]. PLA is susceptible to thermal degradation when exposed to elevated temperatures for a prolonged time. The authors were challenged by the issue of thermal degradation and evaluated several strategies to reduce it. They found that addition of the antioxidant 1010 was the most promising method for the alleviation of the problematic thermal degradation. Ogata et al. have produced submicron fibers from poly-L-lactic acid (PLLA) rods through the laser ME method [29]. A decrease in the fiber diameter was attained by increasing the laser output power, which induced molecular scission and broke down the polymer chains. In a subsequent study, they enhanced the electrical conductivity of PLA rods through a poly(ethylene-*co*-vinyl alcohol) (EVOH) coating or application of pie wedge fiber bundles of PLA/EVOH and obtained PLA nanofibers with average diameters of 400–500 nm [33, 39]. Nano-/microfiber composite and microfibers of PLA were fabricated by Lee and colleagues [58]. Part of the characterization work looked at the mechanical properties of the two typologies of fibers. The PLA/PLA (20/80) nano-/microfiber composite scaffold had higher tensile strength (26.8 gf mm^{-2}) and modulus (2.7 gf mm^{-2}) than those (1.5 gf mm^{-2} and 0.3 gf mm^{-2}) of the PLA microfiber scaffolds. This may be explained by the nanofibrous structure entangled with microfibers. The nanofibers in the nano- /microfiber scaffolds behave like junctions with the microfibers and thus act as physical crosslinks [58].

Another biodegradable polymer, PLGA similar to PLA, has been melt electrospun on a rotating collector, giving fibers between 15 and 28 μm diameter [24]. Thermoplastic polyurethane (TPU) has been successfully melt electrospun for wound-healing applications [25, 65]. It has been reported that skin simulants showed a tensile strength of 18 ± 2 MPa and an elongation at break of 65% ± 5% [25]. Mechanical properties of polyurethane (PU) wound dressing have been characterized by Karchin et al. and they found that the average ultimate tensile strength and elongation of melt electrospun TPU were 36.7 ± 9.9 MPa and 221% ± 95.1% [65]. These values fall between those of collagen and elastin [65], which are the main contributors to the mechanical properties of soft tissue.

Other polymers of industrial interest reported in the literature that have been successfully melt electrospun are Nylon 6, PMMA and PET [30, 38, 56]. A recent publication described ME of poly(vinylidene fluoride), a piezoelectric polymer extensively used in sensors, electroacoustic and electromechanical transducers [31].

4.6 Melt electrospinning writing

MEW involves the addition of a moving collector to an ME setup. This relatively new technology enables accurate deposition of linear fibers, layer by layer to obtain 3D architectures, in a solvent-free approach. Linear fibers can be collected when the translational speed of the collector exceeds the velocity of the jet. Below the optimal translational speed curly fibers are collected (Figure 4.6) and far beyond it, fibers are excessively stretched; the jet can break or fibers may appear with the characteristic bottleneck shape. Deng and coworkers showed that the collector speed is the main parameter in determining the diameter of the fibers [8]. The most common moving collectors are X–Y stages but rotating mandrels or cylindrical collectors have been utilized too [66]. The moving collector is generally software controlled and the deposition pattern and geometry can be designed by the operator using, for example, G code language. Another important feature of MEW is the possibility to tune pore size by altering the programmed distance between adjacent fibers. This aspect is particularly relevant for tissue engineering applications. In order to perform MEW, the temperature of the polymer melt needs to be optimized such that the viscosity is sufficiently low for the flow to be stable [51]. Temperatures above the melting point of the polymer reduce its viscosity. However, temperature should not be that high to cause degradation of the polymer. If the viscosity is reduced too much, the electrospinning jet may undergo bending instability before it deposits onto the collector and the accuracy of the writing is compromised. Moreover, at higher temperatures, fibers may not be sufficiently solidified upon hitting the collector and this will result in flat fibers that fuse with one another. To avoid bending instabilities, the distance between spinneret and collector is minimized. For example, a distance of 30 mm has proved effective for MEW of PCL at an applied voltage of 12 kV and with a collector speed of 8.3×10^{-2} m s^{-1} [51]. With similar settings (applied voltage of 10 kV, collector speed of 2.5×10^{-2} m s^{-1}), Farrugia and her tissue engineering group were able to fabricate a structure with controlled pore size of 46 µm that facilitated good infiltration of fibroblast cells [23]. Increasing the spinneret to collector distance to allow more stretching of the electrospinning jet will cause a corresponding loss in the precision of fiber deposition because the jet experiences bending instabilities. A challenge in building layer-by-layer structures using MEW is the accumulation of residual charge. Hutmacher et al. successfully fabricated structure up to 1 mm thick [51]. For Brown et al. [40] and Hochleitner et al. [67], stacking of PCL fibers on top of one another was limited to 50 layers and then accuracy of deposition was somehow lost. This was related to build up of residual charge within the fiber. The preference of the fibers to stack on top of one another is probably due to smaller distance between the spinneret and the top of the previous layer deposited as compared to the ground and that residual charges are trapped inside the fiber instead of at the surface. Hence, the molten jet preferentially deposits on top of existing

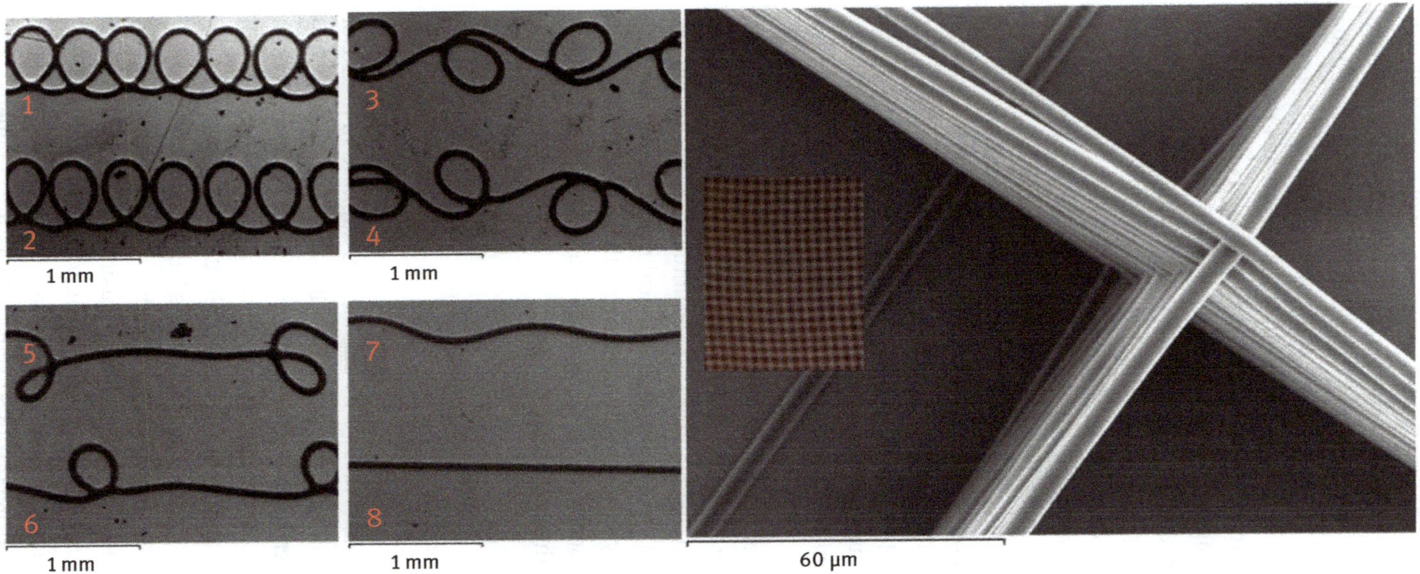

Figure 4.6: (A) Increasing translational speed from line 1 to 8, 3.3, 3.4, 3.8, 4.0, 4.7, 4.8, 6.4, 6.5 m min^{-1}; (B) stacking fibers: the image shows a scaffold of 20 layers, whole 2 × 2 cm scaffold shown in the inset, obtained with PCL 45,000 g mol^{-1}, 80 °C, 5.5 kV, 1 bar air pressure, spinneret to collector distance 6 mm, 24 G stainless steel needle (unpublished micrographs kindly provided by Spraybase®).

deposited fiber up to a limit. In a recent study Hutmacher et al. [68] successfully fabricated high-volume scaffolds with uniform morphologies and fiber diameters of thickness exceeding 7 mm. They overcame the limitation of build-up of residual charge by adjusting the *Z*-axis and increasing the voltage during the MEW process.

The optimum utility of MEW is the production of porous scaffolds on a scale that fills the current research gap between SE and many of the conventional 3D printing techniques. Despite the limitations of ME, the level of thickness and the 3D architectures reached with MEW are far more challenging to achieve with SE [66]. A great example of the versatility of MEW in pore size control and 3D structures comes from a recently published work by Hrynevich et al. [49]. The authors achieved fine control of fiber diameter by varying the pressure driving the polymer melt and collector speed, without changing applied voltage, nozzle size, temperature or collector distance. Remarkably, they were able to change fiber diameter "on the fly," which enabled deposition of multimodal scaffolds with layers of smaller pores connected to layers with larger pores [49]. The combination of small and bigger fibers and layers of different porosity in various multimodal combinations promoted penetration of human mesenchymal stromal cells and supported their proliferation, while in a different application allowed culture of human adipose-derived stem cells.

Mc Coll and colleagues further developed MEW with an adaptation to generate cylindrical scaffolds, particularly relevant for nerve regeneration. They developed a process that uses digitally controlled mandrels and an intuitive software application to print tubular frames. They verified the mathematical model used to develop the software by printing a variety of tubes and comparing them to the theoretical model. Their results enable the user to select the combination of process parameters that achieve a pore size close to their requirements [69].

4.7 Applications

To the best of the author's knowledge, there is no commercial application to date of an ME-derived product. Because of the novelty of this technology, papers published refer to exploratory and characterization work, fabrication in lab scale or work in research and development phase. Achieving comparable fiber diameter to SE will improve the performances of ME and facilitate industrialization. Moreover, the development of multiple Tc dispensing systems such as needleless spinnerets or multinozzle emitters will demonstrate that mass production can be achieved. An advantage offered by ME, which is very attractive for industrialization, is the possibility to process materials in a more environmental friendly manner. Although ME is a process that necessitates heating systems and requires financial input to keep those running, issues and cost associated with solvent recovery are eliminated. The absence of solvents is especially attractive for the life science applications, whereby gaining regulatory approval for a product might be easier given that it is not necessary to prove the removal of toxic solvents.

Nowadays, the biomedical field and in particular tissue engineering are the areas of research where ME and MEW, in particular, show higher potential for future endeavors [70]. As a matter of fact, a significant portion of the ME literature is focused toward the development of fibrous biomaterials onto which cells can be grown into tissues. The inspiring work of Dalton and Hutmacher has established over the last decade that melt electrospun scaffolds of different polymer materials (PCL, PEG/PCL) can be deposited in a controlled manner and can promote cell infiltration, adhesion and growth of functional cells [7, 20, 23, 49, 69, 71]. In their early studies they successfully optimized fiber size and porosity of the fabricated scaffolds that facilitated penetration and growth of fibroblasts. In their later works, they demonstrated infiltration and proliferation of osteoblasts, up to 86 days and human bone marrow-derived mesenchymal stromal cells up to 3 months [66]. Using a different polymer, PU, Sanders and colleagues also cultured fibroblasts for 4 weeks onto melt electrospun scaffolds [65].

Similar results are more challenging to achieve with SE because the low pore sizes associated with the random layering of the submicron fibers is a fundamental issue for solution electrospun mats, which act as a barrier to cell infiltration rather than promoting it. An interesting study published by Mikos et al. explained that fiber diameter must be at least 4 µm to generate a scaffold with a pore size of at least 20 µm in order to promote cell infiltration and growth [72].

Duan et al., using an ME system equipped with an umbrella nozzle, fabricated micro-/nanofiber meshes for mimicking native extracellular matrix structure. They investigated how the fiber alignment and thickness of fabricated scaffolds affected morphology, orientation, proliferation and differentiation of human adipose-derived mesenchymal stem cells. They also co-seeded human umbilical vein endothelial cells on PLLA-aligned fibrous scaffolds to assess the effects of vascularization on the human adipose mesenchymal cells differentiation [47]. They found that the scaffolds promoted cell migration and infiltration into the inner layers and stimulated tissue-like 3D architecture.

Woodruff and Stevens published a manuscript on the use of melt electrospun PCL for bone repair and regeneration. They developed composite scaffolds of PCL-containing Strontium-substituted bioactive glass, a material with the ability to promote osteogenesis. The particles of bioactive glass were homogeneously mixed with PCL melt and then the polymer composite was loaded into a syringe. They were able to reproducibly manufacture composite scaffolds with an interconnected porous structure, which proved to be noncytotoxic in vitro [73].

A remaining challenge in tissue engineering approaches is the in vitro vascularization of engineered constructs or tissues. Groll and coworkers developed a method that enabled the growth of 100 µm thick tissue on melt electrospun scaffolds. This was a great step forward in improving the handling of scaffolds and generation of clinically relevant sample volume [74]. In another recent study, melt electrospun fibers were generated from a triblock copolymer PLA-PEG-PLA blended with bioactive glass particles, which are known for their ability to stimulate bone tissue formation [60].

Aside from tissue engineering, another potential application in the biomedical field is drug delivery. For example, Marosi and team proposed ME for the microencapsulation and controlled release of drugs [75]. The authors were interested in creating a fast release system of a poorly soluble drug. Melt-homogenized drug–polymer mixture was fed into ME equipment. They compared ME to SE and melt extrusion and found that in acidic media, the release profile of the drug from melt electrospun fibers was as good as what evaluated for solution electrospun fibers and better than what was obtained with the grounded melt extrudate or for the free drug. The fast release was achieved because of the high surface area of samples generated with SE and ME and solubility of the polymer in acidic media, with the advantage that no solvents were used in the preparation of the melt electrospun samples [75].

Another study comparing solution electrospun fibers and melt electrospun fibers as drug carriers was conducted by Meng and Lian. They loaded an anticancer drug in PCL fibers and observed that the melt electrospun fibers presented a smooth surface and higher degree of crystallinity that slowed down the undesired burst release of the drug observed for solution electrospun fibers. Furthermore, more drug was loaded into the melt electrospun fibers as the drug has poor solubility in the solvent system used for SE and that caused uneven aggregates of drug to be loaded in the fibers formed with SE [61].

In a recent publication, the authors proposed an interesting application of MEW in cell therapy. The manuscript presents encouraging data on the use of functionalized PCL scaffolds for T-cells expansion. The highly organized 3D structures they manufactured allowed enough space for the cells to grow and exchange nutrients [76].

Other fields of application where ME relevance could expand in the future include filtration, environment and electronics. Melt electrospun fibers can have small enough size to be utilized as nanofiltration membranes [21]. PP is one of the polymers used in the filtration industry and it has been processed using ME [10, 17, 21, 46]. The reproducible fabrication of fibers in the nano-size range could be key for ME to replace SE in the industrial production of filtration material as ME is more environmental friendly and cost effective. The solvents vapor produced with SE are 5–10 times the mass of fibers deposited [77]. Rajabinejad and colleagues reported that ME of used PET plastic bottles into ultrafine fibers is an alternative technique to increase the degradation rate of the polymer and improve recycling hence reducing environmental pollution [26]. Micro-/nano-optical fibers fabricated by means of ME have been described as having applications in optical devices and circuits. They have shown to be promising for applications in optical communication, integrated optics and sensors. Among various active microstructures, polymer-based microfibers are particularly interesting in combination with cost-effective synthesis techniques that do not require hazardous chemicals [78].

ME is not restricted only to polymer. Inorganic materials such as glass may also be melt electrospun if the temperature at the nozzle can reach the desired melting point. Praeger et al. [79] fabricated nanoscale glass fibers by electrospinning molten boron oxide (B_2O_3). They were able to heat B_2O_3 to 850 °C, despite the melting point of the material being 450 °C. They needed such a high temperature to lower the viscosity to 15–26 Pa s and allow ME. With their unique setup, they successfully fabricated glass fibers of about 100 nm diameter.

4.8 Challenges and limitations

The earliest description of ME dates back to 1936 when Charles Norton from MIT was approved a patent describing the phenomenon [80]. The first scientific literature on ME was published many years later, in 1981 by Larrondo and St. John Manley [81]. A second manuscript was only published over 20 years later, in 2003 by Reneker and Rangkupan [82]. The probable cause for such a scarcity of publications is the complexity of the technology. The experimental equipment for ME is more complicated than that of conventional SE. In addition, fibers obtained are generally larger in size due to the low surface charge density of the polymer melt, in comparison to polymer in solution and to the tendency of the polymer to solidify as it leaves the spinneret, as outlined in Section 4.2. The disadvantage of the large fiber diameter produced is a common denominator across the existing solvent-free electrospinning methods, such as supercritical CO_2-assisted electrospinning, anion-curing and UV-curing electrospinning and thermocuring electrospinning [83]. Achieving fibers in the nano-size range is important to maximize surface-to-volume ratio, tunable mechanical properties, flexibility, high porosity and small pore size. The large surface-to-volume ratio is particularly important in areas such as filtration, drug delivery and energy storage. Green and colleagues [77] have modeled downstream heating in ME of polymers with the aim to demonstrate that it is possible to improve the technology. Heating in the ME zone does assist thinning of the fibers. The simulation completed with PLA validated the concept that downstream heating could be used to address the drawbacks of ME. This is attractive from an industrial point of view as it could promote scalable production of submicron fibers in an environmental friendly approach [77].

The fact that the market only offers three commercially available ME instruments [84] highlights the complexity of these devices. The challenges are in sourcing components that can withstand high temperature and materials that are thermally conductive, but electrically insulating to prevent interference of the heating element with the high voltage modulus, which could expose the operator to danger and damage the instrument. The scarcity of commercially available instruments and the difficulty of building home-made system units have a repercussion effect on the growth and spread of this technology. As a result the vast majority of research groups continue to use SE, even for applications where this technique is not the ideal choice.

Another restriction is the use of thermoplastic polymers. Polymers that are thermosensitive or thermally unstable cannot be processed by means of ME. Natural polymers, especially the ideal candidates for medical applications such as collagen, undergo thermal degradation upon heating; therefore, they cannot be processed with ME.

Scale up for mass production constitutes an additional challenge. Multinozzle spinning heads and needleless spinning heads have been developed [44–47]. They are valid evidence demonstrating that multiple Tc can be generated simultaneously. With a spray head composed of 64 nozzles, the authors were able to achieve a mass output of 12.8 mL h^{-1}, which is a noteworthy achievement considering the low throughput of a single spinneret setup.

In 2004 only one publication contained the phrase "melt electrospinning" in the title. This number increased to 45 in 2012 and to 126 at the beginning of 2018 (Web of Science citation indexing service). Novel instrument designs and component materials that will allow the processing of a wider range of commodity polymers, improvement of the reproducibility, smarter heating systems, optimization of nanoscale fabrication and diffusion of commercially available instruments with a focus on high throughput will contribute to the establishment of this exciting and innovative technology.

Abbreviations

EVOH	Poly(ethylene-*co*-vinyl alcohol)
ME	Melt electrospinning
MEW	Melt electrospinning writing
PCL	Polycaprolactone
PE	Polyethylene
PEG	Poly(ethylene glycol)
PET	Poly(ethylene terephthalate)
PDMs	Poly(dimethyl siloxane)
PLA	Poly(lactic acid)
PLGA	Poly(lactic-*co*-glycolic acid)
PLLA	Poly-L-lactic acid
PMMA	Poly(methyl methacrylate)
PP	Polypropylene
i-PP	isotactic polypropylene
a-PP	atactic polypropylene
SE	Solution electrospinning
SO	Sodium oleate
Tc	Taylor cone

References

[1] Brown, TD., Dalton, PD., & Hutmacher, DW. Melt electrospinning today: an opportune time for an emerging polymer process, Prog. Polym. Sci., [Internet] 2015, 56, 116–166. Available from: http://dx.doi.org/10.1016/j.progpolymsci.2016.01.001

[2] Kebarle, P., & Verkerk, UH. Chapter, 1, A brief overview of the mechanisms involved in electrospray mass spectrometry, 2010.

[3] Ramakrishna, S., Fujihara, K., Teo, W-E., & Lim, T-C MZ. An introduction to electrospinning and nanofibres, World Scientific., 2005, 382 p.

[4] Hutmacher, DW., & Dalton, PD. Melt electrospinning, Chem. – An Asian J., 2011, 6(1), 44–56.

[5] Zhmayev, E., Zhou, H., & Joo, YL. Modeling of non-isothermal polymer jets in melt electrospinning, J. Nonnewton. Fluid. Mech., 2008, 153(2–3), 95–108.

[6] Xu, H., Yamamoto, M., & Yamane, H. Melt electrospinning: electrodynamics and spinnability, Polymer (United Kingdom), [Internet] 2017, 132, 206–215. Available from: https://doi.org/10.1016/j.polymer.2017.11.006

[7] Dalton, PD., Joergensen, NT., Groll, J., & Moeller, M. Patterned melt electrospun substrates for tissue engineering, Biomed. Mater., 2008, 3, 3.

[8] He, FL., He, J., Deng, X. et al. Investigation of the effects of melt electrospinning parameters on the direct-writing fiber size using orthogonal design, J. Phys. D Appl. Phys., 2017, 50, 42.

[9] Zhou, H., Green, TB., & Joo, YL. The thermal effects on electrospinning of polylactic acid melts, Polymer. (Guildf), 2006, 47(21), 7497–7505.

[10] Kong, CS., Jo, KJ., Jo, NK., & Kim, HS. Effects of the spin line temperature profile and melt index of poly(propylene) on melt-electrospinning, Polym. Eng. Sci., 2009, 49, 391–396.

[11] Carroll, CP., Zhmayev, E., Kalra, V., & Joo, YL. Nanofibers from electrically driven viscoelastic jets: modeling and experiments, Korea Aust. Rheol. J., 2008, 20(3), 153–164.

[12] Zhmayev, E., Cho, D., & Lak Joo, Y. Electrohydrodynamic quenching in polymer melt electrospinning, Phys. Fluids., 2011, 23, 7.

[13] Hunley, MT., Karikari, AS., McKee, MG. et al. Taking advantage of tailored electrostatics and complementary hydrogen bonding in the design of nanostructures for biomedical applications, Macromol. Symp., 2008, 270(1), 1–7.

[14] Wei, C., & Dong, J. Direct fabrication of high-resolution three-dimensional polymeric scaffolds using electrohydrodynamic hot jet plotting, J. Micromech. Microeng., 2013, 23, 2.

[15] Nayak, R., Kyratzis, IL., Truong, YB., Padhye, R., & Arnold, L. Melt-electrospinning of polypropylene with conductive additives, J. Mater. Sci., 2012, 47(17), 6387–6396.

[16] Lyons, J., Li, C., & Ko, F. Melt-electrospinning part I: processing parameters and geometric properties, Polymer. (Guildf), 2004, 45(22), 7597–7603.

[17] Kadomae, Y., Maruyama, Y., Sugimoto, M., Taniguchi, T., & Koyama, K. Relation between tacticity and fiber diameter in melt-electrospinning of polypropylene, Fibers Polym., 2009, 10(3), 275–279.

[18] Brown, TD., Slotosch, A., Thibaudeau, L. et al. Design and fabrication of tubular scaffolds via direct writing in a melt electrospinning mode, Biointerphases., 2012, 7(1–4), 1–16.

[19] Thibaudeau, L., Taubenberger, A V., Holzapfel, BM. et al. A tissue-engineered humanized xenograft model of human breast cancer metastasis to bone, Dis. Model. Mech., [Internet] 2014, 7(2), 299–309. Available from: http://dmm.biologists.org/cgi/doi/10.1242/dmm.014076

[20] Dalton, PD., Klinkhammer, K., Salber, J., Klee, D., & Möller, M. Direct in vitro electrospinning with polymer melts, Biomacromolecules, 2006, 7(3), 686–690.

[21] Dalton, PD., Grafahrend, D., Klinkhammer, K., Klee, D., & Möller, M. Electrospinning of polymer melts: phenomenological observations, Polymer (Guildf), 2007, 48(23), 6823–6833.

[22] Dalton, PD., Calvet, JL., Mourran, A., Klee, D., & Möller, M. Melt electrospinning of poly-(ethylene glycol-block-ε-caprolactone), Biotechnol. J., 2006, 1(9), 998–1006.
[23] Farrugia, BL., Brown, TD., Upton, Z., Hutmacher, DW., Dalton, PD., & Dargaville, TR. Dermal fibroblast infiltration of poly(ε-caprolactone) scaffolds fabricated by melt electrospinning in a direct writing mode, Biofabrication., 2013, 5, 2.
[24] Kim, SJ., Jang, DH., Park, WH., & Min, BM. Fabrication and characterization of 3-dimensional PLGA nanofiber/microfiber composite scaffolds, Polymer (Guildf)., 2010, 51(6), 1320–1327.
[25] Hacker, C., Karahaliloglu, Z., Seide, G., Denkbas, EB., & Gries, T. Functionally modified, melt-electrospun thermoplastic polyurethane mats for wound-dressing applications, J. Appl. Polym. Sci., 2014, 131(8), 1–12.
[26] Rajabinejad, H., Khajavi, R., Rashidi, A., Mansouri, N., & Yazdanshenas, ME. Recycling of used bottle grade poly ethyleneterephthalate to nanofibers by melt-electrospinning method, Int. J. Environ. Res., 2009, 3(4), 663–670.
[27] Yan, X., Yu, M., Zhang, L-H. et al. A portable electrospinning apparatus based on a small solar cell and a hand generator: design, performance and application, Nanoscale, [Internet] 2016, 8(1), 209–213. Available from: http://xlink.rsc.org/?DOI=C5NR06858D
[28] Qin, C-C., Duan, X-P., Wang, L. et al. Melt electrospinning of poly(lactic acid) and polycaprolactone microfibers by using a hand-operated wimshurst generator, Nanoscale, [Internet] 2015, 7(40), 16611–16615. Available from: http://xlink.rsc.org/?DOI=C5NR05367F
[29] Ogata, N., Yamaguchi, S., Shimada, N. et al. Poly(lactide) nanofibers produced by a melt-electrospinning system with a laser melting device, J. Appl. Polym. Sci., 2007, 104(3), 1640–1645.
[30] Nobuo, Ogata., Naoki, Shimada., Shinji, Yamaguchi., & Koji Nakane, TO. Melt-electrospinning of poly(ethylene terephthalate) and polyalirate, J. Appl. Polym. Sci., 2007, 105(3), 1127–1132.
[31] Asai, H., Kikuchi, M., Shimada, N., & Nakane, K. Effect of melt and solution electrospinning on the formation and structure of poly(vinylidene fluoride) fibres, RSC Adv., [Internet] 2017, 7(29), 17593–17598. Available from: http://xlink.rsc.org/?DOI=C7RA01299C
[32] Ogata, N., Lu, G., Iwata, T., Yamaguchi, K., & Nakane, Ogihara T. Effects of ethylene content of poly(Ethylene-co-vinyl alcohol) on diameter of fibers produced by melt-electrospinning, J. Appl. Polym. Sci., 2007, 104(2), 1368–1375.
[33] Tian, S., Ogata, N., Shimada, N., Nakane, K., Ogihara, T., & Yu, M. Melt electrospinning from poly(L-lactide) rods coated with poly(ethylene-co-vinyl alcohol), J. Appl. Polym. Sci., 2009, 113(2), 1282–1288.
[34] Zhang, L-H., Duan, X-P., Yan, X. et al. Recent advances in melt electrospinning, RSC Adv., [Internet] 2016, 6(58), 53400–53414. Available from: http://xlink.rsc.org/?DOI=C6RA09558E
[35] Wang, XF., & Huang, ZM. Melt-electrospinning of PMMA, Chinese J. Polym. Sci. English Ed, 2010, 28(1), 45–53.
[36] Zhmayev, E., Cho, D., & Joo, YL. Modeling of melt electrospinning for semi-crystalline polymers, Polymer (Guildf), 2010, 51(1), 274–290.
[37] Mota, C., Puppi, D., Gazzarri, M., Bártolo, P., & Chiellini, F. Melt electrospinning writing of three-dimensional star poly(??-caprolactone) scaffolds, Polym. Int., 2013, 62(6), 893–900.
[38] Takasaki, M., Fu, H., Nakata, K., Ohkoshi, Y., & Hirai, T. Ultra-fine fibers produced by laser-electrospinning, Sen-I Gakkaishi., 2008, 64(1), 29–31.
[39] Shimada, N., Ogata, N., Nakane, K., & Ogihara, T. Spot laser melt electrospinning of a fiber bundle composed of poly(lactide)/poly(ethylene-co-vinyl alcohol) pie wedge fibers, J Appl. Polym. Sci., 2012, 125(SUPPL)2.
[40] Brown, TD., Edin, F., Detta, N., Skelton, AD., Hutmacher, DW., & Dalton, PD. Melt electrospinning of poly(ε-caprolactone) scaffolds: phenomenological observations

associated with collection and direct writing, Mater. Sci. Eng. C, [Internet] 2015, 45, 698–708. Available from: http://dx.doi.org/10.1016/j.msec.2014.07.034
[41] Park, SJ., Lee, BK., Na, MH., & Kim, DS. Melt-spun shaped fibers with enhanced surface effects: fiber fabrication, characterization and application to woven scaffolds, Acta Biomater., 2013, 9(8), 7719–7726.
[42] Fang, J., Zhang, L., Sutton, D., Wang, X., & Lin, T. Needleless melt-electrospinning of polypropylene nanofibres, J. Nanomater., 2012, 1–10
[43] Li, X., Liu, H., Wang, J., & Li, C. Preparation and characterization of poly(ε-caprolactone) nonwoven mats via melt electrospinning, Polymer (Guildf)., [Internet] 2012, 53(1), 248–253. Available from: http://dx.doi.org/10.1016/j.polymer.2011.11.008
[44] Li, X., Zhang, Y., Li, H., Chen, H., Ding, Y., & Yang, W. Effect of oriented fiber membrane fabricated via needleless melt electrospinning on water filtration efficiency, Desalination, 2014, 344, 266–273.
[45] Liu, Y., Zhao, F., Zhang, C., Zhang, J., & Yang, W. Solvent-free preparation of poly(lactic acid) fibers by melt electrospinning using an umbrella-like spray head and alleviation of the problematic thermal degradation, J. Serbian Chem. Soc., 2012, 77(8), 1071–1082.
[46] Liu, Yong., Deng, Rongjian., Hao, Mingfeng., & Hua Yan, WY. Orthogonal design study on factors effecting on fibers diameter of melt electrospinning, J. Appl. Polym. Sci., 2012, 125, 2652–2658.
[47] Wu, S., Peng, H., Li, X., Streubel, PN., Liu, Y., & Duan, B. Effect of scaffold morphology and cell co-culture on tenogenic differentiation of HADMSC on centrifugal melt electrospun poly (Llactic acid) fibrous meshes, Biofabrication, 2017, 9, 4.
[48] Singer, JC., Giesa, R., & Schmidt, H-W. Shaping self-assembling small molecules into fibres by melt electrospinning, Soft Matter, [Internet] 2012, 8(39), 9972. Available from: http://xlink.rsc.org/?DOI=c2sm26469b
[49] Hrynevich, A., Elçi, B., Haigh, JN. et al. Dimension-based design of melt electrowritten scaffolds, Small, 2018, 14(22), 1–6.
[50] Malakhov, SN., Khomenko, AY., Belousov, SI. et al. Method of manufacturing nonwovens by electrospinning from polymer melts, Fibre Chem., 2009, 41(6), 355–359.
[51] Brown, TD., Dalton, PD., & Hutmacher, DW. Direct writing by way of melt electrospinning, Adv. Mater., 2011, 23(47), 5651–5657.
[52] Nazari, T., & Garmabi, H. The effects of processing parameters on the morphology of PLA/PEG melt-electrospun fibers, Polym. Int., [Internet] 2017; ;(December). Available from: http://doi.wiley.com/10.1002/pi.5486
[53] Wunner, FM., Maartens, J., Bas, O., Gottschalk, K., De-Juan-Pardo, EM., & Hutmacher, DW. Electrospinning writing with molten poly (ε-caprolactone) from different directions – Examining the effects of gravity, Mater. Lett., [Internet] 2018, 216, 114–118. Available from: https://doi.org/10.1016/j.matlet.2017.12.079
[54] Kim, SJ., Jeong, L., Lee, SJ., Cho, D., & Par, WH. Fabrication and surface modification of melt-electrospun poly(D,L-lactic-co-glycolic acid) microfibers, Fibers Polym., 2013, 14(9), 1491–1496.
[55] Ko, J., Mohtaram, NK., Ahmed, F. et al. Fabrication of poly (ε-caprolactone) microfiber scaffolds with varying topography and mechanical properties for stem cell-based tissue engineering applications, J. Biomater. Sci. Polym. Ed., 2014, 25(1), 1–17.
[56] Brown, TD., Dalton, PD., & Hutmacher, DW. Melt electrospinning today: an opportune time for an emerging polymer process, Prog. Polym. Sci., [Internet] 2015; Available from: http://dx.doi.org/10.1016/j.progpolymsci.2016.01.001
[57] Detta, N., Brown, TD., Edin, FK. et al. Melt electrospinning of polycaprolactone and its blends with poly(ethylene glycol), Polym. Int., 2010, 59(11), 1558–1562.

[58] Lee, SJ., Yoon, Y Il., Park, KE., Lee, SJ., & Park, WH. Fabrication of microfibrous and nano- / microfibrous scaffolds : melt and hybrid electrospinning and surface modification of P. 2013, 2013(March), 2016.
[59] Liu, Y., Chen, Z., He, J., Zhao, F., Liu, Y., & Yuan, H. Effect of polar additives on melt electrospinning of non-polar polypropylene, J. Serbian Chem. Soc., 2014, 79(5), 587–596.
[60] Hochleitner, G., Kessler, M., Schmitz, M., Boccaccini, AR., Teβmar, J., & Groll, J. Melt electrospinning writing of defined scaffolds using polylactide-poly(ethylene glycol) blends with 45S5 bioactive glass particles, Mater. Lett., 2017, 205, 257–260.
[61] Lian, H., & Meng, Z. Melt electrospinning vs. solution electrospinning: a comparative study of drug-loaded poly (ε-caprolactone) fibres, Mater. Sci. Eng. C, [Internet] 2017, 74, 117–123. Available from: http://dx.doi.org/10.1016/j.msec.2017.02.024
[62] Deng, R., Liu, Y., Ding, Y., Xie, P., Luo, L., & Yang, W. Melt electrospinning of low-density polyethylene having a low-melt flow index, J. Appl. Polym. Sci., 2009, 114, 166–175.
[63] Chrzanowska, O., Struszczyk, MH., & Krucinska, I. Small diameter tubular structure design using solvent-free textile techniques, J. Appl. Polym. Sci., 2014, 131(8), 1–14.
[64] Zaiss, S., Brown, TD., Reichert, JC., & Berner, A. Poly(ε-caprolactone) scaffolds fabricated by melt electrospinning for bone tissue engineering, Materials (Basel), 2016, 9(4), 1–15.
[65] Karchin, A., Simonovsky, FI., Ratner, BD., & Sanders, JE. Melt electrospinning of biodegradable polyurethane scaffolds, Acta Biomater., [Internet] 2011, 7(9), 3277–3284. Available from: http://dx.doi.org/10.1016/j.actbio.2011.05.017
[66] Muerza-Cascante, ML., Haylock, D., Hutmacher, DW., & Dalton, PD. Melt electrospinning and its technologization in tissue engineering, Tissue Eng. Part B Rev., [Internet] 2015, 21(2), 187–202. Available from: http://online.liebertpub.com/doi/10.1089/ten.teb.2014.0347
[67] Hochleitner, G., Jüngst, T., Brown, TD. et al. Additive manufacturing of scaffolds with sub-micron filaments via melt electrospinning writing, Biofabrication, [Internet] 2015, 7(3), 35002. Available from: http://dx.doi.org/10.1088/1758-5090/7/3/035002
[68] Wunner, FM., Wille, ML., Noonan, TG. et al. Melt electrospinning writing of highly ordered large volume scaffold architectures, Adv. Mater., 2018, 1706570, 1–6.
[69] McColl, E., Groll, J., Jungst, T., & Dalton, PD. Design and fabrication of melt electrowritten tubes using intuitive software, Mater. Des., [Internet] 2018, 155, 46–58. Available from: https://doi.org/10.1016/j.matdes.2018.05.036
[70] Youssef, A., Hollister, SJ., & Dalton, PD. Additive manufacturing of polymer melts for implantable medical devices and scaffolds, Biofabrication, [Internet] 2017, 9(1), 1–29. Available from: http://dx.doi.org/10.1088/1758-5090/aa5766
[71] Hutmacher, DW., Wagner, F., Martine, LC. et al. Periosteum tissue engineering in an orthotopic in vivo platform, Biomaterials, [Internet] 2017, 121, 193–204. Available from: http://dx.doi.org/10.1016/j.biomaterials.2016.11.016
[72] Pham, QP., Sharma, U., & Mikos, AG. Electrospun poly (ε-caprolactone) microfiber and multilayer nanofiber/microfiber scaffolds: characterization of scaffolds and measurement of cellular infiltration, Biomacromolecules., 2006, 7(10), 2796–2805.
[73] Ren, J., Blackwood, KA., Doustgani, A. et al. Melt-electrospun polycaprolactone strontium-substituted bioactive glass scaffolds for bone regeneration, J. Biomed. Mater. Res. – Part A, 2014, 102(9), 3140–3153.
[74] Bertlein, S., Hikimoto, D., Hochleitner, G. et al. Development of endothelial cell networks in 3D tissues by combination of melt electrospinning writing with cell-accumulation technology, Small, 2018, 14(2), 1–7.

[75] Nagy, ZK., Balogh, A., Dravavolgyi, G., Ferguson, J., Pataki, H., Vajna, B., & Marosi, G. Solvent-free melt electrospinning for preparation of fast dissolving drug delivery system and comparison with solvent-based electrospun and melt extruded systems, J. Pharm. Sci., 2013, 102(2), 508–517.

[76] Delalat, B., Harding, F., Gundsambuu, B. et al. 3D printed lattices as an activation and expansion platform for T cell therapy, Biomaterials, [Internet] 2017, 140, 58–68. Available from: http://dx.doi.org/10.1016/j.biomaterials.2017.05.009

[77] Mayadeo, N., Morikawa, K., Naraghi, M., & Green, MJ. Modeling of downstream heating in melt electrospinning of polymers, J. Polym. Sci. Part B Polym. Phys., 2017, 55(18), 1393–1405.

[78] Yu, X., Feng, M., Zhang, R. et al. Amplified spontaneous emission from a single organic microfiber fabricated by melt electrospinning, Org. Electron Phys. Mater. Appl., 2017, 51, (September): 442–445.

[79] Praeger, M., Saleh, E., Vaughan, A., Stewart, WJ., & Loh, WH. Fabrication of nanoscale glass fibers by electrospinning, Appl. Phys. Lett., 2012, 100(6), 1–4.

[80] Norton, C.L. Method of and apparatus for producing fibrous or filamentary material. 1936;

[81] Larrondo, L., & Manley, RSJ. Electrostatic fiber spinning from polymer melts. I. Experimental observations on fiber formation and properties, J. Polym. Sci. Polym. Phys. Ed., 1981, 19(6), 909–920.

[82] Rangkupan, R., & Reneker, DH. Electrospinning process of molten polypropylene in vacuum, J. Met. Mater. Miner., [Internet] 2003, 12(2), 81–87. Available from: http://www.material.chula.ac.th/journal/V12-2/81-87 Rangkupan.pdf

[83] Zhang, B., Yan, X., He, H-W., Yu, M., Ning, X., & Long, Y-Z. Solvent-free electrospinning: opportunities and challenges, Polym. Chem., [Internet] 2017, 8(2), 333–352. Available from: http://xlink.rsc.org/?DOI=C6PY01898J

[84] Dalton, PD. Melt electrowriting with additive manufacturing principles, Curr. Opin. Biomed. Eng., [Internet] 2017, 2, 49–57. Available from: http://linkinghub.elsevier.com/retrieve/pii/S2468451117300156

Siti Machmudah, Wahyudiono, Hideki Kanda, Motonobu Goto

5 Supercritical fluid-assisted electrospinning

Abstract: Supercritical fluids have been recognized as typical green solvents and possess physical properties that are intermediate to those of gas and liquid states. These properties are continuously tunable with the change of pressures and/or temperatures; therefore, supercritical fluids can be used as a solvent in many distinct areas of applications. Supercritical fluids have also been applied to replace many environmentally hazardous organic solvents currently deployed in industry. Carbon dioxide (CO_2) is the most beneficial supercritical fluid employed in many applications. Supercritical CO_2 has ability to swell and to plasticize polymers; hence the impregnation, the foaming, and the modification of polymeric materials can be performed when supercritical CO_2 is used as a medium in polymer processes. Here, CO_2 at near- and supercritical conditions was used as a medium to generate polymer fibers from a solution of polymer through electrospinning process. This chapter shows that this process is new and offers the possibility that electrospinning process at near- and supercritical CO_2 will be a necessary and beneficial technique for fibers production with hollow interiors.

Keywords: carbon dioxide, electrospinning, fiber, hollow fiber, near-critical, polymer, poly(methyl methacrylate), polyvinylpyrrolidone, supercritical

5.1 Introduction

Supercritical fluids are known as a unique solvent and are used in a wide number of applications. Supercritical fluids based technologies and their associated technological uses are also developed rapidly. In recent years, the applications of these technologies in a wide variety of industrial applications have shown significant progress. The special association of supercritical fluid properties, including liquid-like density and solvating properties as well as gas-like diffusivity and viscosity, makes it an eminent solvent that can be used in different fields such as food, cosmetics, pharmaceutics, materials, chemistry, energy and waste treatment [1–5]. In particular, supercritical fluids have been widely used in extraction and purification processes, as well as in organic synthesis processes. In the polymer reaction

Siti Machmudah, Department of Chemical Engineering, Institut Teknologi Sepuluh Nopember, Kampus ITS Sukolilo, Surabaya, Indonesia
Wahyudiono, Hideki Kanda, Motonobu Goto, Department of Materials Process Engineering, Nagoya University, Furo-cho, Chikusa-ku, Nagoya, Japan

https://doi.org/10.1515/9783110581393-005

engineering, supercritical fluid can be used for a variety of processes, for instance, as solvents in polymer synthesis for polymeric materials modification and as swelling and plasticizing agents or for chemical recycling.

Advantages of using a supercritical fluid are summarized in Table 5.1 [6]. In this table, the advantages of using a supercritical fluid are categorized into environmental, health and safety, process, and chemical. In environmental advantages, the supercritical fluid was allowed to replace many environmentally harmful solvents. Safety and health advantages incorporate the reality that the most necessary supercritical fluids are nontoxic, noncarcinogenic, nonflammable, nonmutagenic, and stable in thermodynamic. The physical properties of supercritical fluids result in process advantages. Because of the high diffusivity, intermediate density, and low viscosity, supercritical fluids mainly can be applied conveniently in semibatch or continuous-flow processes. In terms of chemical advantages, the tunable solvating power of supercritical fluids may provide obvious advantages for chemical synthesis. In addition, the use of supercritical fluids as solvents may also have useful impacts on the chemical reactions because the process is generally cleaner and has high efficiency.

Table 5.1: Advantages of supercritical fluids.

Category	Benefits
Environment	No liquid wastes Not causing crucial toxicity in ecosystem No contribution to smog No harm to ozone layer
Safety and health	Nonflammable Noncarcinogenic Nontoxic
Process	High diffusion rates Density can be tuned No residue of solvents Low viscosity Solvent power can be tuned Easy separation from products
Chemical	High compressibility Inexpensive Local density augmentation High miscible with gases Variable dielectric constant High diffusion rates Transformed cage strength

5.2 Physical properties of supercritical fluid

It was well known that when the substance is at its critical point, the physical properties of its liquid and gas phases congregate. Consequently, only one phase emerges at the critical point, and at this point, there is no different between the gas and liquid phases due to zero vaporization heat at and beyond this point. However, the term "supercritical fluid" generally is defined as any substance located over its critical temperature and pressure. Figure 5.1 depicts the standard phase diagram of a pure substance. The figure illustrates the temperature and pressure regions wherein the substance exists as solid, liquid, and gas. According to their aggregation states, their regions can be clearly distinguished in this diagram. The phases are restricted by the individual phase – transition lines, at which the phases are in equilibrium [7]. These phase boundaries will converge at the triple point where they coexist and are in equilibrium. Shifting up the curve of gas–liquid coexistence, the gas becomes denser with increasing pressure, and the liquid becomes less dense due to the thermal extension. Their densities become identic and the difference between the gas and the liquid vanishes at the critical point. The supercritical fluid region could be found as apart from the three standard entirety states. Table 5.2 lists the properties of some typical supercritical fluid solvents; most of these have been employed with polymers [8].

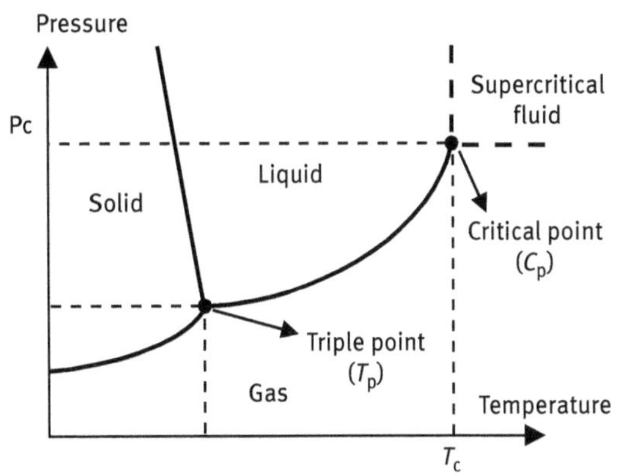

Figure 5.1: General phase diagram of a pure substance.

A supercritical fluid shows a gas-like characteristic as it occupies a chamber and it presumes the form of the chamber. The mobility of supercritical fluid molecules is slightly similar to that of molecules of gas. Besides this, it gives a liquid-like density; therefore, the supercritical fluid exhibits a property of liquid in case of dissolving impact. Table 5.3 lists the general physical properties of a supercritical fluid. The properties are between properties of gas and liquid [9]. Density, viscosity, and diffusivity are the most important supercritical fluid properties. Adjusting the temperature and pressure above the critical points changes these properties

Table 5.2: Critical points of some common supercritical solvents.

Substance	Critical temperature (°C)	Critical pressure (MPa)	Substance	Critical temperature (°C)	Critical pressure (MPa)
Ethylene	9.3	5.06	Ethylene oxide	195.6	7.19
Fluoroform	26.0	4.82	*n*-Pentane	196.6	3.37
Chlorotrifluoromethane	28.9	3.92	Trichlorofluoromethane	198.1	4.41
Carbon dioxide	31.0	7.38	Trifluoroethanol	225.4	4.83
Ethane	32.2	4.88	Acetone	235	4.80
Nitrous oxide	36.5	7.26	Isopropanol	235.2	4.76
Propylene	91.9	4.62	Methanol	240	7.95
Chlorodifluoromethane	96	4.91	Ethanol	243	6.38
Propane	96.7	4.25	Ethyl acetate	257	3.88
1,1,1,2-Tetrafluoroethane	101	4.06	Chloroform	263	5.47
Dimethyl ether	127	5.33	Isooctane	270.7	2.57
Ammonia	132.5	11.28	Cyclohexane	280.3	4.07
1-Cchloro-1, 1-difluoroethane	137	4.05	Benzene	318.6	4.90
n-Butane	152	3.80	Toluene	320.8	4.10
Dichlorofluoromethane	178.5	5.17	p-Xylene	343.1	3.52
Dichlorotrifluoroethane	183	3.68	Water	374.2	22.05

Table 5.3: The physical properties of gas, liquid, and supercritical fluid.

Physical property	Gas	Supercritical fluid	Liquid
Density, ρ (kg m^{-3})	1	100 – 1,000	1,000
Diffusivity, D (m^2 s^{-1})	10^{-6}–10^{-5}	10^{-8}–10^{-7}	10^{-9}
Viscosity, μ (mPa s)	0.01	0.05–0.1	0.5–1

and improves the capability of the supercritical fluid to penetrate and separate the molecules under study from raw materials. As a result, the extraction rates and phase separation can be significantly more rapid than that in common separation processes [10–13]. When the supercritical fluids were introduced in polymers matrix, their mixtures have many characteristics in general with blends of common incompressible solvents and polymers. Hence, a diversity of processes, including polymerization, dissolution and precipitation, swelling, plasticization, impregnation, and extraction, may be performed on polymers in a supercritical fluid medium [14–18]. At a constant temperature, the supercritical fluid density rises with an increase in pressure; conversely, at a constant pressure, it decreases with an increase in temperature. This physical property has significant influence on the dissolving effect of a supercritical fluid; hence, it is an essential parameter for a supercritical fluid when it applied as a solvent medium. Although the viscosity of supercritical fluid is close enough to a viscosity of gas, its diffusivity is intermediate between the two states (liquid and gas). Generally, diffusivity is opposite to pressure and is proportional to temperature. Elevating pressure influences the molecules of supercritical fluid to be closer to each other and results in reducing diffusivity in the substance. The higher diffusivity of a supercritical fluid helps in rapidly proceeding with several processes such as extraction and fractionation. It diffuses more rapidly into a solid matrix and even penetrates the solid matrix. These three main physical properties are associated with each other; the shift in pressure and temperature may influence all of them in diverse associations. In this chapter, we discuss the potential applications of supercritical CO_2 as an alternative solvent medium for the electrospun fibers production from polymer feed solution by using electrospinning process.

5.2.1 Supercritical carbon dioxide

Supercritical fluids have been utilized for isolation of natural products since the end of the 1970s, but for a long time the applications depended only on few products. Such supercritical fluids have the advantage of low viscosity and hence are able to penetrate well solid matrix and result in good mass transfer. As a result, supercritical fluids can be applied to extract active ingredients or analytes from various plants and microbial samples, can be especially useful in the extraction of unknown natural products, and can be used as media for polymerization, dissolution and precipitation, swelling, plasticization, and impregnation in polymeric materials modification. Now, industries are more fascinated toward supercritical technologies and the development of equipment and processes is starting to pave a way [3, 14–22]. This interest is also indicated in a high number of scientific articles dealing with supercritical fluid extraction and polymeric materials modification published in recent years. In addition, since the early 1990s and 1980s, there are a number of patents on the industrial applications of supercritical fluid extraction and polymeric materials modification [3, 19, 23–25].

Many solvents can be used as the solvents' media for supercritical fluid polymeric materials modification; particularly interesting are the substances with low molecular weight having a critical temperature close to ambient temperature (10 °C ≤ T_c ≤ 40 °C) and a critical pressure that is not too high (4.0 ≤ P_c ≤ 6.0 MPa). Because of this, CO_2 is the main supercritical fluid solvent used in polymeric materials modification technique under supercritical conditions. The critical point of CO_2 is at 31.1 °C (T_c) and 7.38 MPa (P_c) [26]. The basic properties and advantages of CO_2 in supercritical polymeric materials modification include that it is inert, relatively nontoxic, nonflammable, inexpensive, easily available, odorless, tasteless, and environment friendly; thus, it does not contaminate products and environment and does not damage the ozone layer. In addition, CO_2 can be easily released from products at ambient conditions; accordingly the next-step operations such as drying and solvent removal can be avoided.

Similar to other pure substances, in the phase diagram of CO_2 with pressure and temperature dependence, the solid, liquid, and gaseous states of CO_2 are also restricted by the curves of melting, sublimation, and evaporation, respectively. The equilibrium of all three states are located at the triple point (T_p). At the critical point (C_p), there is no distinction between liquid or vapor phase, and the supercritical phase possess properties that are frequently reminiscent of both states. At supercritical conditions, CO_2 has a high diffusion rates, very low surface tension, and low viscosity [19, 24]. The diffusion coefficient of supercritical CO_2 is much higher than that of liquid solvents. The supercritical CO_2 viscosity increases quickly with pressure; however, it is still at least an order of magnitude lower than that of liquid solvents. In terms of surface tension, supercritical CO_2 possess zero surface tension; therefore, supercritical CO_2 may penetrate easily into a microporous solid structure. As a result, supercritical CO_2 can swell many polymers when it is applied as a polymer-processing medium. Of course, these properties are helpful for the fabrication of polymer materials with porous structure and metal nanoparticles doped in polymer materials with porous structure [14, 17, 19, 27]. Another important physicochemical property of supercritical CO_2 is its density. The density of supercritical CO_2 can also be easily tuned to the values of liquid-like and the solubility of solid material or liquid in supercritical CO_2 may be orders of magnitude higher than the estimated value from the ideal gas law. Generally, a number of the fluid phase solvent properties including supercritical CO_2 are directly corresponded to the change of their density. Thus, the solvent power of supercritical CO_2 is indeed connected to its polarity, and is also highly affected by its density, which in turn depends on the pressure and temperature [28–30]. As a result, a solute solubility in supercritical CO_2 may be greatly controlled by pressure and temperature, where a higher density conforms mostly to higher solubility.

5.2.2 Interaction of carbon dioxide–polymers

Supercritical CO_2 is considered as an environmentally benign solvent for reaction media in polymer processing and modifying. The sorption of CO_2 may yield in swelling and/or dissolution of a polymers matrix when the supercritical CO_2 was introduced into the polymers matrix. The absorbed CO_2 may also lead to nucleate into bubbles and result in the conformation of foam, or small damages in the polymers matrix, which can remarkably transform the mechanical properties of the polymeric materials when the CO_2 system was depressurized. As a result, supercritical CO_2 can be applied for polymerization, swelling, impregnation, fractionation, purification, and formation of powdered polymers [14, 17, 19, 27, 31]. Nevertheless, because of the potential interactions nature between CO_2 and polymers occur during polymers processing and modifying, the characterization of their interactions is a very important initial step in grasping such interactions, which will give explanation in the appropriate performance and design of extensively polymers processing strategies. These potential interactions may yield in swelling and/or dissolution of a polymers matrix. Therefore, there was a substantial requirement to describe any such contradictory interactions between CO_2 and polymers and to specify the appropriate parameters, whereby a large variation of polymers can be reasonably treated without decay to the design of polymers.

There are several potential interactions between supercritical CO_2 and polymers matrix such as sorption of CO_2 by polymers, swelling of polymers with CO_2, dissolution of polymers in CO_2, dissolution of CO_2 in polymers, plasticization and reducing the glass transition temperature of polymers, crystallization and rise in the melting temperature and the melting enthalpy of polymers, shift of polymers' mechanical properties, shift of polymers' surface properties, and nucleation of gaps within the structures of polymer. However, the interaction between CO_2 and polymers could be categorized into three general application areas: the swelling of the polymers matrix, the dissolution of the supercritical CO_2, and the applications where CO_2 did not interact with the polymers matrix [31, 32]. Scheme 5.2 illustrates the potential interactions between CO_2 and polymers under supercritical conditions and their possible applications. It showed that the sorption and swelling of polymers by CO_2

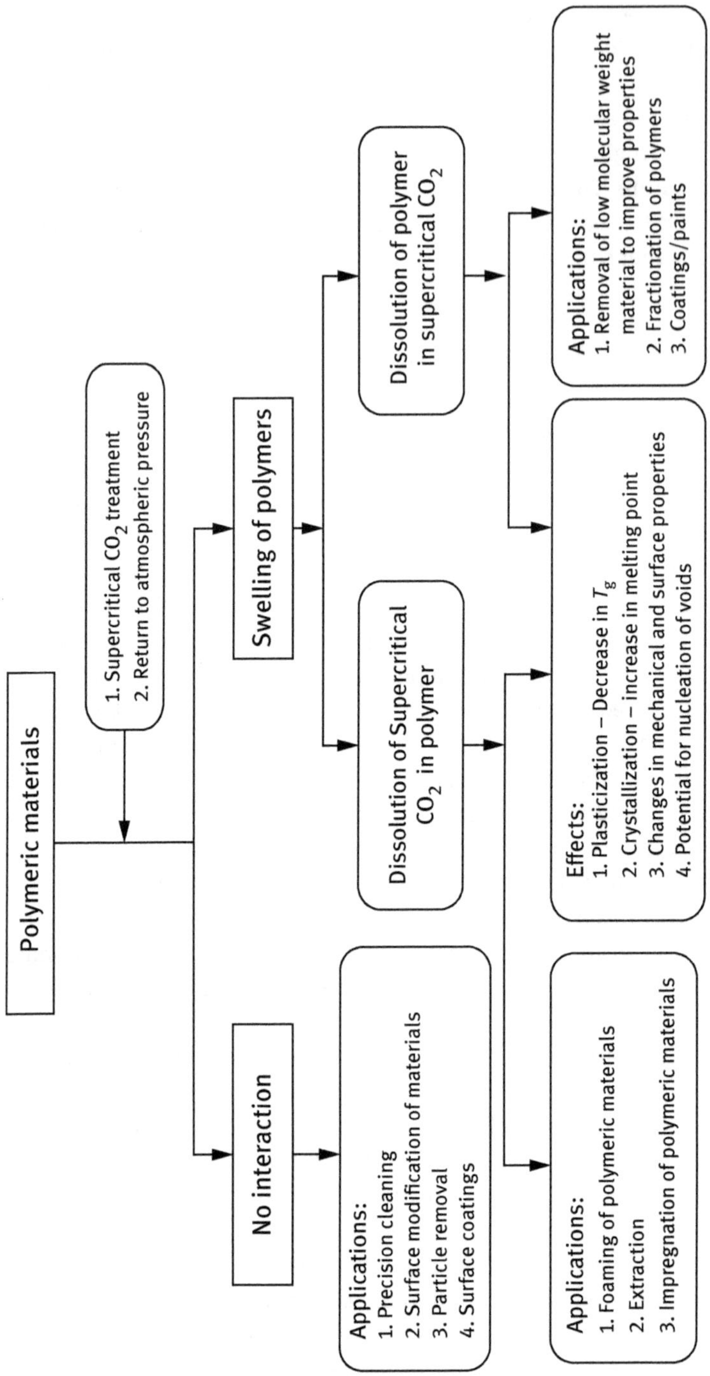

Scheme 5.2: Flow diagram of the interaction between supercritical CO_2 and polymers.

obviously are crucial effects in the design and implementation of polymer processing and modifying due to the important physicochemical properties of CO_2 and polymers matrix such as diffusivity, viscosity, density, glass transition temperature, melting point, compressibility, and expansion may shift at supercritical conditions.

Although supercritical CO_2 possesses the capability to penetrate and to swell many polymers matrix up to several mass percentages, supercritical CO_2 has been known as a solvent with an extremely poor solubility for most polymers in easily reachable conditions. Various techniques developed for the production and/or modification of particles using supercritical technologies exploit the solubility of supercritical CO_2 in a polymer or vice versa. These techniques are gas antisolvent crystallization, rapid expansion of supercritical solutions, precipitation by compressed antisolvent (PCA), supercritical antisolvent micronization, solution-enhanced dispersion by supercritical fluids (SEDS), and particles from gas-saturated solutions (PGSS). Like the supercritical antisolvent precipitation process, here supercritical CO_2 that functionalized as an antisolvent to produce electrospun fibers from polymer feed solution by using electrospinning technique will be described.

5.2.3 Electrospinning

Electrospinning and electrospraying are electrohydrodynamic techniques that can be considered as a branch of fluid mechanics concerned with the electric force effects. Both techniques work on the same principle with very minor and basic differences. The typical equipment of electrospinning and electrospraying comprises of these three main components: a high-voltage power supply to generate electric field, a syringe pump to introduce polymer feed solution via a blunt-ended metal needle, and a grounded products collector to collect electrospun products in nano-microscale. Generally, the generation of nano-microsized electrospun products through electrospinning and electrospraying needs dispersing the polymer as a starting material in a liquid medium and thus both these processes and the shape and size of the obtained electrospun products were influenced by the related solvent properties. In electrospinning, the polymer feed solution was spun into continuous nano-microsized polymer threads by the introduction of high-potential electric field to produce fibers, while electrospraying resulted in nano-microsized particles of the polymer [33, 34]. These techniques have been known as promising techniques, attracting researcher's interest and curiosity due to their possible implementation in food materials and pharmaceutical drugs. In the next sections, the main features of electrospinning technique will be elucidated with more detail.

Electrospinning is a common technique to fabricate polymeric nano-microsized fibers continuously by using electric fields. Compared to other techniques such as template synthesis, drawing, phase separation, and self-assembly, the electrospinning technique is widely accepted as a simple and versatile technique for creating nano-microsized fibers [35–38]. The morphology, size, and porosity of fiber as well as density of overall nanofiber mat can be controlled during electrospinning process. The electrospinning technique is also able to produce both polymers and composite materials with many desired features for high performance in a wide range of industrial applications. In addition, electrospinning technique may be employed to a large variation of polymeric materials; this technique is suitable to scale-up, with a relatively inexpensive and easy arrangement. Hence, the electrospinning technique has the potential for mass production of nano-microsized fibers from various polymers. Table 5.4 lists advantages and disadvantages of each technique for fabrication of polymeric nano-microsized fibers.

Table 5.4: Comparison of different nano-microsized fibers' fabrication techniques.

Fabrication technique	Drawing	Template synthesis	Self-assembly	Phase separation	Electrospinning
Advances	laboratory	laboratory	laboratory	laboratory	laboratory, commercial
Scalability	×	×	×	×	√
Repeatability	√	√	√	√	√
Convenience	√	√	√	×	√
Process Control	×	√	×	×	√

√ = Yes
× = No

In the electrospinning process, a polymer feed solution is introduced at a controlled velocity to keep a polymer droplet at the end of a conductive capillary tube, as well as a metallic needle with a blunt end. The conductive capillary tube is connected to a power supply with high voltage and serves as an electrode to charge the polymer solution. Other electrode, simply grounded, is functionalized as an electrospun fibers collector. The electric field, which is outward from a positive charge and toward a negative point charge, will induce a charge on the face of the polymer solution. When the electric field was increased, a critical value was reached, where the repulsive electrostatic force copes with the polymer solvent surface tension and the charged stream of the polymer solution is spurted from the needle of a conductive capillary tube. The discharged polymer solution stream goes through an instability process of elongation to the grounded target, which leads to long and thin stream . At the same time, the polymer solvent is lost in the evaporation process and the remaining polymer solidifies on the electrospun fibers collector. Hence, it could be said that the electrospinning process possess three stages: (i) initiation and prolongation of the jet throughout a straight line; (ii) instability of bending and further the jet elongation; and (iii) solidification of the jet into nano-microsized fibers [39, 40].

Generally, the generation of nano-microfibers by the electrospinning process is hardly affected by the concentration of the polymer feed solution. The solution of polymer feed used as a precursor for the generation of nano-microsized fibers should have a concentration high enough to form convolution of polymer. Figure 5.2 depicts the SEM photos of the electrospun products fabricated from poly(methyl methacrylate) (PMMA) dissolved in dichloromethane (DCM) at varying concentrations [41]. As shown in the SEM photos, distinct morphologies were produced. It is understood that the polymer solution concentration influences its properties such as viscosity and surface tension [42, 43]. The significant effect of polymer concentrations on the morphologies of the nano-microsized fibers fabricated by electrospinning process is explained as follows. The formation of bead strings and beads was predominantly at 10 and 15 wt% polymer concentrations. At these concentrations, the solutions of PMMA feed might have high surface tension as well as low viscosity. Hence, by using a high voltage in the electrospinning process, the jet during its trajectory was dissociated because of high surface tension, causing the formation of polymer droplets in place of fibers.

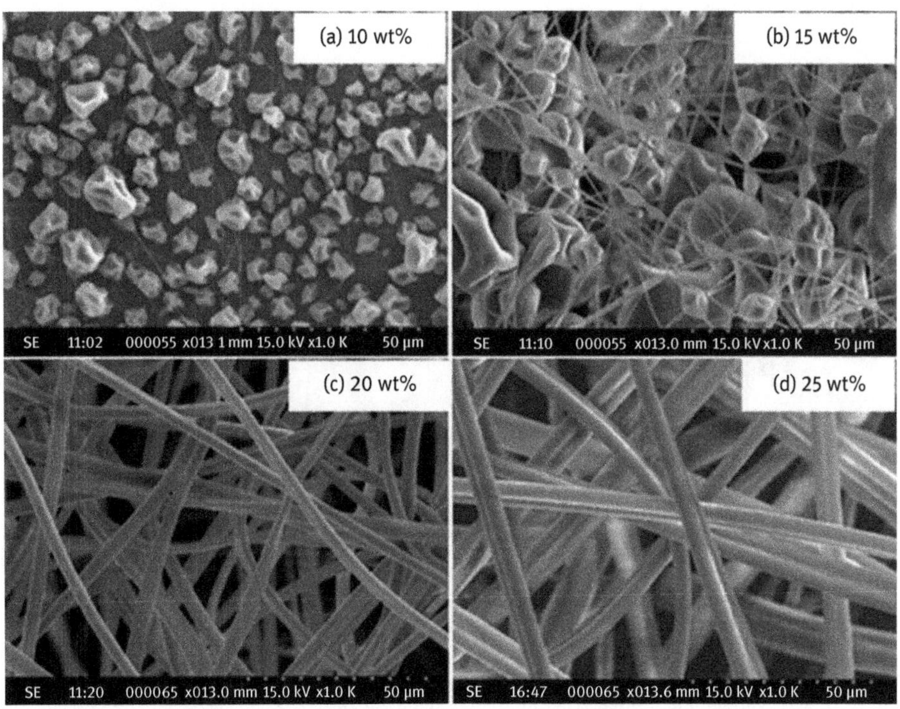

Figure 5.2: SEM photos of electrospun fiber fabricated from solution of PMMA at ambient temperature (adopted from Ref. [41]).

The increasing concentration of PMMA results in a rise in the viscosity as well as an increase in feed solution surface tension. This condition entangles polymeric chains, which afterward results in the production of uniform nano-microsized fibers without beads. Figure 5.2 exhibits that nano-microsized fibers without beads were produced as the concentration of PMMA was raised to 20 and 25 wt%. As reported by Touny et al. [42], the viscosity of the solution increases at higher concentrations, thus decreasing electrostatic repulsion. The decreasing electrostatic repulsion induces a decreasing drawing stress in the jet, and as the result, the generation of fibers in nano-microsized range occurs. In addition, there are several factors that affect the electrospinning process to the production of electrospun fibers: (i) polymer solution parameters as well as elasticity, viscosity, molecular weight, conductivity, and surface tension; (ii) processing parameters as well as applied voltage, flow rate, fiber collector, and distance between the fiber collector and the needle of the syringe; and (iii) environmental parameters such as temperature, humidity, and air flow [44–46].

Electrospinning under pressurized CO_2

Figure 5.3 shows the electrospinning apparatus at near- and supercritical CO_2 [41, 47–50]. The primary apparatus comprises of an autoclave made of a nonconductive

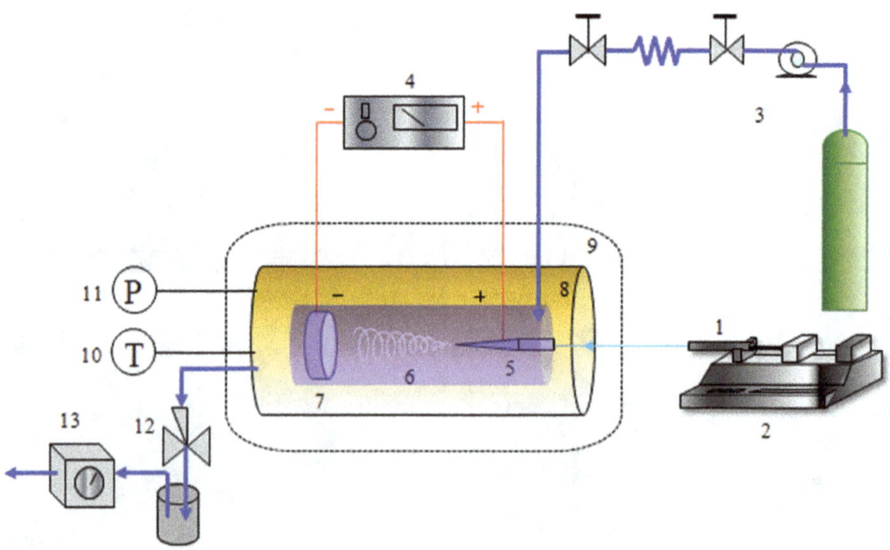

Figure 5.3: Schematic diagram of the electrospinning system. (1. Polymer solution; 2. syringe pump; 3. CO_2 cylinder and preheater; 4. high-voltage power supply; 5. nozzle; 6. polymer fiber; 7. collector wrapped with aluminum foil; 8. PEEK autoclave; 9. heater; 10. temperature monitor; 11. pressure gauge; 12. back pressure regulator (BPR); 13. flow meter).

poly(ether ether ketone) (PEEK) (AKICO PEEK) equipped cartridge heaters combined with an electric fan, a high-pressure pump (JASCO PU-1586), a high-voltage power supply (Matsusada Precision HARb-30P1), a back-pressure regulator (BPR; AKICO HPB-450 SUS-316), a syringe made of stainless steel, and a high-pressure syringe pump (Harvard Apparatus PHD-Ultra 4400). The PEEK autoclave has inner diameter of 6.0 cm, outer diameter of 15.0 cm, and length of 20.0 cm, with two stainless steel flanges. The two flanges were used as an anode electrode connected with a high-voltage power supply and as a cathode electrode wrapped with aluminum foil for collecting electrospun fiber products. The gap between needle and collector was 8 cm. Prior to starting the electrospinning process, the temperature of nonconductive PEEK vessel was increased to certain temperature by cartridge heaters. The temperature electrospinning process was controlled by introducing a thermocouple inside the PEEK chamber and by contacting properly with the PEEK chamber. The distribution of radial temperature was measured by inserting thermocouples (K-type) into the wall of the PEEK vessel. Since a certain temperature was achieved, CO_2 was loaded into the PEEK chamber by a capillary tube made of PEEK by a high-pressure pump to increase the pressure. The pressure was controlled by a BPR. Polymer dissolved in a solvent was introduced by the high-pressure syringe pump equipped with a high-pressure stainless steel syringe into the PEEK chamber at the time when a certain condition was achieved. At the same time, electrostatic force was generated by the high-voltage power supply. By using this apparatus, CO_2 and feed polymer dissolved in a solvent were introduced individually via the nozzle located at the anode electrode of stainless steel flange.

In principle, the electrospinning at near- and supercritical CO_2 conditions is similar to the electrospinning under ambient temperature. At near- and supercritical CO_2 conditions, electrospinning behaviors of polymer feed solution are also governed by a number of parameters. The polymer should be dissolved in a liquid solvent to prepare polymer feed solution for polymer stream formation; hence, the polymer feed solution should possess the electrical conductivity that is especially specified for the polymer type, used solvent, or salt addition. Polymer feed solutions that have high surface tension will need higher applied voltage to commence jetting than those with low surface tension. The lower viscosity of polymer feed solutions frequently results in beads due to the more prominent effect of surface tension than polymer–polymer chain interaction. However, only the dry fiber products will be obtained when the polymer solvent evaporates completely before they deposit on the electrospun fiber products collector. Therefore, the polymer solvent evaporation may play an important role and is one of the challenges to deal when using electrospinning technique for producing nano-microsized fibers from polymer solutions.

There are several important issues, such as drying and solubility, that are involved when near- and supercritical CO_2 is deployed as a polymer synthesis solvent. The benefit of the solubility of the near- and supercritical CO_2 media in the polymers can be utilized to decrease the viscosity of polymer for fiber thinning, with no dissolving or liquefying of the polymer [51, 52]. Hence, the fabricated polymer can be separated from the media of reaction (near- and supercritical CO_2) via modest depressurization, producing a polymer with dry condition. Figure 5.4 exhibits the pictures of the electrospun fiber produced from polyvinylpyrrolidone (PVP) dissolved in DCM for electrospinning process performed at ambient temperature with compressed CO_2. As discussed earlier, one of the defiance's deal using electrospinning fiber fabrication for collecting the fiber is evaporation of polymer solvent. For this issue, the removal of the polymer solvent of DCM takes place before the solution of PVP feed reached the target with a short space between needle and collector wrapped with aluminum foil. The figure shows the difference of electrospun fiber products fabricated by electrospinning in the absence of CO_2 and with compressed CO_2. At ambient temperature, the electrospun fiber products seemed almost moist and the solvent of polymer still appear clearly. It shows that the vaporization rate of solvent was slow. Conversely, the electrospun fiber products fabricated in the media of compressed CO_2 were dry and there was no residual solvent of polymer. It is evident that the solubility of DCM as a solvent of polymer in CO_2 could assist the evaporation process by lowering the

Figure 5.4: Electrospun fiber products fabricated (a) without and (b) with compressed CO_2 (adopted from Ref. [50]).

pressure when the experiment is complete. Thus, the rate of removing polymer solvent could be faster. This result also shows that the controlling pressure and temperature with CO_2 as a solvent is a promising pathway for transition of the thermal behavior of the polymer [41, 47–51].

Nevertheless, similar results were found when the PMMA feed solution at various concentrations were fed in electrospinning process with and without pressurized CO_2. Figure 5.5 shows the SEM photos of electrospun fibers fabricated from PMMA dissolved in DCM at various concentrations with compressed CO_2 (5 MPa). The figures indicate that fiber-free beads were formed at low concentrations of PMMA. The increasing concentration of PMMA caused the production of combined bead fibers. When the concentration of PMMA was increased further, bead-free fibers formed continuously. Interestingly, as shown in Figure 5.2, the nano-microsized fibers without beads were produced at 20 wt% concentration of PMMA and at room temperature and pressure. Conversely, using pressurized CO_2 as an electrospinning medium resulted in beaded fibers and beads at the same concentration of PMMA as shown in Figure 5.5(e) and (f), respectively. It revealed that increasing the solubility of DCM in CO_2 might result in DCM being completely removed from the origin PMMA solution. Moreover, the increasing solubility of DCM in CO_2 might increase the rate of evaporation of DCM from the fiber product. In this case, the affinity of CO_2 was sufficient for dissolving a part of DCM. The affinity of CO_2 may be regulated by altering the temperature and pressure. Generally, the increasing pressure of CO_2 could increase the solubility of DCM, thus increasing the amount of DCM removed [41, 53].

Hollow electrospun products

Recently, there has been a great deal of progress in the potential applications of hollow fibers in the field of energy storage, photonics, and microfluids [54–57]. However, the formation of hollow fibers was still by template processes or using coaxial capillary. For a template process, the formation of hollow fibers is carried out in several steps. Initially fibers are produced by electrospinning, then the fibers are coated with a precursor material prepared by different deposition techniques. Subsequently, the hollow fibers are produced by removing the inner section of electrospun nano-microsized fiber with thermal degradation or selective dissolution. Because the entanglement or imbrication of long and flexible templates inevitably engenders interconnections among the resulting fibers, the template process usually works well for fibers with relatively short structures [58–61]. In the coaxial capillary process, the hollow fibers are fabricated by selectively removing core of core/shell fibers produced from two different solutions spinning simultaneously using a spinneret consisting of two coaxial capillaries [62–66]. There were some difficulties in these processes, such as the choice of preferable solution consisting of a nonsolvent for core to the polymer of shell with the result that a solid film can be instantaneously

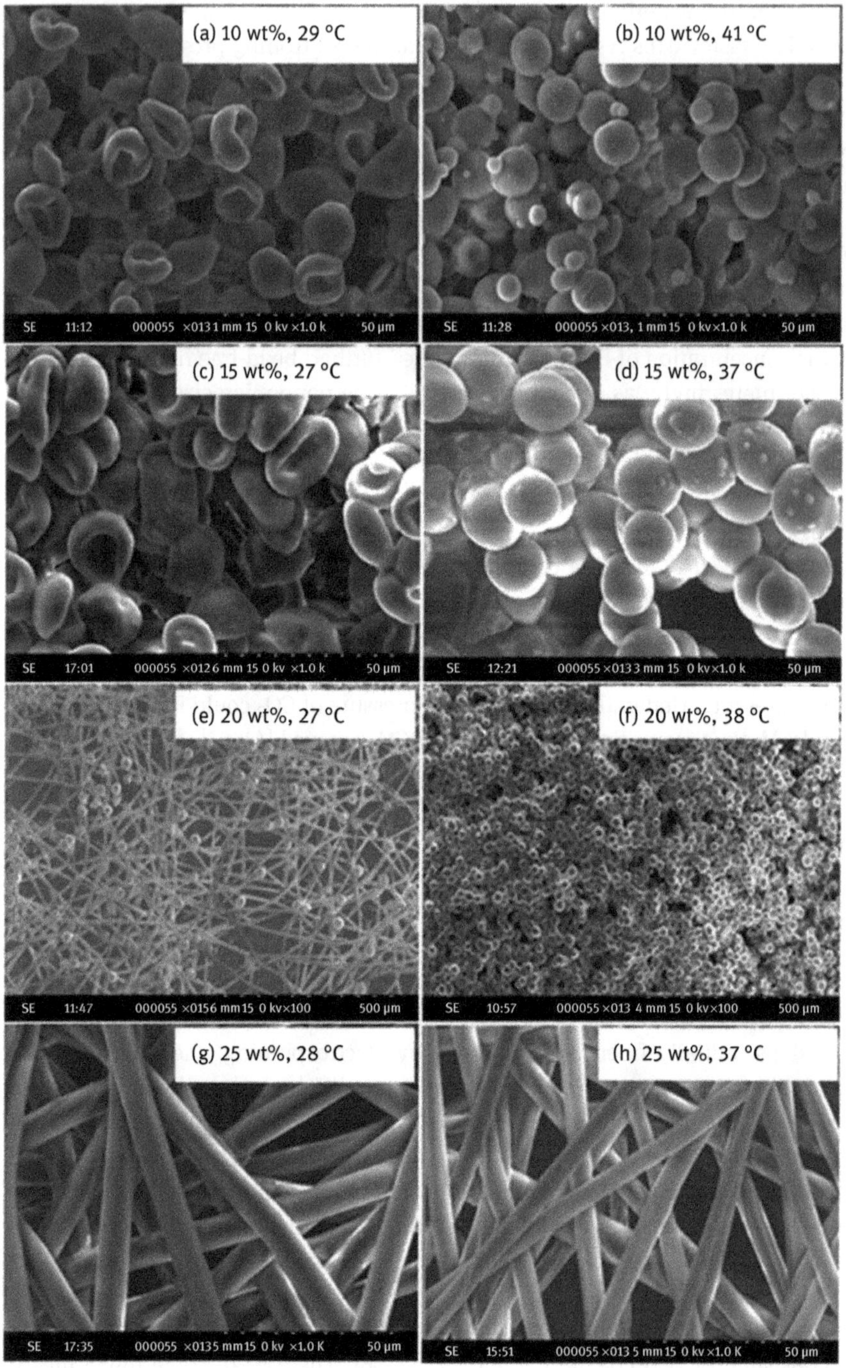

Figure 5.5: SEM photos of electrospun fibers fabricated from PMMA dissolved in DCM with compressed CO_2 (adopted from Ref. [41]).

generated at the interface and how to accurately regulate parameters of electrospinning, and so on. It causes limitation for applied materials to be used as starting materials especially when a water-soluble polymer was incorporated in the inner core. It was also noted that the equipment design and electrohydrodynamic behaviors of these processes were complex.

Figure 5.6 shows the morphology of PVP electrospun products produced by electrospinning at various pressures of CO_2 with temperature at 40 °C. At 3 MPa, the solid-core structure of PVP electrospun products was produced. The same morphology was also found when electrospinning of PVP in DCM was conducted at ambient conditions, with a distance of 20.0 cm distance between the needle to collector [59, 67]. During the electrospinning process, evaporation of the solvent of polymer and the extreme elongation of the solidifying fibers simultaneously control the fabrication of structure within the nano-microsized fibers. During electrospinning, the stream transfers from the needle to the sample collector, where the solid fiber is precipitated, which takes place within 0.1 s. From this time period, the fiber is in solid form. Hence, the time limit for the fabrication of structure is characteristically less than 0.01 s [68, 69]. When CO_2 was introduced at high-pressure conditions, CO_2

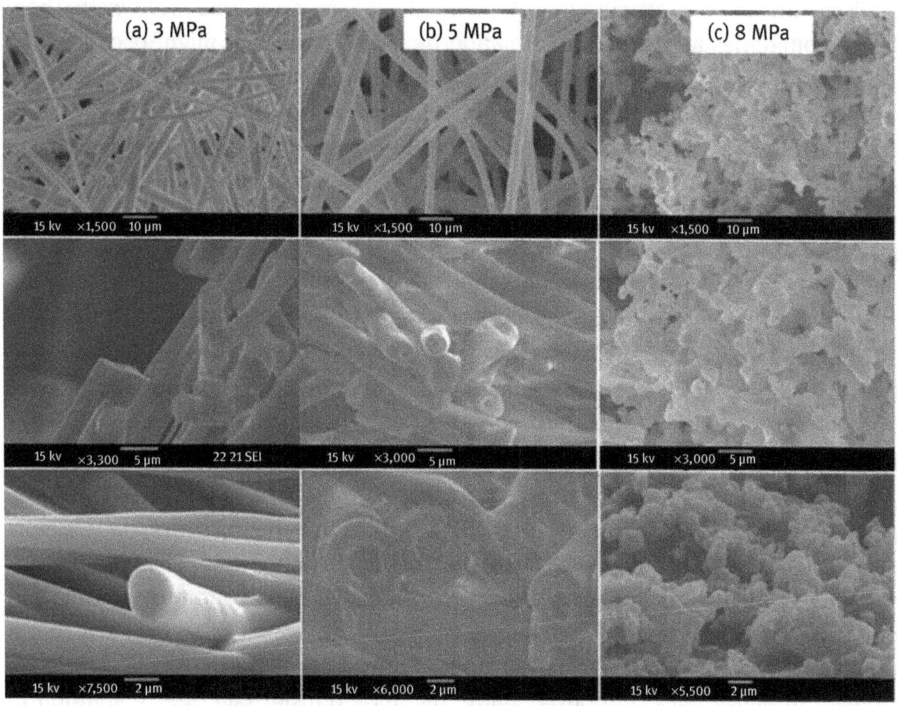

Figure 5.6: SEM photos of electrospun PVP fibers produced by electrospinning at different pressures of CO_2 (adopted from Ref. [50]).

dissolves the DCM present in the solution of PVP–DCM. Tsivintzelis et al. [70] reported the phase behaviors for the system of CO_2–DCM at 35.2, 45.2, and 55.2 °C. The increasing operating pressure resulted in higher solvent power of CO_2 for dissolving polymer solvent of DCM. Therefore, the generation of PVP fiber is attributed well to the phase behavior of the ternary mixture of PVP–DCM–CO_2.

As shown by Natalia and Gary [52], the viscosity of polymer could be adequately reduced by CO_2 to permit quick shear flow sans, demolishing the physical integrality of the fiber products when the formation of fiber was carried out by electrospinning under a CO_2 environment. They estimated the equilibration time between the CO_2 and the polymer sample during electrospinning process using optical changes in the polymer induced by CO_2. They found that equilibrium times were less than 1 min. However, it was still not easy to determine a one-to-one conformity between a dynamic process of fiber spinning and a static condition of phase behavior measurement [71]. The density and thermal diffusivity of CO_2 at 3 MPa and 37 °C were 0.06 (g cm^{-3}) and 3×10^{-5}(cm^2 s^{-1}), respectively. At this condition, CO_2 has the ability to extract DCM solvent from the liquid stream, and to dissolve it into the liquid stream and generate the solution of PVP–DCM to a distinct phase. This leads to spinodal decomposition of phase morphologies within the electrospun fiber products at the initial PVP-rich [51, 67]. As a result, a dramatic increase of the stream surface within milliseconds and the evaporation of DCM will happen on a time scale well. Then, phase boundaries and structure formation by phase separation might occur. As the pressure of CO_2 is increased to 5 MPa, the resultant electrospun fiber products are hollow core fibers, with the diameter almost identical to that obtained at 3 MPa. In the case of excess of CO_2 in the solution of PVP–DCM and the low interfacial tension of the PVP–DCM–CO_2 phase relative to the CO_2–DCM phase, the hollow core electrospun fiber products were generated. It can be seen that increasing the solubility of CO_2 in a DCM-rich phase and the solubility of DCM in a CO_2-rich phase resulted in an increase in CO_2 concentration in the liquid stream of PVP–DCM. Again, the solution of PVP–DCM underwent spinodal decomposition, resulting in netting of PVP–DCM solution accompanied with CO_2-rich bubbles. When the stream transfers to the target, the CO_2-rich bubbles proceed to affiliate and extend irrespective of netting of PVP-rich phase. The bubbles drive PVP-rich phase radially outside contrary to the inner surface of the stream to create a hollow core.

Similarly, a notable difference between the morphology of PVP electrospun prepared at 3, 5, and 8 MPa was observed. The particle in nano-microsized was almost generated at 8 MPa, although the chain among them still occurred. The possible reason was as follows. In this condition, two competing effects are temperature and pressure. They influenced the chemical and physical properties of CO_2, which are usually intermediate states of liquid and gas. Increasing the pressure caused an increase in density of CO_2. Consequently, the mass transfer rate of PVP solution in supercritical CO_2 increased [72]. James and McCarthy [73] reported that the reaction can be performed either in the presence of the supercritical CO_2 solution or by following the disposal of the solution by depressurization. As a common solvent, CO_2 is in gas state at ambient pressure; therefore, the solvent will quickly vanish with the decrease in pressure, causing the electrospun fibers to break apart. Supercritical CO_2 offers some more benefits. For example, it enhances penetration and diffusion rates in solid polymers by adjusting pressure and temperature. Therefore, the conditions of the process as well as pressure and temperature play a significant part in establishing the electrospun fibers structure with electrospinning in a supercritical CO_2 medium. However at higher pressure, the diffusion coefficient of supercritical CO_2 was proportional to the high volumetric expansion of DCM, resulting in rapid diffusion of CO_2 into the solvent as well as rapid solid deposition. Consequently, particles or fibers were not deposited separately; it resulted in larger particle size and denser agglomeration [59]. Yeo and Kiran [74] also suggested that polymer particles and fibers were fabricated under pressurized CO_2 based on the concentrations of polymer, and the variation of morphology was explained by the distinct mechanisms of precipitation. The particles were generated in the range of low polymer concentration due to the mechanisms of nucleation and the growth (at supercritical condition), whereas the fibers were generated at high concentrations of polymer that caused the spinodal decomposition (at near critical condition).

The electrospun products with hollow-structured morphology were also found when 25 wt% solution of PMMA was spun at near-critical conditions of CO_2 (see Figure 5.7). At 4 MPa pressure of CO_2, the electrospun fibers were almost entangled beads or bead strings. In this case, the mass transfer may occur between a PMMA solvent in the droplets and CO_2 as a compressed antisolvent; the separation of phase and PMMA solid precipitation also takes place. The solution of PMMA went through crystallization, deposition or precipitation out of the liquid solution state to be the solid–fluid state, via the two phases state of liquid–solid and liquid–fluid, and all three phases state of solid–liquid–fluid. As the droplets of PMMA solution approach the supersaturation condition, a very fast phase alteration takes place to generate a solid phase due to DCM solvent separation from the droplets, and PMMA deposited on the superficies of the fibers collector in the form of electrospun particles or beads.

The beautiful PMMA electrospun fiber products without beads were obtained when the pressures of CO_2 were increased to 5 and 6 MPa. As in the case of 4 MPa pressurized CO_2, the droplets of PMMA solution also go through the phase segregation and deposition process to generate PMMA in the solid phase. With an increase in CO_2 pressure, the amounts of CO_2 in the system enhance and also affect the composition of CO_2 in the electrospinning vessel. In order to preserve the amount of CO_2 dissolved in the polymer solvent, higher pressures are necessary. In general, the cloud point pressure rises with an increase in the amount of CO_2 in polymer–solvent–CO_2 system [53]. Otherwise, the elevating composition of CO_2 in the system also causes the reduction of dissolving power of the mixed solvent of CO_2 and DCM. Therefore, the electrospun fiber beads free of PMMA are generated when the

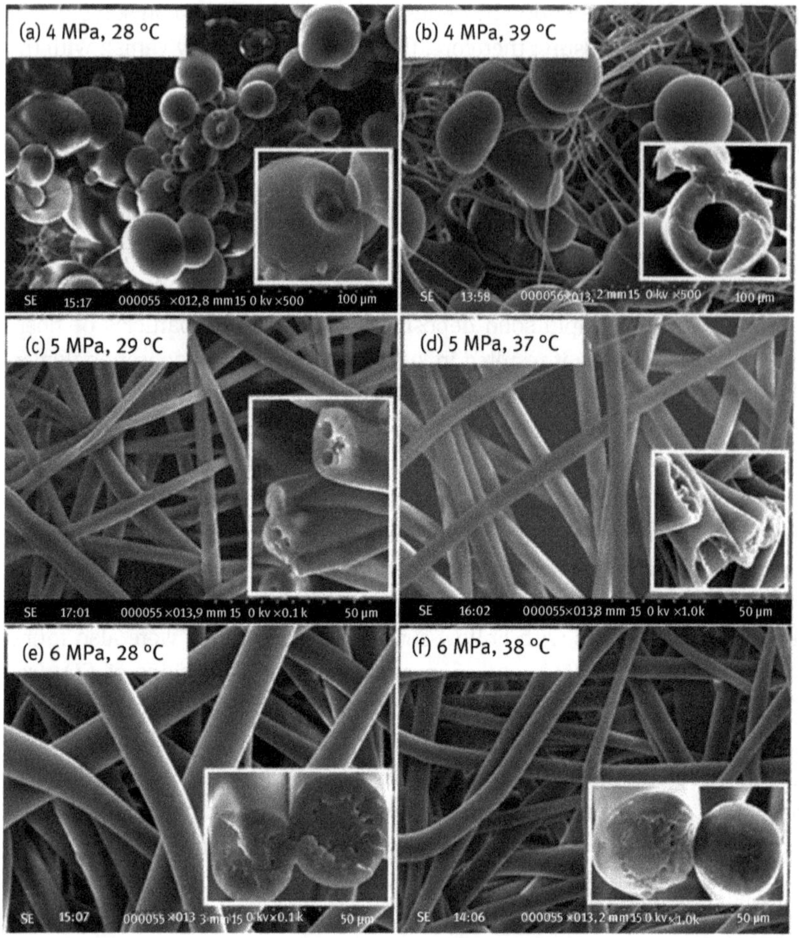

Figure 5.7: SEM photos of PMMA electrospun fabricated by electrospinning under pressurized CO_2 (adopted from Ref. [41]).

electrospinning process was performed at pressures of 5 and 6 MPa. Moreover, at 5 and 6 MPa the lower thermal diffusivity of CO_2 compared to that at 4 MPa also gave a high effect to the mass transfer between DCM as a solvent and CO_2 as an antisolvent. It implies that at these range temperatures the diffusion rate of a species under compressed CO_2 of 5 and 6 MPa may take place slower than that at 4 MPa.

It was well known that when the polymer was dissolved in a liquid organic solvent applied as feed of electrospinning and a compressed CO_2 employed as an antisolvent for the polymer, CO_2 will assimilate quickly into the polymer–liquid organic solvent stream and separate it [50]. Consequently, the products of nascent electrospun were obtained in dry conditions. As depicted in Figure 5.7, as the electrospinning process was done in a compressed CO_2 medium at pressures of 4–6 MPa; the products of electrospun fabricated from the solution of PMMA were also obtained in dry conditions. Interestingly, based on the cross-sectional observation of beads and fiber beads-free electrospun products of PMMA, the hollow morphologies of electrospun products were found. It can be explained that under pressurized CO_2, CO_2 simply dissolved into the PMMA solvent (DCM), and at this condition a large amount of CO_2 could dissolve into DCM-rich liquid phase. Conversely, most of the polymers, such as PMMA, were not dissolved in CO_2 under these conditions. However, very low PMMA solubility in CO_2 resulted in the separation between PMMA and DCM as a PMMA solvent. Because the electrospinning process was carried out by the batch process or at constant pressures, CO_2 dissolves into PMMA–DCM solution and also affects a phase extraction, which occurs by the decomposition in spinodal pattern. The interplay between CO_2 and PMMA–DCM solution causes quick DCM evaporation, and then the phase boundaries and the structure of PMMA electrospun products are formed. After the evaporation of DCM out of the surficies of the PMMA nascent fiber, the formation of beads or fibers on the skin takes place. At the same time, CO_2 also replaces solvent DCM in the PMMA electrospun products and boosts the PMMA expansion. By modest CO_2 depressurization on the completion of experiments, CO_2 was simply separated from the PMMA electrospun fibers or beads, yielding the electrospun products in hollow structures. Hence, it can be seen that the existence of CO_2 in the PMMA–DCM solution and the weak interfacial tension of the PMMA–DCM–CO_2 phase compared to the CO_2–DCM phase lead to a hollow fabrication of electrospun fibers or beads when CO_2 depressurization was applied at the final step of the process.

Glass transition temperature

Generally, polymer is defined as a large molecule, or macromolecule, comprising many repeated subunits of chemicals that are covalently bond. Most of polymers comprised of a combination of crystalline and amorphous regions [75]. The existence of these regions is important for the modification of polymeric materials that have both good strength (contributed largely by the crystalline regions) and some flexibility or softness (contributed largely by the amorphous regions). Hence, polymers may undergo several major thermal transitions by using specific treatments.

It has been known that the interaction of CO_2–polymer in polymer processing at supercritical conditions generated the sorption and expansion of polymers by CO_2, and the diffusion of a supercritical CO_2 into an amorphous polymer is much more pronounced than it is into a crystalline polymer because of better swelling of the amorphous polymer [31]. Other potential interactions may include the following: dissolution, increased diffusion rate, changes in solubility of additives, weight loss, change of mechanical properties, change of surface properties, and nucleation of voids within the polymer structure. However, one of the significant influences on the polymer processing in a supercritical CO_2 medium is the decrease in polymer's glass transition temperature (T_g). It happens when the CO_2 sorption in the solution of polymer is adequate to defeat their normal T_g. T_g is a very important polymer property, resulting in the transformation of polymer from a hard, glassy material to a soft, rubbery material [31, 35]. The temperature may also be interpreted as the border of temperature toward the polymer molecules mobility state. In terms of glassy polymer, the polymer acts in a brittle manner and has temperature lower than its T_g. This state is usually called the *glassy state*. At temperatures above T_g, the state is called the *rubbery state*. The polymer behaves in soft or rubbery manner and the molecular chains of polymer have enough thermal energy to glide.

Figure 5.8 shows the fabricated electrospun fibers obtained from PMMA solution of 25 wt% by the process of electrospinning conducted near the trend line of PMMA glass transition temperature at various CO_2 pressures. This figure shows that the pressure of CO_2 influenced the products of electrospun morphologies at these conditions. The hollow structure of electrospun beads string showed a glass-like brittle behavior at pressure of 4.5 MPa and temperature of 28 °C. When the CO_2 pressure increased to 5 MPa, the electrospun fibers in hollow morphology exhibited glass–rubber-like characteristics. At 29 °C and pressures of 5.5 and 6 MPa, the electrospun products in hollow morphology showed rubber-like characteristics. These results indicated that the morphologies of electrospun products, such as crack and hollow, and its thermal behavior were hardly influenced by CO_2 pressure. When pressurized CO_2 penetrates into the cavity among the chains of PMMA to change the PMMA chain mobility and free volume, and the T_g of PMMA was also altered [76–79]. The influence of high-pressurized CO_2 on the PMMA structure determined by Fourier-transformed infrared spectroscopy at temperatures of 20–80 °C was investigated by Noto et al. [78]. They observed that the increase in CO_2 pressure caused changes in PMMA spectra, where the vibrational band at 1,680 cm^{-1} vanishes with the increase in pressure of CO_2.

Effect of depressurization time

Since supercritical CO_2 has unique properties, as well as the association of liquid-like densities or high solvent power with gas-like viscosities or high diffusion rates, it is usually employed as a porogen to generate pores inside a matrix of polymer. Porous polymer foams can be formed when a polymer is plasticized with saturated supercritical CO_2, with decreasing CO_2 pressure at a constant temperature. It has been reported that the foaming process of polymers using supercritical CO_2 could be divided into three steps [79–81]: (i) the polymer is saturated by the supercritical CO_2 at high pressure; (ii) the combined polymer–CO_2 is extinguished into a supersaturated state by either depressurization or by an increase in temperature; and (iii) nucleation and growth of CO_2 cells dispersed as the polymer sample evolves.

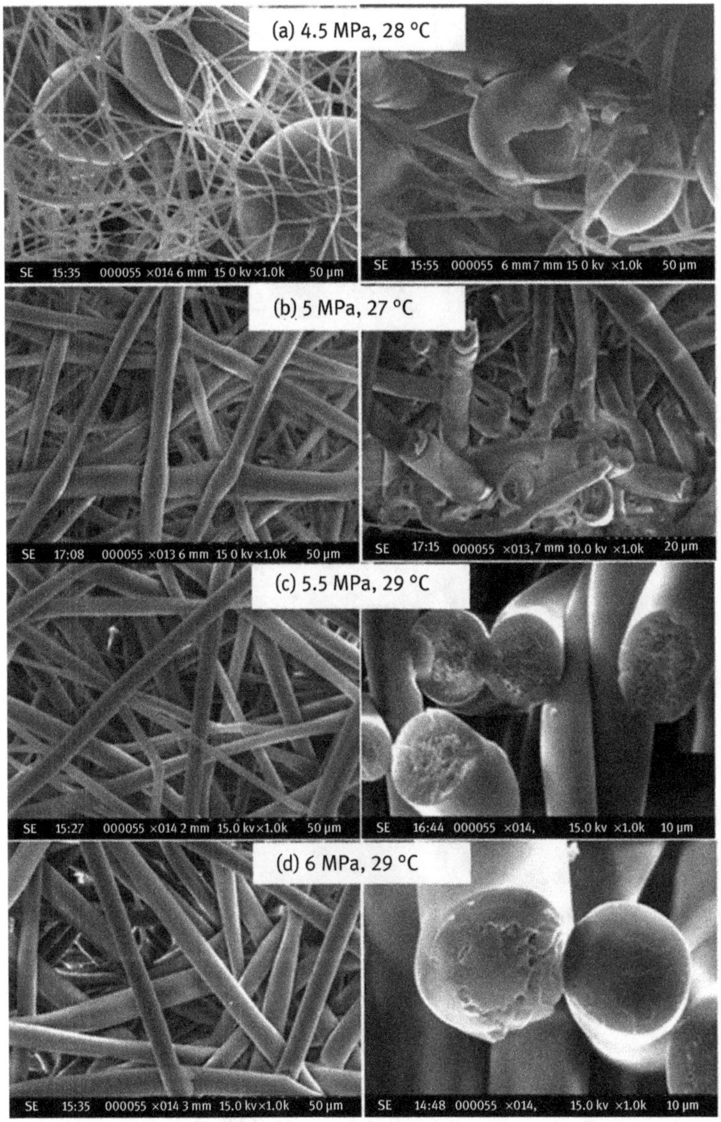

Figure 5.8: SEM photos of electrospun fibers fabricated at near glass transition temperature condition (adopted from Ref. [41]).

Figure 5.9 exhibits the SEM photos of the electrospun fibers fabricated from PMMA solution of 25 wt% at varying temperatures of 28–29 °C and varying pressures of 5–6 MPa, with depressurization time of 5–20 s. The results show that hollow morphology of electrospun fibers was generated at these operating conditions. Basically, the foaming process of polymer in supercritical CO_2 can be considered as the

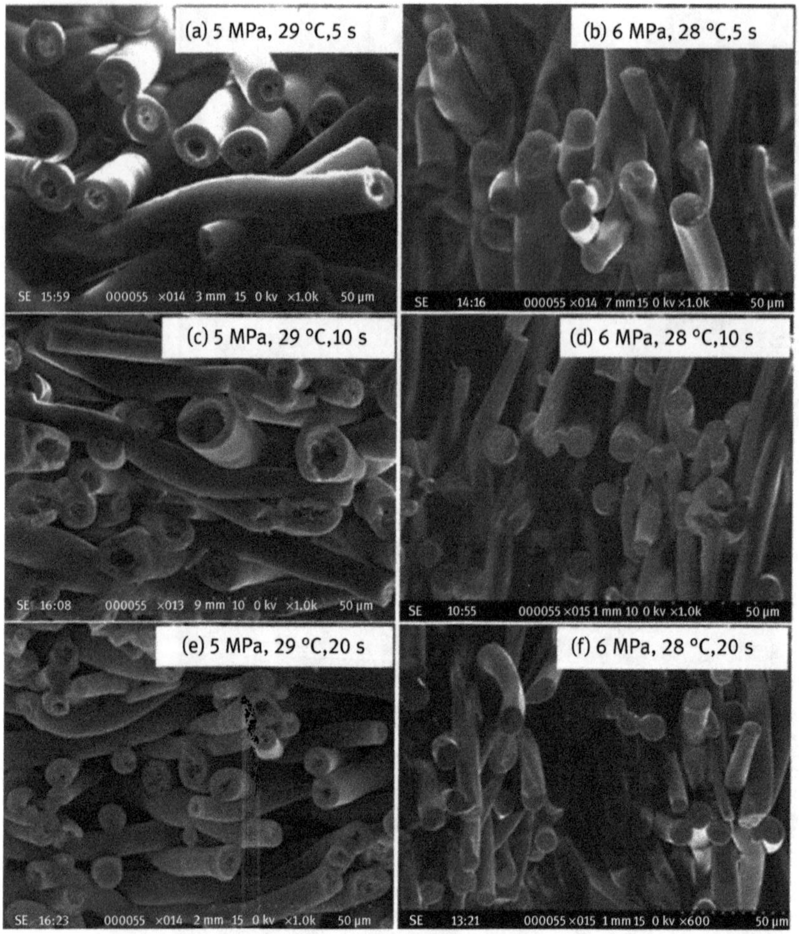

Figure 5.9: SEM photos of electrospun fibers fabricated at varying depressurization processes (adopted from Ref. [41]).

saturation of a polymer at a certain temperature and CO_2 pressure, followed by a fast depressurization step. During the depressurization process, CO_2 in the polymer becomes oversaturated. Consequently, phase separation and nucleation occur, and subsequently the generated nuclei grow into cells, resulting in the generation of small pores [78]. Generally, the increasing depressurization rate resulted in a decrease in porosity and mean pore size. It exhibited that the variation of depressurization time did not significantly affect the morphology of electrospun fibers at the same pressure as that in the electrospinning process. Hence, the hollow sizes were not determined.

As depicted in Figure 5.9, single-hollow morphology of electrospun fibers was primarily obtained as the electrospinning process was carried out at a pressure of 5 MPa, temperature of 29 °C, and depressurization times of 5–20 s. On the contrary, multihollow morphology of electrospun fibers was produced at a pressure of 6 MPa, temperature of 28 °C, and depressurization times of 5–20 s. These phenomena were explained by the saturation of CO_2 into a polymer that takes place either in glassy or in rubbery state. The polymer–CO_2 thermodynamic stable condition could be attained by altering the electrospinning pressure and/or temperature. In the rubbery state, which is at a temperature close to an amorphous polymer T_g, the nucleation and pore growth take place. Once the porous morphology is formed, it is stabilized as CO_2 breaks away from the matrix of polymer in a glassy state. In accordance with PMMA–CO_2 system T_g line [82], at 5 MPa and 29 °C, the system of electrospinning is in a glassy state; however, it shifts to a rubbery state when the process was performed at a pressure of 6 MPa and temperature of 28 °C. Because of this, the electrospun fibers obtained had distinct hollow morphologies. Moreover, the deformation that occurs during the depressurization process in the rubbery state was restored by the shape memory polymer layer because of the shape memory effect.

5.3 Conclusion and future direction

Electrospinning has been considered as a very simple and versatile technique to generate polymeric materials with high-functional and high-performance nano-microsized fibers. The process can be applied to a wide range of polymer and biopolymer solutions that can be spun directly. The electrospinning process has been accomplished by other methods such as drawing, template synthesis, self-assembly, and phase separation for producing of nano-microsized fibers from polymer solutions. Supercritical CO_2 is the most general application for supercritical fluid technologies, which offer important advantages over other organic solvent technology. It is considered as an environmental friendly technique since by using this technique the consumption of organic solvents can be reduced. When the supercritical CO_2 was applied to the electrospinning process, the organic solvent used as a solvent medium can be removed quickly and completely from the electrospun fiber products because of the high solubility of this solvent in the supercritical CO_2. Other advantages of using supercritical CO_2 for the production of nano-microsized fibers from polymer solutions are a low operating temperature and a low operating pressure. This chapter briefly described the benefits of using near- and supercritical CO_2 as an electrospinning process environment. At ambient temperature, the products of electrospun fiber looked nearly wet, with the solvent of polymer still obviously apparent. On the contrary, the electrospun fiber products formed were dry with no existing the solvent of polymer as the electrospinning process was performed in near- and supercritical CO_2. In near-critical conditions, when the CO_2 pressure is 4–6 MPa, the resultant electrospun products have hollow core fibers and the diameter is almost identical to that obtained at 3 MPa. The polymer particles in nano-microscales were generated at nearly supercritical condition, even though the chain among them still takes place and the varying time for depressurization did not have prominent influence on the electrospun fibers morphology at the same pressure of the electrospinning process. Nevertheless, this work demonstrated that electrospinning in pressurized CO_2 will be an absolutely necessary and worthwhile method for the formation of organic polymer with the hollow interior structures.

Abbreviations

BPR	Back-pressure regulator
CO_2	Carbon dioxide
C_p	Critical point
DCM	Dichloromethane
FT-IR	Fourier-transformed infrared
GAS	Gas antisolvent
T_g	Glass transition temperature
PGSS	Particles from gas-saturated solutions
PEEK	Poly(ether ether ketone)
PMMA	Poly(methyl methacrylate)
PVP	Polyvinylpyrrolidone
PCA	Precipitation by compressed antisolvent
RESS	Rapid expansion of supercritical solutions
SEM	Scanning electron microscope
SEDS	Solution-enhanced dispersion by supercritical fluids
T_p	Triple point

References

[1] Perrut, M. Supercritical fluid applications: industrial developments and economic issues, Ind. Eng. Chem. Res., 2000, 39, 4531–4535.

[2] Cansell, F., Aymonier, C., & Serani, A. Review on materials science and supercritical fluids, Curr. Opin. Solid State Mater. Sci., 2003, 7, 331–340.

[3] Brunner, G. Applications of supercritical fluids, Annu. Rev. Chem. Biomol. Eng., 2010, 1, 321–342.

[4] Knez, Z., Markocic, E., Leitgeb, M., Primozic, M., Knez Hrncic, M., & Skerget, M. Industrial applications of supercritical fluids: a review, Energy, 2014, 77, 235–243.

[5] Mezzomo, N., & Ferreira, SRS. Carotenoids functionality, sources, and processing by supercritical technology: a review, J. Chem., 2016, Article ID 3164312, 16 pages. http://dx.doi.org/10.1155/2016/3164312

[6] Wu, T., & Han, B. Supercritical Carbon Dioxide (CO_2) as Green Solvent. PT Anastas & JB. Zimmerman. eds. Innovations in Green Chemistry and Green Engineering: Selected Entries from the Encyclopedia of Sustainability Science and Technology, Springer Science & Business Media, New York, 2012, 297–326.

[7] Haynes, WM., Lide, DR., & Bruno, TJ. CRC Handbook of Chemistry and Physics, 95th Edition, CRC Press, Taylor & Francis Group, Boca Raton, FL, 2014.

[8] Shine, AD. Polymers and Supercritical Fluids. JE. Marked. Physical Properties of Polymers Handbook, Vol. 1076, New York, Springer, 2007, 319–338.

[9] Pereda, S., Bottini, SB., & Brignole, EA. Fundamentals of Supercritical Fluid Technology. JL. Martinez. ed. Supercritical Fluid Extraction of Nutraceuticals and Bioactive Compounds, CRC Press, Taylor & Francis Group, Boca Raton, FL, 2008, 1–23.

[10] Lang, Q., & Wai, CM. Supercritical fluid extraction in herbal and natural product studies–a practical review, Talanta, 2001, 53, 771–782.

[11] McKenzie, LC., Thompson, JE., Sullivan, R., & Hutchison, JE. Green chemical processing in the teaching laboratory: a convenient liquid CO_2 extraction of natural products, Green Chem, 2004, 6, 355–358.

[12] Herrero, M., Mendiola, JA., Cifuentes, A., & Ibanez, E. Supercritical fluid extraction: recent advances and applications, J. Chromatogr. A, 2010, 1217, 2495–2511.

[13] Da Silva, RPFF., Rocha-Santos, TAP., & Duarte, AC. Supercritical fluid extraction of bioactive compounds, TrAC Trends Anal. Chem., 2016, 76, 40–51.

[14] Kazarian, SG. Polymer processing with supercritical fluids, Polym. Sci. Ser. C, 2000, 42, 78–101.

[15] Reverchon, E., Adami, R., Cardea, S., & Della Porta, G. Supercritical fluids processing of polymers for pharmaceutical and medical applications, J. Supercrit. Fluids, 2009, 47, 484–492.

[16] Kiran, E. Supercritical fluids and polymers–The year in review–2014, J. Supercritical Fluids, 2016, 110, 126–153.

[17] Tsivintzelis, I., Sanxaridou, G., Pavlidou, E., & Panayiotou, C. Foaming of polymers with supercritical fluids: a thermodynamic investigation, J. Supercritical Fluids, 2016, 110, 240–250.

[18] Lee, JK., Yao, SX., Li, G., Jun, MBG., & Lee, PC. Measurement methods for solubility and diffusivity of gases and supercritical fluids in polymers and its applications, J. Polym. Rev., 2017, 57, 695–747.

[19] Beckman, EJ. Supercritical and near-critical CO_2 in green chemical synthesis and processing, J. Supercrit. Fluids, 2004, 28, 121–191.

[20] Brunner, G. Supercritical fluids: technology and application to food processing, J. Food Eng., 2005, 67, 21–33.

[21] Conde-Hernandez, LA., Espinosa-Victoria, JR., & Trejo, A. CO_2-supercritical extraction, hydrodistillation and steam distillation of essential oil of rosemary (Rosmarinus officinalis), J. Food Eng., 2017, 200, 81–86.

[22] Michalak, I., Chojnacka, K., & Saeid, A. Plant growth biostimulants, dietary feed supplements and cosmetics formulated with supercritical CO_2 algal extracts, Molecules, 2017, 22(66), doi:10.3390/molecules22010066

[23] Fukushima, Y. Application of supercritical fluids, R&D Rev. Toyota CRDL, 1999, 35, 1–9.

[24] Tomasko, DL., Li, H., Liu, D., Han, X., Wingert, MJ., Lee, LJ., & Koelling, KW. A review of CO2 applications in the processing of polymers, Ind. Eng. Chem. Res., 2003, 42, 6431–6456.

[25] Schutz, E. Supercritical fluids and applications: a patent review, Chem. Eng. Technol., 2007, 30, 685–688.

[26] Span, R., & Wagner, W. A new equation of state for carbon dioxide covering the fluid region from the triple point temperature to 1100 K at pressures up to 800 MPa, J. Phys. Chem. Ref. Data, 1996, 25, 1509–1596.

[27] Cooper, AI. Polymer synthesis and processing using supercritical carbon dioxide, J. Mater. Chem., 2000, 10, 207–234.

[28] Rindfleisch, F., DiNoia, TP., & McHugh, MA. Solubility of polymers and copolymers in supercritical CO_2, J. Phys. Chem., 1996, 100, 15581–15587.

[29] Yang, C., Brazhkin, VV., Dove, MT., & Trachenko, K. Frenkel line and solubility maximum in supercritical fluids, Phys. Rev. E, 2015, 91, 012112.

[30] Ingrosso, F., & Ruiz-Lopez, MF. Modeling solvation in supercritical CO_2, Chem. Phys. Chem., 2017, 18, 1–14.

[31] Kemmere, MF., & Meyer, T. Supercritical Carbon Dioxide in Polymer Reaction Engineering, Wiley-VCH Verlag GmbH & Co. KGaA, Weinheim, Germany, 2005.

[32] Shieh, YT., Su, JH., Manivannan, G., Lee, PHC., Sawan, SP., & Spall, WD. Interaction of supercritical carbon dioxide with polymers. I. Crystalline polymers, J. Appl. Polym. Sci., 1996, 59, 695–707.

[33] Kessick, R., Fenn, J., & Tepper, G. The use of AC potentials in electrospraying and electrospinning processes, Polymer, 2004, 45, 2981–2984.

[34] Bhushani, JA., & Anandharamakrishnan, C. Electrospinning and electrospraying techniques: potential food based applications, Trends Food Sci. Technol., 2014, 38, 21–33.

[35] Ramakrishna, S., Fujihara, K., Teo, WE., Lim, TC., & Ma, Z. An Introduction to Electrospinning and Nanofibers, World Scientific Publishing, Singapore, 2005.

[36] Shi, X., Zhou, W., Ma, D., Ma, Q., Bridges, D., Ma, Y., & Hu, A. Electrospinning of nanofibers and their applications for energy devices, J. Nanomater, 2015, 16(ID), 140716. http://dx.doi.org/10.1155/2015/140716

[37] Silva, PES., Vistulo de Abreu, F., & Godinho, MH. Shaping helical electrospun filaments: a review. Soft Matter, 2017, Advance Article, DOI: 10.1039/C7SM01280B.

[38] Zhang, B., Yan, X., He, HW., Yu, M., Ning, X., & Long, YZ. Solvent-free electrospinning: opportunities and challenges, Polym. Chem., 2017, 8, 333–352.

[39] Subbiah, T., Bhat, GS., Tock, RW., Parameswaran, S., & Ramkumar, SS. Electrospinning of nanofibres, J. Appl. Polym. Sci., 2005, 96, 557–569.

[40] Agarwal, P., Mishra, PK., & Srivastava, P. Statistical optimization of the electrospinning process for chitosan polylactide nanofabrication using response surface methodology, J. Mater. Sci., 2012, 47, 4262–4269.

[41] Wahyudiono, Okamoto K., Machmudah, S., Kanda, H., & Goto, M. Generation of multihollow structured poly (methyl methacrylate) fibers by electrospinning under pressurized CO_2, Polym. Eng. Sci., 2016, 56, 752–759.

[42] Touny, AH., Lawrence, JG., Jones, AD., & Bhaduri, SB. Effect of electrospinning parameters on the characterization of PLA/HNT nanocomposite fibers, J. Mater. Res., 2010, 25, 857–865.

[43] Bosworth, LA., & Downes, S. Acetone, a sustainable solvent for electrospinning poly (ε-Caprolactone) fibres: effect of varying parameters and solution concentrations on fibre diameter, J. Polym. Environ., 2012, 20, 879–886.

[44] Li, Z., & Wang, C. One-Dimensional Nanostructures: Electrospinning Technique and Unique Nanofibers, 9, New York, NY, USA, Springer, 2013.

[45] Hu, X., Liu, S., Zhou, G., Huang, Y., Xie, Z., & Jing, X. Electrospinning of polymeric nanofibers for drug delivery applications, J. Controlled Release, 2014, 185, 12–21.

[46] Okutan, N., Terzi, P., & Altay, F. Affecting parameters on electrospinning process and characterization of electrospun gelatin nanofibers, Food Hydrocolloids, 2014, 39, 19–26.

[47] Wahyudiono, Murakami K., Machmudah, S., Sasaki, M., & Goto, M. Production of nanofibers by electrospinning under pressurized CO_2, High Pressure Res., 2012, 32, 54–59.

[48] Wahyudiono, Murakami K., Machmudah, S., Sasaki, M., & Goto, M. A dry process for polymer nano-microfibers prepared by electrospinning under pressurized CO_2, Jpn, J. Appl. Phys., 2012, 51(8S1), 08HF07.

[49] Wahyudiono, Machmudah S., Murakami, K., Okubayashi, S., & Goto, M. Generation of PVP fibers by electrospinning in one-step process under high-pressure CO_2, Int. J. Ind. Chem., 2013, 4, 27.
[50] Wahyudiono, Machmudah S., Kanda, H., Okubayashi, S., & Goto, M. Formation of PVP hollow fibers by electrospinning in one-step process at sub and supercritical CO_2, Chem. Eng. Process Process Intensif., 2014, 77, 1–6.
[51] Shen, Z., Thompson, BE., & McHugh, MA. Electrospinning in near-critical CO_2, Macromolecules, 2006, 39, 8553–8555.
[52] Natalia, L., & Gary, T. Supercritical CO_2-assisted electrospinning, J. Supercrit Fluids, 2004, 31, 329–333.
[53] Shin, MS., Lee, JH., & Kim, H. Phase behavior of the poly(vinyl pyrrolidone) + dichloromethane + supercritical carbon dioxide system, Fluid Phase Equilib., 2008, 272, 42–46.
[54] Li, D., McCann, JT., Xia, Y., & Marquez, M. Electrospinning: a simple and versatile technique for producing ceramic nanofibers and nanotubes, J. Am. Chem. Soc., 2006, 89, 1861–1869.
[55] Gu, Y., Chen, D., & Jiao, M. Synthesis and electrochemical properties of nanostructured LiCoO2 fibers as cathode materials for lithium ion batteries, J. Phys. Chem. B, 2005, 109, 17901–17906.
[56] Song, MY., Ahn, YR., Jo, SM., Kim, DY., & Ahn, JP. TiO_2 single-crystalline nanorod electrode for quasi-solid-state dye-sensitized solar cells, Appl. Phys. Lett., 2005, 87, 1131131?3.
[57] Gouma, PI. Nanostructured polymorphic oxides for advanced chemo sensor, Rev. Adv. Mater. Sci., 2003, 5, 147–154.
[58] McCann, JT., Marquez, M., & Xia, YN. Highly porous fibers by electrospinning into a cryogenic liquid, J. Am. Chem. Soc., 2006, 128, 1436–1437.
[59] Bognitzki, M., Frese, T., Steinhart, M., Greiner, A., Wendorff, JH., Schaper, A., & Hellwig, M. Preparation of fibers with nanoscaled morphologies: electrospinning of polymer blends, Polym. Eng. Sci., 2001, 41, 982–989.
[60] Li, D., & Xia, YN. Direct fabrication of composite and ceramic hollow nano fibers by electrospinning, Nano. Lett., 2004, 4, 933–938.
[61] Loscertales, IG., Barrero, A., Marquez, M., Spretz, R., Velarde-Ortiz, R., & Larsen, G. Electrically forced coaxial nano jets for one-step hollow nanofiber design, J. Am. Chem. Soc., 2004, 126, 5376–5377.
[62] McCann, JT., Li, D., & Xia, YN. Electrospinning of nanofibers with core-sheath, hollow, or porous structures, J. Mater. Chem., 2005, 15, 735–738.
[63] Zhan, SH., Chen, DR., Jiao, XL., & Tao, CH. Long TiO_2 hollow fibers with mesoporous walls: solgel combined electrospun fabrication and photocatalytic properties, J. Phys. Chem. B, 2008, 110, 11199–11204.
[64] Zussman, E., Yarin, AL., Bazilevsky, AV., Avrahami, R., & Feldman, M. Electrospun polyaniline/poly(methyl methacrylate) -derived turbo stratic carbon micro-/nanotubes, Adv. Mater., 2006, 18, 348–353.
[65] Ismail, AF., & Yean, LP. Review on the development of defect-free and ultrathin-skinned asymmetric membranes for gas separation through manipulation of phase inversion and rheological factors, J. Appl. Polym. Sci., 2003, 88, 442–451.
[66] Chen, JT., Chen, WL., & Fan, PW. Hierarchical structures by wetting porous templates with electrospun polymer fibers, ACS Macro. Lett., 2012, 1, 41–46.
[67] Bognitzki, M., Hou, H., Ishaque, M., Frese, T., Hellwig, M., Schwarte, C., Schaper, A., Wendorff, JH., & Greiner, A. Polymer, metal, and hybrid nano- and mesotubes by coating degradable polymer template fibers (TUFT Process), Adv. Mater., 2000, 12, 637–640.

[68] Reneker, DH., Yarin, AL., Zussman, E., & Xu, H. Electrospinning of nanofibers from polymer solutions and melts, Adv. Appl. Mech., 2007, 41, 43–195.

[69] Greiner, A., & Wendorff, JH. Electrospinning a fascinating method for the preparation of ultrathin fibers, Angew Chem. Int. Ed., 2007, 46, 5670–6703.

[70] Tsivintzelis, I., Missopolinou, D., Kalogiannis, K., & Panayiotou, C. Phase compositions and saturated densities for the binary systems of carbon dioxide with ethanol and dichloromethane, Fluid Phase Equilib., 2004, 224, 89–96.

[71] Liu, J., Shen, Z., Lee, SH., Marquez, M., & McHugh, MA. Electrospinning in compressed carbon dioxide: hollow or open-cell fiber formation with a single nozzle configuration, J. Supercrit. Fluids, 2010, 53, 142–150.

[72] Werling, JO., & Debenedetti, PG. Numerical modelling of mass transfer in the supercritical antisolvent process, J. Supercrit. Fluids, 1999, 16, 167–181.

[73] James, JW., & McCarthy, TJ. Polymerization in supercritical fluid-swollen polymers: a new route to polymer blends, Macromolecules, 1994, 27, 4845–4847.

[74] Yeo, SD., & Kiran, E. Formation of polymer particles with supercritical fluids: a review, J. Supercrit Fluids, 2005, 4, 287–308.

[75] Alegria, A., & Colmenero, J. Dielectric relaxation of polymers: segmental dynamics under structural constraints, Soft Matter, 2016, 12, 7709–7725.

[76] Li, XK., Lu, H., Cao, GP., Qian, YH., Chen, LH., Zhang, RH., Liu, HL., & Shi, YH. Experimental study of the synergistic plasticizing effect of carbon dioxide and ibuprofen on the glass transition temperature of poly(methyl methacrylate), Ind. Eng. Chem. Res., 2014, 53, 5873–5885.

[77] Noto, VD., Vezzu, K., Giffin, GA., Conti, F., & Bertucco, A. Effect of high pressure CO_2 on the structure of PMMA: a FT-IR study, J. Phys. Chem. B., 2011, 115, 13519–13525.

[78] White, LJ., Hutter, V., Tai, H., Howdle, SM., & Shakesheff, KM. The effect of processing variables on morphological and mechanical properties of supercritical CO_2 foamed scaffolds for tissue engineering, Acta Biomater., 2012, 8, 61–71.

[79] Guo, H., & Kumar, V. Solid-state poly(methyl methacrylate) (PMMA) nano foams. Part I: low-temperature CO_2 sorption, diffusion, and the depression in PMMA glass transition, Polymer, 2015, 57, 157–163.

[80] Reverchon, E., & Cardea, S. Production of controlled polymeric foams by supercritical CO_2, J. Supercrit. Fluids, 2007, 40, 144–152.

[81] Jacobs, LJM., Kemmere, MF., Keurentjes, JTF., Mantelis, ChA., & Meyer, Th. Temperature-induced morphology control in the polymer-foaming process, AIChE J., 2007, 53, 2651–2658.

[82] Handa, YP., & Zhang, Z. A new technique for measuring retrograde vitrification in polymer–gas systems and for making ultramicrocellular foams from the retrograde phase, J. Polym. Sci. Part B: Polym. Phys., 2000, 38, 716–725.

Razan Badran, Ryan Gharios and Ali R. Tehrani-Bagha

6 Water-based electrospinning

Abstract: Water-based electrospinning offers an attractive alternative to organic solvent–rich electrospinning methods. This is to satisfy both increasing environmental concerns and ensure a higher degree of occupational safety as the process becomes ubiquitous and increasingly scalable in both academic and industrial settings. This chapter presents the fundamentals of water-based electrospinning, as well as an overview of water-soluble (WS) polymers and their applications. The discussion is then focused on the most relevant and widely used synthetic and natural-based polymers. The difficulties related to their electrospinning and the strategies to overcome them are then discussed by referring to several research papers. The most important applications of WS electrospun meshes are also highlighted accordingly.

Keywords: drug delivery, electrospinning, nanofibrous mesh, polymer blending, tissue engineering, water-soluble polymers

6.1 Introduction

Electrospinning is an attractive and promising method for the production of nonwoven meshes. These meshes, which consist of submicron fibers, are formed by the uniaxial stretching of a polymeric solution through the application of a high voltage. Prior to electrospinning, the polymer should dissolve in a good solvent. At low concentration/viscosity and under a high applied voltage, droplets of the polymer solution are ejected from the tip of the nozzle toward the collector, which results in the formation of submicron fibers. Polymer chain entanglement, essential for successful electrospinning, increases with concentration. Electrospun beadless fibers can only be produced within a certain range of concentration and viscosity. Moreover, the applied voltage, which is needed to overcome the viscous and surface tension forces of the polymer solution, relies significantly on the properties of both polymer and solvent used [1]. The obtained electrospun meshes can then be used for a wide range of applications, which span from common applications in filters [2] to more advanced applications in tissue engineering and drug delivery [3–6].

From a purely environmental standpoint, the electrospinning of water-soluble (WS) polymers in water offers a green processing alternative to other organic solvent–rich methods. More importantly, however, is the toxicity of certain organic solvents typically used. In fact, for a solvent to be used in electrospinning, it must

Razan Badran, Ryan Gharios, Ali R. Tehrani-Bagha, Department of Chemical and Petroleum Engineering, American University of Beirut, Beirut, Lebanon

https://doi.org/10.1515/9783110581393-006

dissolve the polymer chain, and be adequately volatile to evaporate within the required distance separating the nozzle from the collector. This necessitates the use of organic solvents with higher toxicities, which in turn precludes the usage of this technique in environments where occupational safety is key, such as the pharmaceutical and food industries [7].

Water is very desirable solvent for various applications as it is abundant, nontoxic, nonflammable, biocompatible, and inexpensive. However, the electrospinning of WS polymers from their aqueous solutions is a challenging task as (a) water has a high surface tension and dielectric constant, and (b) some WS polymers have very high molecular weights, crystalline structures, and rigid molecular conformations [8]. To overcome these issues, researchers have adopted various strategies such as polymer modification (by functionalization, grafting, or hydrolysis), blending with other WS polymers, or addition of spinning aids such as surfactants, cosolvents, and inorganic salts/acids to the polymer solution. These strategies will be discussed in more details in this chapter.

WS electrospun meshes have recently found promising prospective applications in nanomedicine [8]. However, their applicability can be further diversified by reducing their water solubility through either chemical treatment or cross-linking [9–12]. Another viable strategy is to produce organic–inorganic hybrid fibers and calcinate the resulting mesh at high temperatures to obtain pure inorganic nanofibers [13]. This latter approach, has the potential to prepare reinforced clay aerogel composites for various industrial applications [14].

6.2 Water-soluble (WS) polymers

Based on their origin, WS polymers can be classified into two main categories: natural-based and synthetic. Other classifications can also be based on molecular weight (e.g., low, medium, and high), functional side groups (e.g., nonionic, cationic, and anionic), or polymer chain structure (e.g., linear, branched, and cross-linked). The reader is directed to the following reviews covering WS polymer classification [15–17]. It should be noted that the water solubility of these polymers does not in any way impart them biodegradability. On the contrary, polymers with a carbon-chain backbone generally do not biodegrade, with one notable exception being polyvinyl alcohol (PVA) [15].

The physicochemical properties of WS polymers are directly linked to their chemical structure, charge, molecular weight, and other characteristics [18–20]. They are typically used industrially as thickeners, gelling agents, flocculants, emulsifiers, and emulsion stabilizers.

Common WS polymers used as thickeners are polysaccharides such as xanthan gum, pectin, alginates, and cellulose gums. Thickeners and gel-forming polymers have a wide range of applications in the food industry [21, 22], oil recovery [23], cleaning products [24], personal care products [25–27], textile printing [28, 29], paper coating [30], as well as building materials [31].

Examples of commonly used polymeric flocculants are dextran, carboxymethyl cellulose (CMC), sodium acrylate, quaternized cationic acrylate esters, and polyethylene oxide (PEO). These are used in a host of applications such as water injection systems [32], oily water clarification/wastewater treatment [33], and paper-making [34].

Some WS polymers such as hydroxypropyl methylcellulose (HPMC), methylcellulose (MC), and gelatin can emulsify and stabilize oil droplets in water at relatively low concentrations via adsorption to the oil–water interface and reduction of the interfacial tension, or an increase in the solution viscosity or alternatively by steric stabilization [35]. The effectiveness of WS polymers as emulsifiers depends directly on their aforementioned properties. The resulting emulsion should be able to tolerate slight changes in pH, temperature, ionic strength, and solvent composition [36].

6.3 Electrospinning of WS polymers

Water is a good solvent as it is nonflammable, nontoxic, abundant, and inexpensive. Despite these promising attributes, water is not a suitable solvent for electrospinning due to its high surface tension and dielectric constant. This makes electrospinning from aqueous solutions a highly challenging task, especially when WS polymers are used. For example, the electrospinning of WS natural-based polymers is facilitated more or less exclusively by blending with a synthetic WS polymer. The following sections deal with the successful water-based electrospinning of the most relevant WS polymers in order to provide more in-depth coverage of the topic.

6.3.1 Electrospinning of synthetic polymers

The majority of WS synthetic polymers can be electrospun from their aqueous solutions without the addition of another polymer, surfactant, or salt. For this reason, this category is presented before that of WS natural-based polymers. As such, the electrospinning of the most commonly used WS synthetic polymers [i.e., PEO, PVA, and polyvinyl pyrrolidone (PVP)] are presented:

Polyethylene oxide

PEO is a biocompatible polymer that has been clinically approved by the Food and Drug Administration (FDA) and other regulatory agencies. Therefore, it is extensively investigated and used for biomedical applications [8]. Aqueous PEO solutions can be electrospun successfully with or without blending with a second polymer. Its electrospinning is optimized by varying the preparation methods and the process parameters. The effects of many important parameters such as polymer concentration, applied voltage (*V*), tip-to-collector distance (TCD), and feed rate (FR) on the distribution of fiber diameters are summarized in Table 6.1. The optimization of these parameters is in fact crucial for successful electrospinning [37, 38].

Changing the solution concentration alters the viscosity of the solution and thus significantly affects the resulting fiber morphology. Thin beaded fibers are

Table 6.1: Electrospinning conditions of PEO in aqueous solutions.

Polymer solution	Electrospinning conditions	Fiber diameter	Reference
PEO in water MW 400,000 g mol^{-1}	[PEO] = 4–8 wt% V = 4–10 kV TCD = 12.5 cm FR = 0.2–1.5 mL h^{-1}	150–350 nm As a function of both [PEO] and applied V	[42]
PEO in water MW 600,000 g mol^{-1}	[PEO] = 5% (w/v), 0.1 g carvedilol, and 0.02 mL of polysorbat 80 V = 28 kV, TCD = 15 cm FR = 0.2–1 mL h^{-1}	144–345 nm Increases by increasing the FR under constant V and TCD	[43]
PEO in water MW 400,000 g mol^{-1} PEO 1:1 keratin	[PEO] = 5, 7 wt% [polymer] = 7, 10 wt% V = 15–30 kV TCD = 20 cm FR = 0.01–0.05 mL h^{-1}	150–270 nm Decreases as voltage increases	[44]
PEO in water MW 600 kDa HASE MW 100–200 kDa	[PEO] = 3–4 wt% [HASE] = 0.1–0.2 wt% V = 7–13 kV, TCD = 15 cm FR = 1 mL h^{-1}	4% PEO → $\bar{D}$ = 230 nm 3% PEO + 0.2% HASE →$\bar{D}$ = 180 nm 3% PEO + 0.1% HASE (with NaCl) →$\bar{D}$ = 120 nm	[45]
PEO in water MW 300,000 g mol^{-1} Surfactant: F127	[PEO] = 12 wt% [F127] = 5 wt% V = 22 kV, TCD = 16 cm FR = 0.2 mL h^{-1}	114–562 nm $\bar{D}$ = 305 ± 129 nm without lactase $\bar{D}$ = 292 ± 55 nm with lactase	[46]

HASE, hydrophobically modified alkali-soluble emulsion, which is a copolymer of methacrylic acid, ethyl acrylate, and macro-monomer, with molar ratio (43.57/56.21/0.22); F127, polyethylene oxide/polypropylene oxide block copolymer (EO_{100}–PO_{65}–EO_{100}).

obtained at low concentrations, whereas thicker and more uniform fibers are obtained at high concentrations. For example, it has been shown that the average diameter of PEO electrospun fibers increases from 100 to 400 nm as the polymer concentration increases from 3 to 8 wt%. In addition, beadless PEO fibers can be obtained at PEO concentrations above 8 wt% [39]. Electrospun beadless fibers can only be produced within a certain range of concentration and viscosity, and there is a critical polymer concentration beyond which no electrospinning is possible. Thus, the polymer concentration is an essential design parameter to keep in mind when approaching nanofiber electrospinning [37]. Different agitation rates can also lead to widely differing outcomes. Ultrasonication leads to severe material changes, shaking the solution favors aggregate formation, and stirring the solution uniformly over time usually results in uniform PEO fibers with an average diameter of 350 nm [39].

The effect of the solvent evaporation rate on the morphology of PEO electrospun fibers is determined by varying the relative humidity of the electrospinning chamber. At lower relative humidity values, the charged jet solidifies faster. This in turn results in the formation of larger fiber diameters due to an overall faster evaporation of water [40].

The addition of a cationic polyelectrolyte (polyallylamine hydrochloride) and an anionic polyelectrolyte (polyacrylic acid sodium salt) to a 7 wt% aqueous PEO solution significantly increases the conductivity and reduces the average fiber diameter of the produced meshes. In fact, the addition of these polyelectrolytes to the solution even at concentrations as low as 0.1 wt% can significantly increase the charge density of the polymeric jet. Consequently, the jet is stretched more strongly toward the collector, and this is due to interchain repulsion when under the same applied voltage. It should also be noted that neither the surface tension nor the viscosity of the PEO aqueous solution were noticeably affected by addition of the polyelectrolytes (within the 0.1–4 wt% range) [41].

Polyvinyl alcohol

PVA is a biodegradable, biocompatible, and nontoxic synthetic polymer. Due to the presence of hydroxyl side groups, strong hydrogen bonding is established between neighboring polymer chains, which significantly enhances the mechanical strength of the resulting electrospun mesh.

The effect of various electrospinning parameters such as solution concentration, feed rate (FR), applied voltage (V), TCD, ionic salt addition, molecular weight, degree of hydrolysis (DH), pH, surfactant addition, and type of collector has been studied and summarized in Table 6.2.

PVA solutions with various molecular weights have been successfully electrospun even in the absence of any other additives. By the addition of surfactants, the surface tension of the polymer solution decreases, thereby significantly facilitating

Table 6.2: Electrospinning conditions of PVA in aqueous solutions.

Polymer solution	Electrospinning conditions	Fiber diameter	Reference
PVA in water MW 145,000 g mol^{-1} DH 99%	[PVA] = 10 wt% V = 55 kV, TCD = 12 cm FR = not mentioned	350–700 nm	[11]
PVA in water MW 72,000 g mol^{-1} DH 97.5–99.5%	[PVA] = 10 wt% V = 15 kV, TCD = 15 cm FR = 1 mL h^{-1}	130 ± 23 nm Beadless	[50]
PVA MW 31000 Da	[PVA] = 20 wt% TCD = 15 cm, V = 10–35 kV	100–130 nm ~300 nm by encapsulation of 33 wt% Donepezil HCl	[51]
PVA in water TL in 2-propanol 3:1 water	[PVA] = 10 wt%; [TL] = 40 wt% 5g TL + 10 g PVA V = 25 kV, TCD = 10 cm FR = 0.9 mL h^{-1}	200–350 nm TiO_2 nanofibers obtained by calcination	[52]
PVA in water MW 88,000 g mol^{-1} DH 88% MA as a cross-linker	[PVA] = 8 wt% PVA: MA (mole 20:1 mol) V = 20 kV, TCD = 20 cm FR: NM	PVA → 193 nm PVA: MA → 238 nm Less than 5 wt% mass loss for PVA: MA	[53]
PVA in water MW 146–186,000 g mol^{-1}, DH 99%	[PVA] = 10% w/w V = 11 kV, TCD = 13 cm F.R. = 0.06–0.3 mL h^{-1}	70–300 nm $\bar{D}$ = 150 nm	[54]
PVA MW 78,000 g mol^{-1} DH 99.7% +HfO_2 NPs 3–5 nm	[PVA] = 10 wt% V = 12 kV, TCD = 13 cm F.R = 0.54 mL h^{-1}	100–1200 nm Depending on PVA:HfO_2 mass ratio and calcination conditions	[13]

TL, titanium lactate; MA, maleic anhydride; HfO_2, hafnium dioxide; NPs, nanoparticles.

the electrospinning process. The addition of anionic and cationic surfactants in particular has also been shown to effectively increase solution viscosity in small increments [47]. The effect of solution pH on the electrospinning process has also been investigated. pH is an especially important parameter affecting the solution conductivity, which is an essential consideration prior to performing electrospinning. For example, as pH increases from a value of 2 to 12.9, the fibers become straighter and finer. In fact, beaded fibers are formed under acidic conditions due to a noncontinuous electrospinning process, a problem that is readily solved by shifting the medium to more alkaline conditions [48]. It has also been reported that the conductivity of the PVA solution is greatly increased by the introduction of nano-diamonds. By incorporating nano-diamonds into the solution, not only does the conductivity increase, but the electrospun meshes also gain enhanced mechanical properties, such as higher tensile strength, Young's modulus, and toughness [49].

The effects of the DH of PVA (DH = 88%, 96%, 99.9%) and its concentration (8–12 wt%) on the polymeric solution properties and the average fiber diameter have been investigated [55]. The solution viscosity increases with increasing DH at high concentrations (10–12 wt%). In fact, in aqueous solution, the aforementioned hydrogen bonds between the polar hydroxyl groups in the PVA molecules greatly influence the solution rheological properties. Furthermore, the surface tension of the solution also increases mainly by increasing the DH and, comparatively, remains mostly unaffected by increasing the polymer concentration except at DH = 99.9%, most probably due to the formation of gelling structures [55]. The fiber diameter increases with the DH as shown in Figure 6.1. It seems that there is a direct relation between the surface tension of PVA solution and the average fiber diameter of the resulting PVA mesh. By increasing the DH, the viscosity of PVA solution increases and leads to the formation of more uniform fibers.

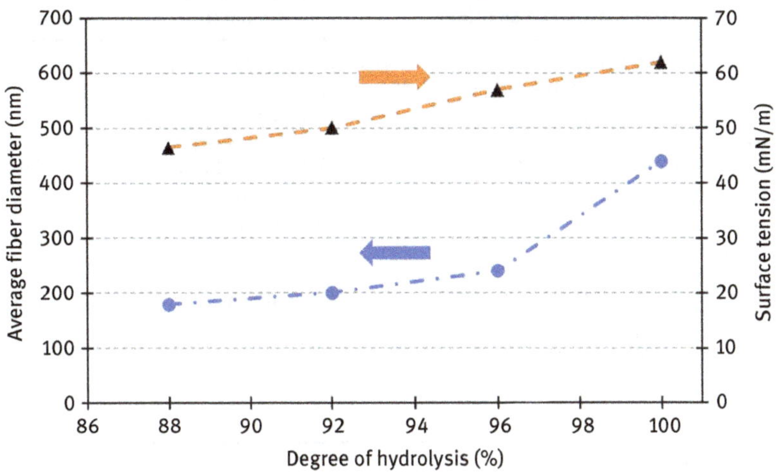

Figure 6.1: Variation in electrospun fiber diameter with the degree of hydrolysis of PVA. The electrospinning conditions: [PVA] =8 wt%, V =10 kV, TCD of 15 cm. Adapted with permission from Ref. [55].

Polyvinyl pyrrolidone

PVP is a WS polymer characterized by low chemical toxicity, a high degree of biocompatibility, and a capability to interact with a wide range of hydrophilic materials. It also possesses excellent capping properties since the carbonyl groups of the polymeric chain interact readily with neighboring metal ions. To exploit this, many researchers are investigating the incorporation of inorganic materials such as silver, gold, cadmium sulfide, and $CaCO_3$ into PVP nanofibers [56].

Table 6.3 shows the electrospinning conditions of various PVP solutions using water as the main solvent. The average diameter of the obtained fibers is 150 ± 34 nm immediately after electrospinning and 169 ± 64 nm 2 months after electrospinning. Electrospun PVP mats have also been tested for their ability to preserve encapsulated viruses (e.g., bacteriophage T7) [57]. The addition of another WS polymer such as dextran to PVP solution can reduce the charge density and result in the production of finer fibers. Some additives, however, might negate this trend. For example, one study has reported that the presence of 200 mM ibuprofen sodium salt or acetylsalicylic acid in the PVP aqueous solution had in fact resulted in larger fiber diameters (300–650 nm) [58].

Table 6.3: Electrospinning conditions of PVP in aqueous solutions.

Polymer solution	Electrospinning conditions	Fiber diameter	Reference
PVP in water/buffer MW 130,000	[PVP] = 15 wt% V = 25 kV, TCD = 12 cm FR = 0.5 mL h^{-1}	100–200 nm Water→$\bar{D}$ = 150 ± 37 nm buffer→$\bar{D}$ = 138 ± 65 nm	[57]
PVP in water MW 40,000 Dextran MW = 10,000, 40,000	[PVP] = 45 wt%, [Dextran] = 2 wt% TCD = 15 cm, V = 15 kV FR = 15 μL h^{-1}	PVP→$\bar{D}$ = 462 ± 67 nm PVP + dextran →$\bar{D}$ = 350 ± 100 nm	[58]
PVP in water MW 360,000 Drugs: helicid, mannitol	[PVP] = 9%, [helicid] = 5%, w/v [SDS] = 0.5%, [mannitol] = 2% (w/v) V = 15 kV, TCD = 12 cm FR = 2 mL h^{-1}	400–600 nm	[59]

SDS, sodium dodecyl sulfate.

6.3.2 Electrospinning of natural-based WS polymers

The most commonly used natural-based WS polymers in electrospinning research are alginate, cellulose derivatives, chitosan derivatives, collagen, dextran, and gelatin. The electrospinning of these polymers is often challenging due to a variety of reasons that will be discussed later. Blending with synthetic WS polymers is a viable method to overcome the electrospinning problems [12], and we have presented several examples to highlight this approach.

Alginate

Electrospinning of pure alginate, a WS anionic polysaccharide, in water is greatly hindered by its morphology and solution properties. In fact, alginate is formed of highly rigid and extended worm-like molecular chains, which prevent the formation of chain entanglements in aqueous solution. In addition, the high surface tension and electrical conductivity of alginate aqueous solutions often impede fiber stretching, ultimately leading to unsuccessful electrospinning [60].

The aforementioned difficulties have been overcome through the use of the following strategies:

- Addition of a miscible cosolvent such as glycerol [61], Triton-X100 [62], and dimethyl sulfoxide (DMSO) [60]
- Blending of the alginate with other nonionic synthetic polymers such as PEO, PVA, or PVP [60]
- Addition of salts such as NaCl and $AgNO_3$ to regulate solution conductivity [62]

- Effect of miscible solvents and surfactants on the alginate solution

Addition of a strong polar cosolvent such as glycerol to aqueous alginate solutions has facilitated the production of beadless fibers with an average diameter of 200 nm. In fact, adding glycerol to the polymeric solution decreased both the surface tension (from 61.4 to 37.5 mN m^{-1}) and the electrical conductivity of the solution (from 4,180 to 193 µS cm^{-1}). Interestingly, the viscosity of the polymeric solution increased from 22.2 to 701.3 Pa.s, which was attributed to the facilitated alginate chain entanglement in the presence of glycerol. As a result of these changes, the spinnability of the polymer solution was significantly enhanced [61].

Addition of surfactants such as Triton-X100 lowers the surface tension of the polymer aqueous solution (from 63 ± 2 to 29 ± 2 mN m^{-1}), thereby suppressing beads formation with little to no effect on the onset of fiber formation [62]. This strategy also enables researchers to work with higher concentrations of alginate. In fact, electrospinning of 70–85 wt% alginate solution was made possible by the addition of only 1 wt% Triton-X100 [60]. However, due to the cytotoxicity of Triton-X100, it is advisable to replace it for future purposes with FDA-approved nonionic surfactants such as Pluronic F127 [62] or natural surfactants such as lecithin [63].

- Effect of polymer blends on the alginate solution

Polymer blending enhances solution properties and facilitates the formation of composite nanofibers. As mentioned previously, the rigid molecular chains in alginate typically prevent the formation of chain entanglements in aqueous solution [64]. The presence of synthetic WS polymers such as PEO can solve this issue by increasing the relaxation time and viscosity of the solution and decreasing both its electrical conductivity and surface tension [60]. This shows that the use of high-molecular-weight polymers is required to induce sufficient entanglement and enhance fiber formation.

For biomedical applications, the use of low-molecular-weight polymer chains is especially desirable since these have a faster rate of in vivo degradation than their high-molecular-weight counterparts. This can be achieved by blending low-molecular-weight alginate with both PEO and a set of surfactants to facilitate processability [62, 65, 66].

– Effect of adding salts to the alginate solution

For proper electrospinning, the electrical conductivity of alginate solution should be well controlled. This can be achieved by the addition of salts such as NaCl or $AgNO_3$ [62]. It should be noted that at very high electrical conductivity, significant disruption during fiber formation can occur, which leads to defects in the resulting architecture and is thus highly undesirable [67].

In the electrospinning of alginate in a glycerol 2:1 water mixture, the average fiber diameter increased from 120 to 300 nm by increasing the concentration from 1.6% to 2.4% (w/v). The solution properties of 2% (w/v) alginate solution at various glycerol to water ratios v/v were also found to change noticeably upon investigation. In fact, successfully electrospun fibers were only obtained at a glycerol/water volume ratio of 2:1, highlighting the importance of the control of properties such as polymer solution viscosity, conductivity, and surface tension on subsequent electrospinnability [61].

Cellulose derivatives

The electrospinning of cellulose in aqueous solution imposes several difficulties as cellulose cannot be dissolved in pure water. However, cellulose can be dissolved at low concentrations in aqueous solutions containing N-methyl-morpholine N-oxide (NMMO) or NaOH/urea. The resulting problem is that ionic liquids such as NMMO have very low evaporation rates, which makes them unsuitable for use as solvents in electrospinning [68]. Furthermore, solutions such as NaOH/urea/water can also perturb electrospinning due to the presence of large amounts of Na^+ and OH^- ions [69]. Electrospinning of cellulose from aqueous NMMO (50%) solution is only possible for a concentration of around 4 wt%, at a voltage of 15 kV, and at room temperature. These conditions lead to formation of fiber diameters in the range of 90–250 nm. The effects of other process parameters such as needle diameter and rotation speed on the subsequent morphology have also been studied [70]. It turns out that the average fiber diameter decreases with a decrease in needle diameter, but remains independent of the collector rotational speed.

A range of water-soluble cellulose derivatives such as CMC and HPMC have been electrospun successfully [71]. The influence of several parameters such as molecular weight, degree of substitution, and substitution pattern on the structure of the fibers has been systematically studied. Both molecular weight and degree of substitution have no effect on fiber morphology. However, the substitution pattern is a crucial parameter. For example, bulky substitutional groups distinctly lead to the formation of inhomogeneous and coarse nonwoven meshes with spherical beads [71]. The various electrospinning parameters of different cellulose derivatives are summarized in Table 6.4.

Table 6.4: Electrospinning conditions of cellulose and cellulose derivatives in aqueous solution.

Polymeric solution	Electrospinning conditions	Fiber diameter	Reference
α-Cellulose from mercerized cellulose pulp in water 1:1 NMMO	[Cellulose] = 1, 2, 4 wt% V = 9–10 kV TCD = 15 cm T = 80–130 °C	200–400 nm for 2 wt% mercerized and raw cellulose at 90–100 °C	[72]
Cellulose + PVA Cellulose + HMPEG Solvent: [NaOH] = 7 wt% [Urea] = 12 wt% Water 81 wt%	[Cellulose] = [PVA] = 4 wt% [Cellulose/HMPEG]= 4 wt%/3 wt% V = 25 kV TCD = 15 cm	2 μm for cellulose 1:1 PVA 400 nm for cellulose 8:2 HMPEG	[69]
CNF + PVA in water	[CNF] = 1–5 wt% [PVA] = 10 wt% V = 16 kV, TCD = 19 cm	1 wt% CNF →102 nm 3–4 wt% CNF →85 nm 5 wt% CNF →80 nm	[73]
CMC 1:1 PEO in water	[CMC] = [PEO] = 4 wt% V = 35 kV TCD = 20 cm	200–250 nm	[71]
HPMC in water 1:1 ethanol	[HPMC] = 2.86 wt% V = 35 kV, TCD = 20 cm	128 nm	[71]

NMMO, *N*-methyl-morpholine-*N*-oxide; HMPEG, high-molecular-weight polyethylene glycol; CNF, cellulose nanofibrils; CMC, carboxymethyl cellulose; HPMC, hydroxypropyl methylcellulose.

Chitosan and their derivatives

Chitosan is produced by the deacetylation of chitin via hydrolysis under alkaline conditions. Chitin, the second most abundant polysaccharide after cellulose, is composed of *N*-acetyl-D-glucosamine and can be extracted from natural sources (e.g., shrimp and other crustaceans) [1]. The solubility of chitosan in organic solvents is limited due to its high crystallinity and high intra-/intermolecular hydrogen bonding. Chitosan is soluble in acidic media primarily due to the presence of amine groups in its chemical structure. In order to improve its solubility across a wider range of pH values, chitosan could be reacted with monochloroacetic acid to produce carboxylated chitosan or, alternatively, could be grafted with a nonionic polyethylene glycol [74, 75].

The electrospinning of chitin derivative carboxymethyl chitin (CMCtn) alone yielded globular drop-like depositions on the collector rather than fibrillary structures due to its excessively high viscosity and conductivity. However, electrospinning of CMCtn blended with PVA was successful at $V = 10$ kV and TCD = 8 cm. In fact, upon the addition of PVA in solution, both surface tension and conductivity decreased, and fiber formation conditions were enhanced [76].

The electrospinning of chitosan alone in pure water has not yet been reported. This is attributed to its polycationic nature, rigid structure, strong intra-/intermolecular hydrogen bonding, and high viscosity even at relatively low concentrations [77, 78]. By increasing acetic acid concentration up to 90% in the polymer solution, the surface tension decreases noticeably down to 31.5 mN m^{-1} and the viscosity slightly increases. Both of these trends present more favorable electrospinning conditions. It was found that the molecular weight of chitosan and its concentration, the applied voltage, and the acetic acid concentration are among the most important factors affecting the process. A beadless nanofibrous electrospun chitosan mesh was successfully produced from 7 wt% chitosan (MW 106,000 g mol^{-1}) in 90% aqueous acetic + 10% water at a V/TCD ratio of 3–5 kV cm^{-1} [79].

To overcome the difficulties associated with the electrospinning of chitosan in pure water, researchers have tried the polymer blend strategy. A chitosan solution in formic acid (7 wt%) was mixed with an aqueous PVA solution (9 wt%) at various volume ratios. Beadless nanofibrous meshes were only achieved when the PVA solution volume ratio was above a threshold of 50% [80]. The mixture of chitosan and PEO (4 wt%) in water + 3 wt% acetic acid has also been successfully electrospun under the following reported conditions ($V = 21$ kV, TCD = 18 cm, FR = 0.5 mL h^{-1}) [81].

The effect of metal ions on the morphology of chitosan electrospun mats was investigated as well. Several metal ion–rich chitosan/PEO solutions (3 wt% chitosan, 5 wt% PEO, and 0.5 wt% acetic acid) were electrospun under similar conditions ($V = 12$ kV, TCD = 8 cm, FR = 0.2 mL h^{-1}). The electrospinning of chitosan/PEO (3:7) with 0.4 wt% NaCl (or KCl) produced nanofibers accompanied by a recrystallization of inorganic salts. Comparatively, the electrospinning of the same solution but with 0.4–1 wt% $CaCl_2$ or 0.4–1.2 wt% $FeCl_3$ led to uniform fibers with an average diameter of 200 nm. In general, adding metal ions enhances chain entanglement and increases solution conductivity, which leads to overall improved spinnability. However, it is also important to note that fiber integrity is negatively influenced by the presence of excessive Fe^{3+} [82].

Carboxymethyl chitosan (CMCS) can be produced by reacting chitosan with 40 wt% NaOH and monochloroacetic acid [83]. Although CMCS with a degree of substitution above 0.73 is completely WS, its electrospinning at a wide range of concentrations produces only droplets. Neither adding a nonionic surfactant nor varying the applied voltage resulted in successful fiber production. This is attributed to insufficient entanglement of CMCS rigid chains. As such, the polymer blend strategy was found to be the most preferable method for the electrospinning of chitosan and its derivatives. The results for blending with four different synthetic WS polymers are summarized in Table 6.5. Similarly, the electrospinning conditions of other WS chitosan derivatives are presented in Table 6.6.

Table 6.5: Electrospinning of carboxymethyl chitosan blended with four different synthetic WS polymers using water as a solvent. From the Ref. [83].

Polymers	Electrospinning conditions	Average fiber diameter
CMCS 3:7 PEO	V = 10–15 kV, TCD = 17–20 cm [Polymer] = 7 wt%	300 nm
CMCS 9:1 PAAm	V = 10–15 kV, TCD = 17–20 cm [Polymer]=18 wt%	300 nm
CMCS 5:5 PAA	V = 10–15 kV, TCD = 17–20 cm [Polymer] = 10 wt%	100 nm
CMCS <50% PVA 50–80%	V = 10–15 kV, TCD = 17–20 cm [Polymer] = 8.5 wt%	CMCS 50% →170 nm CMCS 20% →210 nm

CMCS, carboxymethyl chitosan; PEO, polyethylene oxide; PAAm, polyacrylamide; PAA, polyacrylic acid; PVA, polyvinyl alcohol.

Table 6.6: Electrospinning conditions of chitosan derivatives in aqueous solution.

Polymer/s	Electrospinning conditions	Fiber diameter	Reference
QCh + PVP Various ratios	[Polymer] = 20 wt% V/TCD = 1.6–2.8 kV cm^{-1}	1,500–2,800 nm	[84]
CECh + PAAm Various ratios	[Polymer] = 3 wt% V/TCD = 0.7–1.6 kV cm^{-1}	50–260 nm	[85]
HTCC + PVA Various ratios	[Polymer] = 1% (w/v) V = 10, 15, 20 kV TCD = 15 cm FR = 1.2 mL h^{-1}	200–600 nm Decreases with decreasing [PVA] and increasing V	[86]
PVA + COS + MMT	[Polymer] = 12.5 wt% V = 15 kV, TCD = 15 cm	200–700 nm	[87]
Chitosan + PEO Water + 3% acetic acid	[Polymer] = 3–5 wt% Chitosan:PEO = 1:1 V = 21 kV, TCD = 18 cm FR = 0.5 mL h^{-1}	77–124 nm Depending on the load of metronidazole	[81]

QCh, quaternized chitosan; CECh, *N*-carboxyethyl chitosan; HTCC, *N*-[(2-hydroxy-3-trimethylammonium)propyl] chitosan; COS, chitosan oligosaccharide; MMT, montmorillonite clay.

Collagen

Collagen is one of the most abundant proteins present in the mammalian extracellular matrix (ECM). It is biocompatible, antigenic, and nonimmunogenic, and promotes cellular attachment and growth [88]. There are multiple types of collagen characterized by minor genetic variations. Types I, II, III, and IV have all been reported to form fibers upon electrospinning [89, 90]. This highlights that electrospinning has the potential to precisely engineer and tune fibrous scaffolds that effectively mimic the natural collagen ECM [88].

The electrospinning of collagen alone in acidic aqueous solution cannot be accomplished due to low solution conductivity and viscosity. Blending with synthetic WS polymers such as PEO overcomes these obstacles and facilitates fiber formation. Uniform fibers with diameters in the range of 151–278 nm were produced upon electrospinning of a blend solution of 5 wt% collagen, 3–5 wt% PEO, and 34 mM NaCl under the following conditions: V = 16 kV, FR = 0.8 mL h^{-1}, TCD = 20 cm [91]. Similarly, electrospun collagen nanofibers with diameters in the range of 100–150 nm were produced using 2 wt% collagen–PEO, 34 mM NaCl at V = 18 kV, FR = 100 μL min^{-1}, and a TCD = 20–30 cm [92]. A mixture of ethanol/water 1:1 was also used for the electrospinning of collagen at an initial concertation of 15 wt%. The experimental conditions were as follows: V = 20 kV, FR = 1 mL h^{-1}, and TCD = 15 cm [93].

Collagen loses its mechanical integrity upon hydration in electrospun form [89]. To enhance its mechanical properties, the resulting electrospun collagen mesh can be cross-linked with glutaraldehyde and 1-ethyl-3-(3-dimethylaminopropyl) carbodiimide hydrochloride [89]. Another viable method to improve the mechanical stability of the collagen electrospun structure is through the incorporation of biodegradable synthetic polymers or proteins into the structure. This strengthens the structure of all while maintaining a high level of bioactivity [89].

Dextran

Dextran is a WS, biodegradable, and inexpensive polysaccharide that can be used for electrospinning. Dextran (MW = 64,000–76,000 Da) was electrospun from its aqueous solution at various concentrations (41.2–56.5 w/w%) under the following conditions: V = 9–21 kV, TCD = 15 cm, and FR = 0.25 mL h^{-1}. The fiber diameter increased from 290 nm at a concentration of 0.7 mg L^{-1} to 1,950 nm at 1.3 mg L^{-1}. However, beadless fibers were only obtained for concentrations above 0.9 mg L^{-1} [94].

Some organic solvents such as DMSO, DMF, and THF are miscible with water and can be used as cosolvents for the electrospinning of WS polymers. For example, a dextran solution in a water/DMSO solvent 60:40 has been used for electrospinning, resulting in uniform nanofibers of 150 nm diameter under optimized conditions ([Dextran] = 0.75 g mL^{-1}, FR = 1.2 mL h^{-1}, V = 25 kV, TCD = 15 cm) [95].

Gelatin

Gelatin is derived from collagen by controlled hydrolysis. It is utilized in various applications such as food gel formation, capsules, sizing agents, and coloring agents in soft drinks [96]. It can also enhance wound healing and tissue regeneration, since cells are able to adhere to the surface of gelatin porous mats and proliferate [97]. Gelatin is soluble in warm water as well as various organic solvents such as hexafluoroisopropanol, trifluoroethanol, and formic acid.

When gelatin is dissolved in water at room temperature, a three-dimensional interconnected network of chains forms, thus rendering its electrospinning difficult [98]. Gelatin is a temperature-sensitive polymer such that its viscosity and electrical conductivity vary significantly with changes in temperature. In fact, when temperature is increased, the viscosity of aqueous gelatin solution decreases due to the breakup of hydrogen bonding established between neighboring chains and the rise of the gelatin molecular activity in the solution. The electrical conductivity of the solution increases as well since ionization of gelatin molecules takes place at elevated temperatures. Gelatin is hardly, if at all, electrospun at temperatures below 30 °C – this process becoming markedly easier, however, at higher temperatures (35–50 °C). Nevertheless, very high temperatures could also lead to electrospinning failure by inducing gelatin degradation and forming an unstable jet during the process [98].

Another strategy to ensure stable electrospinning of gelatin solution is polymer blending. Several gelatin-based blends are reported to form uniform mats. These include polymers such as polyaniline, PVA, and PEO. In fact, adding a polymer enhances the mechanical properties of the ensuing mesh. Upon electrospinning of a blend solution of PVA + gelatin, increasing the gelatin concentration decreases the solution viscosity, which in turn leads to a decrease in fiber diameter. Figure 6.2 shows the variation of fiber diameter, viscosity, conductivity, and surface tension with the PVA/gelatin ratio.

The electrospinnability of various concentrations of gelatin, water, and acetic acid is demonstrated in Figure 6.3. The mixtures with very high viscosity, which typically end up forming gels, cannot be used for electrospinning. The same goes for the mixtures with very low viscosity, as barely any chain entanglement occurs, a fact that results in electrospraying rather than electrospinning. The beaded fibers region is only achieved as the chain entanglement density of solution increases [99].

In order to make the resulting electrospun gelatin mesh water insoluble for various applications, it can be cross-linked with glutaraldehyde [97], 1-ethyl-3-(dimethyl-aminopropyl) carbodiimide hydrochloride, and *N*-hydroxyl succinimide [98].

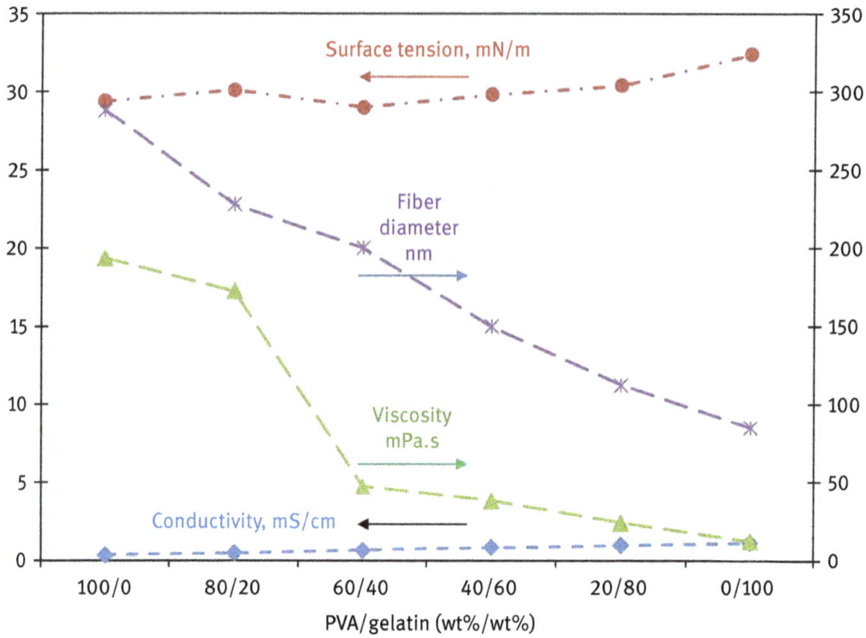

Figure 6.2: Variation of fiber diameter (nm), viscosity (mPa s), conductivity (mS cm^{-1}), and surface tension (mN m^{-1}) with PVA/gelatin ratio (electrospun conditions: TCD = 10 cm, FR = 1 mL h^{-1}, V = 10–25 kV). From Ref. [97].

6.4 Applications of electrospun WS polymers

Despite a large number of applications for WS polymers (e.g., thickeners, flocculants, and emulsifiers), the applications of WS electrospun meshes are mainly limited to drug delivery and tissue engineering [3–6]. From a biological standpoint, many human tissues and organs are found in nanofibrous forms, with examples including bone, skin, denture, and collagen [100]. As such, interactions of cells with their submicron-scale surrounding topography necessitates the formation of biomimetic nanostructures, so that phenomena such as cellular adhesion, proliferation, migration, as well as gene expression could be better modeled, understood, and ultimately exploited [101, 102]. Porous nanofibrous scaffolds can be prepared by electrospinning, and the fiber diameter and pore size can be engineered and tuned by varying the electrospinning parameters. Nanofibrous electrospun scaffolds with high surface area-to-volume ratio can mimic the native ECM morphology and be used for tissue engineering [103].

The goal of drug delivery is to provide proper vehicles for therapeutics (especially novel macromolecular drugs such as proteins and peptides) to reach their target sites in a controlled manner. The design of drug carriers is very important as

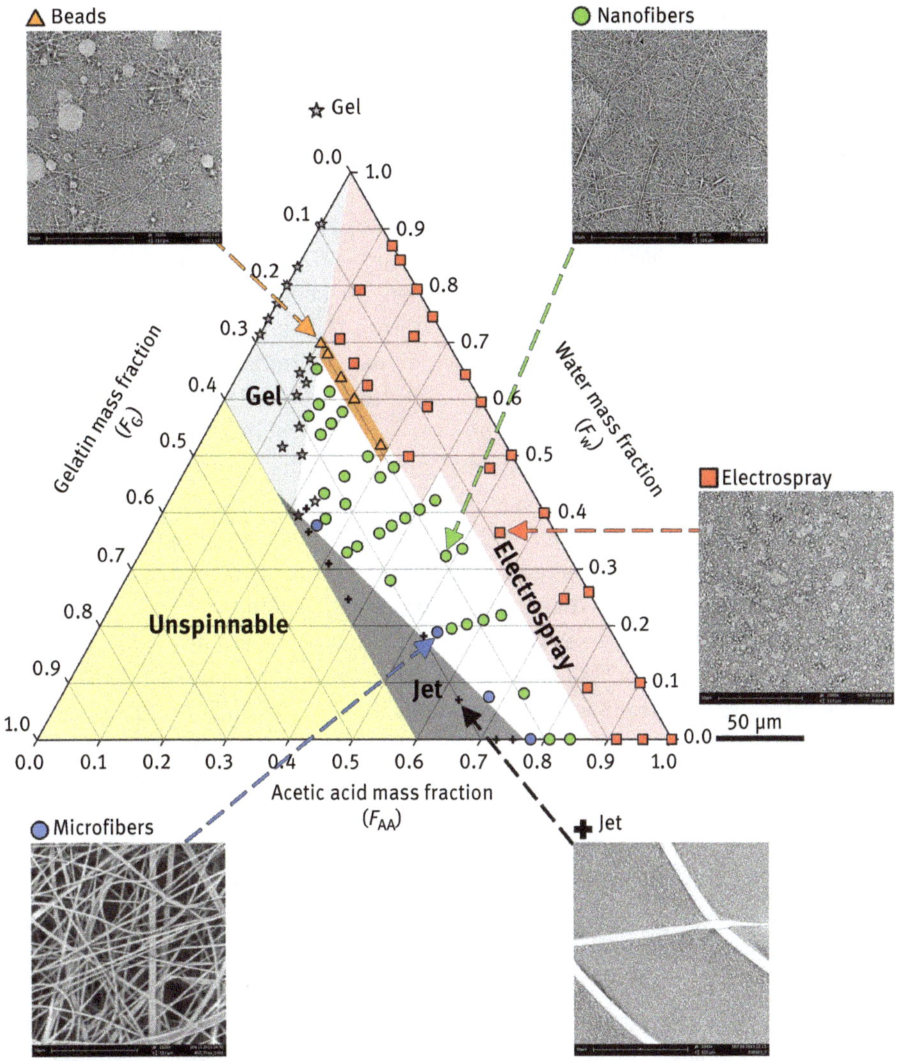

Figure 6.3: Ternary diagram showing the electrospinnability domains based on the gelatin (G), acetic acid (AA), and water (w) mass fractions. Color-coding has been used to show different regions (e.g., unspinnable, jet, gel, electrospraying, and electrospinning) along with representative SEM images of each domain. Adapted with permission from Ref. [99]. Copyright (2014) American Chemical Society.

they can be highly toxic when delivered systemically to off-target tissues and cells. Electrospinning boosts research efforts in this area as it allows the formation of nanofibers with an inherently high surface area-to-volume ratio, which then translates into more desirable properties for the nanocarrier, namely, increased drug encapsulation efficiency, better bioavailability, and enhanced mass diffusion properties [104, 105]. A variety of drugs have been successfully incorporated into electrospun fibers, in addition to novel therapeutics such as proteins, DNA, RNAi, various growth factors, as well as cells for gene therapy [106].

For electrospinning applications, a judicious choice of polymer is essential for a successful run of experiments. The polymer chemical structure plays an important role on drug loading as well as release behavior from nanofibers. In fact, an essential feature of most novel therapies is their hydrophobicity, which precludes their direct intravenous injection in the bloodstream, or direct use in oral delivery. Chemical modification such as pH adjustment to ensure drug solubilization in water works only to a limited extent as the drug is bound to recrystallize in the blood buffer system, which leads to a very high risk of thrombogenicity/blood coagulation. Coupling the drug with other cosolutes to enhance delivery also works but only up to a limited extent, as the introduction of cosolutes in the bloodstream might introduce complications of its own due to their potential inherent toxicity [107]. WS polymers, that are typically nonimmunogenic and nonthrombogenic, have found important applications in drug encapsulation, solubilization, and delivery. Many resulting microstructures, ranging from microspheres, to micelles to polymer–drug conjugates (where separation occurs through spacer molecules) have led to significant increases in drug loading and bioavailability [108]. The ultimate goal of drug delivery systems is to break down the process of delivering therapeutics into a simple feedback-control system, which responds to environmental stimuli coming from surrounding physiologic conditions, in a way which enhances patient compliance and quality of life. This entails moving away from treatments that are invasive and more toward minimally or preferably noninvasive treatments such as oral delivery [109]. WS electrospun meshes have been used for delivery of a number of drugs as shown in Table 6.7. The electrospinning conditions along with the release rate of each drug have also been presented in this table to better elucidate the structure–functionality dependence in WS polymeric systems. A survey of the existing literature shows that PVA is the most widely used for these applications. In fact, in addition to its desirable characteristics such as its hydrophilicity, nonimmunogenicity, biocompatibility, and excellent adhesive properties, it has the added distinction of being biodegradable (under both aerobic and anaerobic conditions), making it significantly more environmentally friendly to use than other WS nonbiodegradable synthetic polymers [110]. Moreover, through an inter-report comparison, it is shown that polymers with higher molecular weights typically have lower rates of solubility in aqueous media which in turn leads to slower drug release rates. Conversely, however, the motivation behind the electrospinning of membranes with ever-finer fiber diameters is the formation of a larger surface area with the surrounding medium. This normally implies a much more significant dissolution rate. Overall, there is no “one size fits all” solution to the question of a best electrospun WS mesh design – the specifics of the release profile are tailored with respect to the needs of each application.

Table 6.7: Some examples of water-based electrospinning of WS nanofibrous mesh for drug delivery.

Polymer	ES conditions	Drug-release profile and additional notes	Reference
[PVA] = 8% (w/v), [chitosan] = 0.6%(w/v)	V = 15 kV TCD = 10 cm FR = 0.1–0.3 mL h^{-1} RH = 40%	Drug: tetracycline 80% release in 2 h – A first-order model was observed Has potential usage as antimicrobial wound dressing	[111]
CS 3:1 PEO [polymer] = 4% in water 1:1 AA	V = 21 kV; TCD = 10 cm FR = 1 mL h^{-1} T = 25–35 °C RH = 35–42%	Drugs: hydrocortisone 82% within 12 h and imipenem/cilastatin 20% in 12 h The CS-PEO nanofibrous mats contained 3% ZnO NPs, 2% hydrocortisone, 3% imipenem/cilastatin-loaded ZnO NPs	[112]
CS 1:1 PEO [polymer] = 4% in 3% AA (w/v)	V = 21 kV TCD = 18 cm FR = 0.5 mL h^{-1} T = 30 °C RH = 35%	Drug: metronizadole 60% release in 10 min, and 95% release in 2 h The release kinetics depends on the available fluid at the therapeutic site	[81]

ES, electrospun; CS, chitosan; AA, acetic acid; EtOH, ethanol; NPs, nanoparticles.

Polysaccharides, as mentioned earlier in the chapter, are endowed with desirable properties such as biocompatibility, biodegradability, hydrophilicity, and enhanced cell adhesion as compared to synthetic polymers. Sodium alginate (SA), for instance, is extensively used as polymeric scaffold for cell culture [113], and drug carrier for a wide range of therapeutics, such as proteins [114]. Polymer blending has the advantage of meeting the chain entanglement requirement without greatly increasing the solution viscosity, which might itself lead to complications in the electrospinning process, such as the formation of polymer beads.

Core–shell blend electrospun meshes were obtained by electrospinning for the colon delivery of bovine serum albumin (BSA) as a model substance for protein-release behavior. The core and shell consisted of chitosan/PVA 50:50 and SA/PEO 80:20, respectively. Pluronic F127, an FDA-approved nonionic surfactant, was used to facilitate the electrospinning, and a mesh with fiber diameter in the 200–600 nm range was ultimately obtained under optimum conditions (V = 17–18 kV, TCD = 15–16 cm, FR = 0.25–0.3 mL h^{-1}). SA as the shell layer resists the acidic conditions imposed by the surrounding media, and chitosan is insoluble in the small intestine. The results showed that less than 4% BSA is released in simulated gastric fluid (pH 1.2) after 2 h, whereas 15% BSA was released in simulated intestinal fluid (pH 6.8) after 4 h, and around 75% of BSA was released in simulated colonic fluid (pH 7.4) after 16 h. Thus, the core–shell electrospun matrix has the potential to be used for the targeted delivery of BSA to the colon over a 24 h period [114].

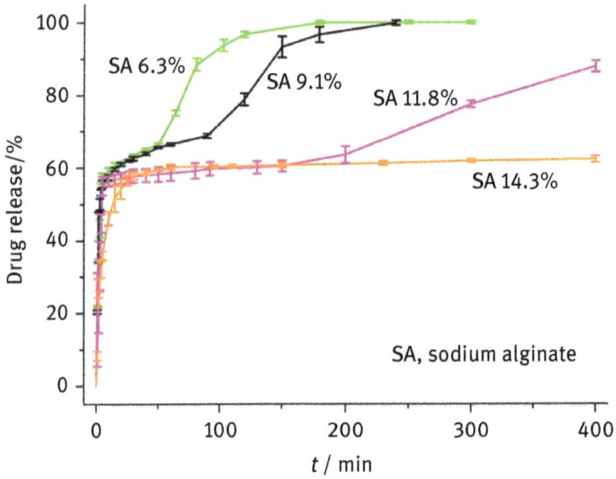

Figure 6.4: Ibuprofen release from electrospun PEO–sodium alginate (SA) fibers with varied SA contents at pH 3. Adapted with permission from Ref. [115].

The release of ibuprofen from electrospun PEO–SA blend nanofibers can be seen in Figure 6.4. The drug release decreases when SA concentration is increased in the composition since it has very limited solubility at low pH values, whereas PEO is more readily soluble in water under acidic conditions. As shown in the following figure, the release has a two-stage pulsatile release profile and the system offers effective protection to the drug against fast release in the acidic conditions of the stomach [115].

Some papers report very fast drug-release profiles from WS electrospun meshes. For example, the release of Donepezil HCl from PVA mesh is less than 1 min, while the release of the same drug from commercial tablets or casted films with the same composition is 30 times slower. This has been attributed to the larger surface area-to-volume area of the electrospun mesh [51]. Similar behavior has been reported for the release of ondansetron from a PVA electrospun mesh [116] as well as the release of ibuprofen and acetylsalicylic acid from a PVP–dextran electrospun mesh [58]. In order to reduce and better control this release rate, the PVA or PVP can be blended with natural-based polymers or alternatively cross-linked to reduce water solubility.

PVA-MC polymeric solutions containing $AgNO_3$ and methacrylate groups were all electrospun from their aqueous solutions. The resulting meshes (fiber diameter = 300–500 nm) were cross-linked under UV irradiation and annealed at 100 °C for the reduction of $AgNO_3$ to Ag nanoparticles. The electrospun meshes showed antimicrobial activity against *Escherichia coli* due to the presence of Ag nanoparticles [117].

WS electrospun meshes for tissue engineering application feature less often in the literature since the WS polymers that are most widely used for regenerative medicine (such as chitosan) are relatively difficult to be electrospun. This difficulty is mainly overcome in most papers through the use of blending, typically with WS synthetic polymers such as PEO [118] or PVA [119]. In fact, for an electrospinning process to be successful, the polymer solution used has to be within a certain range for both concentration and viscosity, outside of which beading might occur or fibers might not be formed in the first place. Blending the polymer solution with even small amounts of synthetic polymer additives induces chain entanglement (thereby increasing solution viscosity significantly) without having to greatly increase the total polymer solution concentration. This approach is used to render the otherwise difficult electrospinning of polymers such as chitosan and cellulose (and their respective derivatives) easier, without having to use organic solvents and staying within the realm of benign aqueous media. Also, in the case of some applications such as cell culture, the mechanical properties of the electrospun matrix may be found to be lacking in some respects. This can be overcome by the incorporation of additives into the system. For example, the addition of graphene oxide into the chitosan/PEO matrix increased the tensile strength of the mesh [118]. Moreover, the polymer ratios in the blend to be electrospun as well as the overall eventual solute concentration typically have to be optimized in each study for biocompatibility. In fact, finding the most appropriate environment confers a better scaffolding for phenomena such as cell growth and adhesion to occur [83].

6.5 Conclusion

Water-based electrospinning is a green process for the production of nanofibrous meshes, and it is well-suited for advanced applications such as drug delivery and tissue engineering. It can also be used for producing reinforced clay aerogel by calcination.

Due to the high surface tension and dielectric constant of water, researchers have employed different strategies to overcome the electrospinning difficulties by addition of other spinning aids (e.g., surfactants, cosolvents, and inorganic salts/acids). They have also reduced the water solubility of the resulting electrospun meshes by cross-linking using various chemical reagents.

List of abbreviations

AA	Acetic acid
Ag	Silver
Au	Gold
BSA	Bovine serum albumin
CMC	Carboxymethyl cellulose

CMCtn	Carboxymethyl chitin
CMCS	Carboxymethyl chitosan
DH	Degree of hydrolysis
DMSO	Dimethyl sulfoxide
ECM	Extracellular matrix
EtOH	Ethanol
F127	Polyethylene oxide/polypropylene oxide block copolymer (EO_{100}–PO_{65}–EO_{100})
FR	Feed rate
HPMC	Hydroxypropyl methylcellulose
MC	Methylcellulose
NMMO	*N*-methyl-morpholine *N*-oxide
PEO	Polyethylene oxide
PVA	Polyvinyl alcohol
PVP	Polyvinyl pyrrolidone
SA	Sodium alginate
SDS	Sodium dodecyl sulfate
TCD	Tip-to-collector distance
V	Voltage
WS	Water soluble

References

[1] Elsabee, M.Z., Naguib, H.F., & Morsi, R.E. Chitosan based nanofibers review, Mat. Sci. Eng. C-Mater., 2012, 32, 1711–1726.

[2] Maksoud, F.J., Lameh, M., Fayyad, S., Ismail, N., Tehrani-Bagha, A.R., Ghaddar, N., & Ghali, K. Electrospun waterproof breathable membrane with a high level of aerosol filtration, J. Appl. Polym .Sci., 2018, 135, 45660.

[3] Dai, X.Y., Nie, W., Wang, Y.C., Shen, Y., Li, Y., & Gan, S.J. Electrospun emodin polyvinylpyrrolidone blended nanofibrous membrane: a novel medicated biomaterial for drug delivery and accelerated wound healing, J. Mater. Sci-Mater. M, 2012, 23, 2709–2716.

[4] Immich, A.P.S., Arias, M.L., Carreras, N., Boemo, R.L., & Tornero, J.A. Drug delivery systems using sandwich configurations of electrospun poly(lactic acid) nanofiber membranes and ibuprofen, Mat .Sci. Eng. C-Mater., 2013, 33, 4002–4008.

[5] Fernandes, J.G., Correia, D.M., Botelho, G., Padrao, J., Dourado, F., Ribeiro, C., Lanceros-Mendez, S., & Sencadas, V. PHB-PEO electrospun fiber membranes containing chlorhexidine for drug delivery applications, Polym .Test, 2014, 34, 64–71.

[6] Dai, X., Hopfner, U., Zheng, H., Liu, J., Sudhop, S., Prein, C., Shen, Y., Machens, H.G., & Schilling, A. Electrospun curcumin-gelatin blended nanofibrous membrane: A novel medicated biomaterial for drug delivery and accelerated wound healing, Tissue Eng. Pt A, 2015, 21, S351–S351.

[7] Nagy, Z.K., Balogh, A., Demuth, B., Pataki, H., Vigh, T., Szabo, B., Molnar, K., Schmidt, B.T., Horak, P., Marosi, G., Verreck, G., Van Assche, I., & Brewster, M.E. High speed electrospinning for scaled-up production of amorphous solid dispersion of itraconazole, Int. J. Pharmaceut., 2015, 480, 137–142.

[8] Krishnan, R., Sundarrajan, S., & Ramakrishna, S. Green processing of nanofibers for regenerative medicine, Macromol. Mater. Eng., 2013, 298, 1034–1058.

[9] Jiang, S.H., Hou, H.Q., Agarwal, S., & Greiner, A. Polyimide nanofibers by "Green" electrospinning via aqueous solution for filtration applications, ACS Sustain .Chem. Eng., 2016, 4, 4797–4804.
[10] Yi, F., Guo, Z.X., Hu, P., Fang, Z.X., Yu, J., & Li, Q. Mimetics of eggshell membrane protein fibers by electrospinning, Macromol. Rapid Comm., 2004, 25, 1038–1043.
[11] Zeng, J., Hou, H.Q., Wendorff, J.H., & Greiner, A. Photo-induced solid-state crosslinking of electrospun poly(vinyl alcohol) fibers, Macromol. Rapid Comm., 2005, 26, 1557–1562.
[12] Liu, Y.R., Bolger, B., Cahill, P.A., & McGuinness, G.B. Water resistance of photocrosslinked polyvinyl alcohol based fibers, Mater .Lett., 2009, 63, 419–421.
[13] Cho, D., Bae, W.J., Joo, Y.L., Ober, C.K., & Frey, M.W. Properties of PVA/HfO2 hybrid electrospun fibers and calcined inorganic HfO2 fibers, J. Phys. Chem. C., 2011, 115, 5535–5544.
[14] Madyan, O.A., Fan, M., Feo, L., & Hui, D. Enhancing mechanical properties of clay aerogel composites: An overview, Compos .Part B-Eng., 2016, 98, 314–329.
[15] Swift, G., Creamer, M., Wei, X., & Yocom, K.M. Water-soluble biodegradable polymers: Synthetic or natural-based raw materials?, Macromol. Symp., 1998, 130, 379–391.
[16] Huang, S.J., Seery, T.A.P., & Swift, G. Biomedical applications of water-soluble polymers and hydrogels – Preface, J. Macromol. Sci. Pure., 1999, A36, Ix-Ix.
[17] Kadajji, V.G., & Betageri, G.V. Water soluble polymers for pharmaceutical applications, Polym.-Basel, 2011, 3, 1972–2009.
[18] P.M, V., Bayraktar, O., & Picó, G. Polyelectrolyte: Thermodynamics and Rheology, Springer, Switzerland, 2014.
[19] Swift, G. A Review of water-soluble polymers in the environment – biodegradable or recalcitrant, Abstr. Pap. Am. Chem S, 1990, 200, 187-Pmse.
[20] Zhang, L.M. New water-soluble cellulosic polymers: A review, Macromol .Mater .Eng., 2001, 286, 267–275.
[21] Imeson, A. Thickening and gelling agents for food, 2nd, Blackie Academic & Professional, London, 1997.
[22] Caessens, P.W., Gruppen, H., Slangen, C.J. Visser, S., & Voragen, A.G. Functionality of β-casein peptides: Importance of amphipathicity for emulsion-stabilizing properties, J. Agri. Food Chem., 1999, 47, 1856–1862.
[23] Wang, Y.F., Lai, C.L., Wang, X.W., Liu, Y., Hu, H.W., Guo, Y.J., Ma, K.K., Fei, B., & Xin, J.H. Beads-on-string structured nanofibers for smart and reversible oil/water separation with outstanding antifouling property, Acs .Appl. Mater. Inter., 2016, 8, 25612–25620.
[24] Miskiel, F.J., & Solanki, Y., Acidic cleaning compositions containing xanthan gum, Google Patents, 1999.
[25] Gilbert, L., Loisel, V., Savary, G., Grisel, M., & Picard, C. Stretching properties of xanthan, carob, modified guar and celluloses in cosmetic emulsions, Carbohydr. Polym., 2013, 93, 644–650.
[26] Himes, R., Lee, S., McMenigall, K., & Russell-Jones, G.J. Reduction in inflammation in the footpad of carrageenan treated mice following the topical administration of anti-TNF molecules formulated in a micro-emulsion, J. Controlled Release, 2010, 145, 210–213.
[27] Wen, J., Zhang, Q., Zhu, D., & Zhang, W. Performance study on particle size variables for nano multiple emulsions, J. Dispersion Sci. Technol., 2017, 38, 801–806.
[28] Abdel-Halim, E.S., Emam, H.E., & El-Rafie, M.H. Utilization of hydroxypropyl cellulose and poly(acrylic acid)-hydroxypropyl cellulose composite as thickeners for textile printing, Carbohydr. Polym., 2008, 74, 938–941.
[29] NPCS Board of Consultants & Engineers, Handbook on textile auxiliaries, dyes and dye intermediates technology, Asia Pacific Business Press, India, 2009.

[30] Comaposada, J., Gou, P., Marcos, B., & Arnau, J. Physical properties of sodium alginate solutions and edible wet calcium alginate coatings, LWT – Food Sci. Technol., 2015, 64, 212–219.

[31] Li, J.H., Wei, Z., & Wu, C. Preparation and properties of novel building materials at low temperature, Mater. Des., 2015, 67, 464–468.

[32] Rabiee, A. Acrylamide-based anionic polyelectrolytes and their applications: A survey, J. Vinyl Add. Technol., 2010, 16, 111–119.

[33] Henze, M., Harremoes, P., la Cour Jansen, J., & Arvin, E. Wastewater treatment: biological and chemical processes, Springer Science & Business Media, Germany, 2001.

[34] Fujimoto, N. Aqueous dispersion of a water-soluble anionic polymer as flocculants, Kinoshi Kenkyu Kaishi/Annals of the High Performance Paper Society, Japan, 2002, 59–65.

[35] Zhang, M., Yang, B.X., Liu, W., & Li, S.M. Influence of hydroxypropyl methylcellulose, methylcellulose, gelatin, poloxamer 407 and poloxamer 188 on the formation and stability of soybean oil-in-water emulsions, Asian J. Pharm. Sci., 2017, 12, 521–531.

[36] Williams, P.A. Handbook-of-Industrial-Water-Soluble-Polymers, Blackwell Publishing Ltd, UK, 2007.

[37] Jacobs, V., Anandjiwala, R.D., & Maaza, M. The influence of electrospinning parameters on the structural morphology and diameter of electrospun nanofibers, J. Appl .Polym. Sci., 2010, 115, 3130–3136.

[38] Chowdhury, M., & Stylios, G.K. Analysis of the effect of experimental parameters on the morphology of electrospun polyethylene oxide nanofibres and on their thermal properties, J. Text. Inst., 2012, 103, 124–138.

[39] Peer, P., & Filip, P. Nanofibrous web quality in dependence on the preparation of poly (ethylene oxide) aqueous solutions, J. Text. I., 2017, 108, 2021–2026.

[40] Tripatanasuwan, S., Zhong, Z.X., & Reneker, D.H. Effect of evaporation and solidification of the charged jet in electrospinning of poly(ethylene oxide) aqueous solution, Polymer, 2007, 48, 5742–5746.

[41] Son, W.K., Youk, J.H., Lee, T.S., & Park, W.H. The effects of solution properties and polyelectrolyte on electrospinning of ultrafine poly(ethylene oxide) fibers, Polymer, 2004, 45, 2959–2966.

[42] Call, C.C., & Schwartz, C.J. The use of weber number to predict morphology in the electrospinning of poly(ethylene oxide) nanofibers, J. Appl .Polym .Sci., 2011, 120, 880–885.

[43] Krstic, M., Radojevic, M., Stojanovic, D., Radojevic, V., Uskokovic, P., & Ibric, S. Formulation and characterization of nanofibers and films with carvedilol prepared by electrospinning and solution casting method, Eur. J. Pharm. Sci., 2017, 101, 160–166.

[44] Aluigi, A., Varesano, A., Montarsolo, A., Vineis, C., Ferrero, F., Mazzuchetti, G., & Tonin, C. Electrospinning of keratin/poly(ethylene oxide) blend nanofibers, J. Appl. Polym .Sci., 2007, 104, 863–870.

[45] Talwar, S., Hinestroza, J., Pourdeyhimi, B., & Khan, S.A. Associative polymer facilitated electrospinning of nanofibers, Macromolecules., 2008, 41, 4275–4283.

[46] Wong, D.E., Dai, M.H., Talbert, J.N., Nugen, S.R., & Goddard, J.M. Biocatalytic polymer nanofibers for stabilization and delivery of enzymes, J. Mol. Catal. B-Enzym., 2014, 110, 16–22.

[47] Jia, L., & Qin, X.-h. The effect of different surfactants on the electrospinning poly(vinyl alcohol) (PVA) nanofibers, J. Therm. Anal. Calorim., 2012, 112, 595–605.

[48] Son, W.K., Youk, J.H., Lee, T.S., & Park, W.H. Effect of pH on electrospinning of poly(vinyl alcohol), Mater. Lett., 2005, 59, 1571–1575.

[49] Wang, Z.L., Cai, N., Zhao, D.P., Xu, J., Dai, Q., Xue, Y.A., Luo, X.G., Yang, Y., & Yu, F.Q. Mechanical reinforcement of electrospun water-soluble polymer nanofibers using nanodiamonds, Polym. Compos., 2013, 34, 1735–1744.

[50] Taepaiboon, P., Rungsardthong, U., & Supaphol, P. Drug-loaded electrospun mats of poly (vinyl alcohol) fibres and their release characteristics of four model drugs, Nanotechnology, 2006, 17, 2317–2329.

[51] Nagy, Z.K., Nyul, K., Wagner, I., Molnar, K., & Marosi, G. Electrospun water soluble polymer mat for ultrafast release of Donepezil HCl, Express .Polym. Lett., 2010, 4, 763–772.

[52] Nakane, K., Yasuda, K., Ogihara, T., Ogata, N., & Yamaguchi, S. Formation of poly(vinyl alcohol)-titanium lactate hybrid nanofibers and properties of TiO_2 nanofibers obtained by calcination of the hybrids, J. Appl .Polym .Sci., 2007, 104, 1232–1235.

[53] Yang, E.L., Qin, X.H., & Wang, S.Y. Electrospun crosslinked polyvinyl alcohol membrane, Mater. Lett., 2008, 62, 3555–3557.

[54] Lopez-Rubio, A., Sanchez, E., Sanz, Y., & Lagaron, J.M. Encapsulation of living Bifidobacteria in ultrathin PVOH electrospun fibers, Biomacromolecules, 2009, 10, 2823–2829.

[55] Park, J.-C., Ito, T., Kim, K.-O., Kim, K.-W., Kim, B.-S., Khil, M.-S., Kim, H.-Y., & Kim, I.-S. Electrospun poly(vinyl alcohol) nanofibers: effects of degree of hydrolysis and enhanced water stability, Polym. J., 2010, 42, 273.

[56] Zhang, L.J., Bai, J., Zhang, Y.F., Li, C.P., & Zhang, J.B. A novel approach to prepare $CaCO_3$/ polyvinylpyrrolidone (PVP) composite nanofibers, J. Macromol. Sci. B., 2012, 51, 2133–2140.

[57] Dai, M.H., Senecal, A., & Nugen, S.R. Electrospun water-soluble polymer nanofibers for the dehydration and storage of sensitive reagents, Nanotechnology, 2014, 25, 225101.

[58] Maslakci, N.N., Ulusoy, S., Uygun, E., Cevikbas, H., Oksuz, L., Can, H.K., & Oksuz, A.U. Ibuprofen and acetylsalicylic acid loaded electrospun PVP-dextran nanofiber mats for biomedical applications, Polym. Bull., 2017, 74, 3283–3299.

[59] Yu, D.G., Gao, L.D., White, K., Branford-White, C., Lu, W.Y., & Zhu, L.M. Multicomponent amorphous nanofibers electrospun from hot aqueous solutions of a poorly soluble drug, Pharm. Res-Dordr., 2010, 27, 2466–2477.

[60] Saquing, C.D., Tang, C., Monian, B., Bonino, C.A., Manasco, J.L., Alsberg, E., & Khan, S.A. Alginate–polyethylene oxide blend nanofibers and the role of the carrier polymer in electrospinning, Ind. Eng. Chem. Res., 2013, 52, 8692–8704.

[61] Nie, H.R., He, A.H., Zheng, J.F., Xu, S.S., Li, J.X., & Han, C.C. Effects of chain conformation and entanglement on the electrospinning of pure alginate, Biomacromolecules, 2008, 9, 1362–1365.

[62] Bonino, C.A., Krebs, M.D., Saquing, C.D., Jeong, S.I., Shearer, K.L., Alsberg, E., & Khan, S.A. Electrospinning alginate-based nanofibers: From blends to crosslinked low molecular weight alginate-only systems, Carbohydr. Polym., 2011, 85, 111–119.

[63] Park, S.A., Park, K.E., & Kim, W.D. Preparation of sodium alginate/poly(ethylene oxide) blend nanofibers with lecithin, Macromol. Res., 2010, 18, 891–896.

[64] Nie, H., He, A., Wu, W., Zheng, J., Xu, S., Li, J., & Han, C.C. Effect of poly(ethylene oxide) with different molecular weights on the electrospinnability of sodium alginate, Polymer, 2009, 50, 4926–4934.

[65] Safi, S., Morshed, M., Hosseini-Ravandi, S.A., & Ghiaci, M. Study of electrospinning of sodium alginate, blended solutions of sodium alginate/poly(vinyl alcohol) and sodium alginate/poly(ethylene oxide), J. Appl. Polym. Sci., 2007, 104, 3245–3255.

[66] Lee, Y.J., Shin, D.S., Kwon, O.W., Park, W.H., Choi, H.G., Lee, Y.R., Han, S.S., Noh, S.K., & Lyoo, W.S. Preparation of atactic poly(vinyl alcohol)/sodium alginate blend nanowebs by electrospinning, J. Appl. Polym. Sci., 2007, 106, 1337–1342.

[67] Li, Z., & Wang, C. Effects of Working Parameters on Electrospinning, Springer Berlin Heidelberg, Berlin, Heidelberg, 2013, 15–28.

[68] Frey, M.W. Electrospinning cellulose and cellulose derivatives, Polym. Rev., 2008, 48, 378–391.

[69] Qi, H., Sui, X., Yuan, J., Wei, Y., & Zhang, L. Electrospinning of cellulose-based fibers from NaOH/urea aqueous system, Macromol. Mater. Eng., 2010, 295, 695–700.

[70] Khil, M.S., Kim, H.Y., Kang, Y.S., Bang, H.J., Lee, D.R., & Doo, J.K. Preparation of electrospun oxidized cellulose mats and their in vitro degradation behavior, Macromol. Res., 2005, 13, 62–67.
[71] Frenot, A., Henriksson, M.W., & Walkenström, P. Electrospinning of cellulose-based nanofibers, J. Appl. Polym. Sci., 2007, 103, 1473–1482.
[72] Kulpinski, P. Cellulose nanofibers prepared by the N-methylmorpholine-N-oxide method, J. Appl. Polym .Sci., 2005, 98, 1855–1859.
[73] Park, B.-D., Um, I.C., Lee, S.-Y., & Dufresne, A. Preparation and characterization of cellulose nanofibril/polyvinyl alcohol composite nanofibers by electrospinning, J. Korean Wood Sci. Technol., 2014, 42, 119–129.
[74] Casettari, L., Vllasaliu, D., Castagnino, E., Stolnik, S., Howdle, S., & Illum, L. PEGylated chitosan derivatives: Synthesis, characterizations and pharmaceutical applications, Prog. Polym. Sci., 2012, 37, 659–685.
[75] Sugimoto, M., Morimoto, M., Sashiwa, H., Saimoto, H., & Shigemasa, Y. Preparation and characterization of water-soluble chitin and chitosan derivatives, Carbohydr. Polym., 1998, 36, 49–59.
[76] Shalumon, K.T., Binulal, N.S., Selvamurugan, N., Nair, S.V., Menon, D., Furuike, T., Tamura, H., & Jayakumar, R. Electrospinning of carboxymethyl chitin/poly(vinyl alcohol) nanofibrous scaffolds for tissue engineering applications, Carbohydr. Polym., 2009, 77, 863–869.
[77] Singh, B.K., & Dutta, P.K., Chitin, chitosan, and silk fibroin electrospun nanofibrous scaffolds: A prospective approach for regenerative medicine, in Chtin and Chitosan for Regenerative medicine (Ed.), 2015, pp. 151–189.
[78] Pakravan, M., Heuzey, M.C., & Ajji, A. A fundamental study of chitosan/PEO electrospinning, Polymer, 2011, 52, 4813–4824.
[79] Geng, X.Y., Kwon, O.H., & Jang, J.H. Electrospinning of chitosan dissolved in concentrated acetic acid solution, Biomaterials, 2005, 26, 5427–5432.
[80] Ohkawa, K., Cha, D.I., Kim, H., Nishida, A., & Yamamoto, H. Electrospinning of chitosan, Macromol. Rapid. Comm., 2004, 25, 1600–1605.
[81] Zupancic, S., Potrc, T., Baumgartner, S., Kocbek, P., & Kristl, J. Formulation and evaluation of chitosan/polyethylene oxide nanofibers loaded with metronidazole for local infections, Eur. J. Pharm. Sci., 2016, 95, 152–160.
[82] Su, P., Wang, C., Yang, X., Chen, X., Gao, C., Feng, X.-X., Chen, J.-Y., Ye, J., & Gou, Z. Electrospinning of chitosan nanofibers: The favorable effect of metal ions, Carbohydr. Polym., 2011, 84, 239–246.
[83] Du, J., & Hsieh, Y.L. Nanofibrous membranes from aqueous electrospinning of carboxymethyl chitosan, Nanotechnology, 2008, 19, 125707.
[84] Ignatova, M., Manolova, N., & Rashkov, I. Novel antibacterial fibers of quaternized chitosan and poly(vinyl pyrrolidone) prepared by electrospinning, Eur. Polym. J., 2007, 43, 1112–1122.
[85] Mincheva, R., Manolova, N., Paneva, D., & Rashkov, I. Preparation of polyelectrolyte-containing nanofibers by electrospinning in the presence of a non-ionogenic water-soluble polymer, J. Bioact. Compat. Polym., 2005, 20, 419–435.
[86] Alipour, S.M., Nouri, M., Mokhtari, J., & Bahrami, S.H. Electrospinning of poly(vinyl alcohol)-water-soluble quaternized chitosan derivative blend, Carbohydr. Res., 2009, 344, 2496–2501.
[87] Park, J.H., Lee, H.W., Chae, D.K., Oh, W., Yun, J.D., Deng, Y., & Yeum, J.H. Electrospinning and characterization of poly(vinyl alcohol)/chitosan oligosaccharide/clay nanocomposite nanofibers in aqueous solutions, Colloid. Polym. Sci., 2009, 287, 943–950.
[88] Fullana, M.J., & Wnek, G.E. Electrospun collagen and its applications in regenerative medicine, Drug. Deliv. Transl. Res., 2012, 2, 313–322.

[89] Sell, S.A., McClure, M.J., Garg, K., Wolfe, P.S., & Bowlin, G.L. Electrospinning of collagen/biopolymers for regenerative medicine and cardiovascular tissue engineering, Adv. Drug. Deliv. Rev., 2009, 61, 1007–1019.
[90] Matthews, J.A., Boland, E.D., Wnek, G.E., Simpson, D.G., & Bowlin, G.L. Electrospinning of collagen type II: A feasibility study, J. Bioact. Compat. Polym., 2003, 18, 125–134.
[91] Chen, L., Zhu, C., Fan, D., Liu, B., Ma, X., Duan, Z., & Zhou, Y. A human-like collagen/chitosan electrospun nanofibrous scaffold from aqueous solution: Electrospun mechanism and biocompatibility, J. Biomed. Mater. Res. – Part A, 2011, 99(A), 395–409.
[92] Huang, L., Nagapudi, K., Apkarian, R. P., & Chaikof, E.L. Engineered collagen–PEO nanofibers and fabrics, J. Biomater. Sci. Polym. Ed., 2001, 12, 979–993.
[93] Jiang, Q., Reddy, N., Zhang, S., Roscioli, N., & Yang, Y. Water-stable electrospun collagen fibers from a non-toxic solvent and crosslinking system, J. Biomed. Mater. Res. Part A, 2013, 101A, 1237–1247.
[94] Ritcharoen, W., Thaiying, Y., Saejeng, Y., Jangchud, I., Rangkupan, R., Meechaisue, C., & Supaphol, P. Electrospun dextran fibrous membranes, Cellulose, 2008, 15, 435–444.
[95] Jiang, H.L., Fang, D.F., Hsiao, B.S., Chu, B., & Chen, W.L. Optimization and characterization of dextran membranes prepared by electrospinning, Biomacromolecules, 2004, 5, 326–333.
[96] Righi, T.M., Almeida, R.S., & d'Ávila, M.A. Electrospinning of Gelatin/PEO blends: Influence of process parameters in the nanofiber properties, Macromol. Symp., 2012, 319, 230–234.
[97] Yang, C., Wu, X.M., Zhao, Y.H., Xu, L., & Wei, S.C. Nanofibrous scaffold prepared by electrospinning of poly(vinyl alcohol)/gelatin aqueous solutions, J. Appl. Polym .Sci., 2011, 121, 3047–3055.
[98] Zhang, S., Huang, Y., Yang, X., Mei, F., Ma, Q., Chen, G., Ryu, S., & Deng, X. Gelatin nanofibrous membrane fabricated by electrospinning of aqueous gelatin solution for guided tissue regeneration, J. Biomed. Mater. Res. A., 2009, 90, 671–679.
[99] Erencia, M., Cano, F., Tornero, J.A., Macanás, J., & Carrillo, F. Resolving the electrospinnability zones and diameter prediction for the electrospinning of the gelatin/water/acetic acid system, Langmuir, 2014, 30, 7198–7205.
[100] Qi, H., Hu, P., Xu, J., & Wang, A. Encapsulation of drug reservoirs in fibers by emulsion electrospinning: Morphology characterization and preliminary release assessment, Biomacromolecules, 2006, 7, 2327–2330.
[101] Kishan, A.P., & Cosgriff-Hernandez, E.M. Recent advancements in electrospinning design for tissue engineering applications: A review, J. Biomed. Mater. Res. A., 2017, 105, 2892–2905.
[102] Rajzer, I., Menaszek, E., & Castano, O. Electrospun polymer scaffolds modified with drugs for tissue engineering, Mater. Sci. Eng. C Mater. Biol. Appl., 2017, 77, 493–499.
[103] Pham, Q.P., Sharma, U., & Mikos, A.G. Electrospinning of polymeric nanofibers for tissue engineering applications: A review, Tissue. Eng., 2006, 12, 1197–1211.
[104] Thakkar, S., & Misra, M. Electrospun polymeric nanofibers: New horizons in drug delivery, Eur. J. Pharm. Sci., 2017, 107, 148–167.
[105] Singh, A., Rath, G., & Goyal, A.K. Nanofibers: An effective tool for controlled and sustained drug delivery, Curr. Drug. Deliv., 2017, 15, 155–166.
[106] Goke, K., Lorenz, T., Repanas, A., Schneider, F., Steiner, D., Baumann, K., Bunjes, H., Dietzel, A., Finke, J.H., Glasmacher, B., & Kwade, A. Novel strategies for the formulation and processing of poorly water-soluble drugs, Eur. J .Pharm. Biopharm., 2017, 126, 40–56.
[107] Dragojevic, S., Ryu, J.S., & Raucher, D. Polymer-based prodrugs: improving tumor targeting and the solubility of small molecule drugs in cancer therapy, Molecules, 2015, 20, 21750–21769.

[108] Luo, Y., & Prestwich, G.D. Cancer-targeted polymeric drugs, Curr. Cancer. Drug .Tar., 2002, 2, 209–226.

[109] Ignatious, F., Sun, L.H., Lee, C.P., & Baldoni, J. Electrospun nanofibers in oral drug delivery, Pharm. Res-Dordr., 2010, 27, 576–588.

[110] Ben Halima, N. Poly(vinyl alcohol): review of its promising applications and insights into biodegradation, RSC Adv., 2016, 6, 39823–39832.

[111] Alavarse, A.C., de Oliveira Silva, F.W., Colque, J.T., Da Silva, V.M., Prieto, T., Venancio, E.C., & Bonvent, J.J. Tetracycline hydrochloride-loaded electrospun nanofibers mats based on PVA and chitosan for wound dressing, Mater. Sci. Eng. C Mater. Biol. Appl., 2017, 77, 271–281.

[112] Fazli, Y., Shariatinia, Z., Kohsari, I., Azadmehr, A., & Pourmortazavi, S.M. A novel chitosan-polyethylene oxide nanofibrous mat designed for controlled co-release of hydrocortisone and imipenem/cilastatin drugs, Int. J. Pharmaceut., 2016, 513, 636–647.

[113] Yang, J.M., Yang, J.H., Tsou, S.C., Ding, C.H., Hsu, C.C., Yang, K.C., Yang, C.C., Chen, K.S., Chen, S.W., & Wang, J.S. Cell proliferation on PVA/sodium alginate and PVA/poly(gamma-glutamic acid) electrospun fiber, Mat. Sci. Eng. C-Mater., 2016, 66, 170–177.

[114] Wen, P., Wen, Y., Huang, X., Zong, M.H., & Wu, H. Preparation and characterization of protein-loaded electrospun fiber mat and its release kinetics, J. Agr. Food. Chem., 2017, 65, 4786–4796.

[115] Kaassis, A.Y.A., Young, N., Sano, N., Merchant, H.A., Yu, D.G., Chatterton, N.P., & Williams, G.R. Pulsatile drug release from electrospun poly(ethylene oxide)-sodium alginate blend nanofibres, J. Mater. Chem. B., 2014, 2, 1400–1407.

[116] Vuddanda, P.R., Mathew, A.P., & Velaga, S. Electrospun nanofiber mats for ultrafast release of ondansetron, React. Funct. Polym., 2016, 99, 65–72.

[117] Mahanta, N., & Valiyaveettil, S. In situ preparation of silver nanoparticles on biocompatible methacrylated poly(vinyl alcohol) and cellulose based polymeric nanofibers, RSC Adv., 2012, 2, 11389–11396.

[118] Cao, L., Zhang, F., Wang, Q., & Wu, X. Fabrication of chitosan/graphene oxide polymer nanofiber and its biocompatibility for cartilage tissue engineering, Mater. Sci. Eng. C Mater. Biol. Appl., 2017, 79, 697–701.

[119] Kheradmandi, M., Vasheghani-Farahani, E., Ghiaseddin, A., & Ganji, F. Skeletal muscle regeneration via engineered tissue culture over electrospun nanofibrous chitosan/PVA scaffold, J. Biomed. Mater. Res. – Part A, 2016, 104, 1720–1727.

Funda Cengiz Çallioğlu, Hülya Kesici Güler

7 Natural nanofibers and applications

Abstract: In this chapter, natural nanofibers and application areas are given in details in terms of green electrospinning approach for clean, non-toxic, eco-friendly and safe production. Generally, water soluble polymers, pure natural polymers, or synthetic/natural polymer blends have been used as a raw material of nanofibers. Firstly; emulsion and suspension electrospinning methods which are useful for green production are explained and supported with previous literature studies. Then, natural polymer based nanofibers such as chitin/chitosan, cellulose, gelatin, alginate, hyaluronic acid, fibrin, collagen etc. are described in terms of production methods and end product properties. On the other hand; natural/synthetic polymer mixture of nanofibers production and applications, suitable and green solvent selection, usage are also researched and underlined importance for healthy process. Lastly, application areas of natural nanofibers under main titles of biomedical, filtration and food industry are discussed and focused on importance for green process. In addition, some important medical applications of natural polymer based nanofibers; drug delivery, controlled release, wound dressing, tissue engineering are also explained and interpreted with literature. The vital key point of this chapter is about natural polymer based nanofibers and green solvents which are environmentally-friendly materials are main components of green electrospinning. In the near future; it is expected that green electrospinning approach will be more indispensable for eco-friendly, healthy, clean and safe industrial process steps.

Keywords: biomedical applications, controlled released, drug delivery, emulsion electrospinning, food industry, green electrospinning, natural nanofibers, natural polymers, suspension electrospinning, tissue engineering, wound dressing

7.1 Introduction

Electrospun nanofibers have become very attractive in recent years. However, some harmful polymer/solvent systems can be a serious problem for some application areas. Electrospinning is a novel, efficient, flexible, and commonly used method to obtain nanofibrous surfaces. During electrospinning, harmful organic solvents such as formic acid, dimethylformamide, dichloromethane, and chloroform are used to produce nanofibers. Natural nanofibers are critically important for some application

Funda Cengiz Çallioğlu, Hülya Kesici Güler, Textile Engineering Department, Süleyman Demirel University/Engineering Faculty, Isparta, Turkey

https://doi.org/10.1515/9783110581393-007

areas (drug delivery, water filtration, artificial tissue, wound dressing, cosmetics, etc.). Because of these reasons, green electrospinning is a nonhazardous and environmentally friendly approach. Also, natural and soft solvents (distilled water, acetic acid, etc.) are used to obtain nanofibers via this method.

The key point of this chapter is to show the importance of green electrospinning in terms of polymers, solvents, and application areas, as well as providing details. First, the green electrospinning approach is explained; then, polymers and solvents that are used for green products are given, and compared with the literature. Lastly, details of the main application areas and importance of green products are explained.

7.2 Green electrospinning

It is well known from the literature that electrospinning is a very attractive method for the production of polymeric nanofibers. There are some advantages of this method, such as using a wide range of polymers, low expense, effectiveness, simplicity, and versatility [1–4].

Nonwoven fibrous materials made of electrospun nanofibers offer many superior functional characteristics, such as high specific surface area ($m^2\ g^{-1}$), high porosity, small and controllable pore size, and very small fiber diameter [5–7].

However, there are also some disadvantages of electrospinning in terms of green production (unsuitable polymers, solvents, etc.). The most important disadvantage of this method is the use of harmful, toxic, and harsh solvents (chloroform, dichloromethane, etc.). These organic solvents are used for insoluble polymers in distilled water or green solvents. The drawbacks of organic solvents are that they are expensive, toxic, and non-eco-friendly. Also, organic solvents are not preferable for certain applications for various medical areas such as wound dressing, tissue engineering, and drug delivery [3, 8, 9].

Green electrospinning is a type of electrospinning that is clean and safe [10], and is vitally important for some application areas (medical, food, etc.). The aim of green electrospinning is to utilize nontoxic, cheap, eco-friendly, mild, green solvents, and materials [2, 11].

Green electrospinning is a new approach, and there are few studies in the literature about this method. Besides solvents, suitable polymer selection is also very important for green electrospinning, and therefore final product properties and application areas. According to the literature, the polymers that have been used in studies about green electrospinning are polycaprolactone/poly(ethylene oxide) (PEO) [12], poly(vinyl alcohol) (PVA)/polyvinylpyrrolidone (PVP) [13], collagen [11], silk fibroin/PEO [10], silk fibroin [9, 14], polyimide [1], PEO/silk fibroin [15], poly(*N*-isopropylacrylamide-*co*-methacrylic acid-*co*-vinylpyrrolidone) [8], PVA/poly(acrylic acid) (PAA) [3], poly(hexamethylene adipate)/PEO block copolymers [16], and polystyrene [4].

In addition, application areas are also important for green electrospinning. In particular, medicine, agriculture [12, 16], bioengineering [13], biomedical applications [11], drug carriers for tissue engineering [10], cosmetics for the personal skin care field [14], filtration [1], tissue regeneration and wound healing [3, 9, 15], and drug delivery systems [8] are very common application areas.

7.2.1 Emulsion electrospinning

An emulsion term has been defined as a dispersion of two or more immiscible liquids in between each other. There are two phases in emulsion electrospinning: the continuous and dispersed phases. The general rule is that droplets are distributed evenly in the continuous phase [17].

Emulsion electrospinning is also a new approach to producing nanofibers with green treatments. In the literature, there are three possible approaches to loading bioactive molecules into fibers during electrospinning, such as the methods of blending directly with polymer solution, coaxial, and emulsion electrospinning. It has been observed that there are some difficulties with the first two approaches, such as the fact that the incorporated drugs or bioactive proteins in the blending electrospun fibers show a burst release phenomenon in the first release point. The reason for this burst release phenomenon may be the inclination of drugs or bioactive proteins placed on the surface of the fibers under the strong electrostatic force experienced during the electrospinning process. Coaxial electrospinning also has some difficulties, such as the need for special apparatus, and to adjust the conditions for the expected results [18, 19].

Emulsion electrospinning is a more simple and effective method than these two approaches. Therefore, this method has quickly increased in popularity in recent years. In the literature, Sanders et al. [20] called this method two-phase electrospinning, instead of emulsion electrospinning. Two-phase electrospinning builds up the fundamentals of emulsion electrospinning. Emulsion electrospinning has some favorable characteristics, for instance, incorporation of bioactive molecules into nanofibers, the sustained release of bioactive molecules from nanofibers, and also its appropriateness for hydrophilic components (proteins and DNA) that are sensitive to organic solvents [17]. Also, it is possible to incorporate hydrophilic and hydrophobic drugs into nanofibers using emulsion electrospinning as a dispersed phase. Similarly, aqueous or lyophilic solutions can be used as a continuous phase. Besides these approaches, oil-in-water and water-in-oil emulsions can also be prepared via the electrospinning method [19, 21]. In addition, another advantage of emulsion electrospinning in terms of process is that it is not necessary to use special apparatus, and conventional electrospinning can be used for this aim [22]. A representation of emulsion electrospinning is given in Figure 7.1.

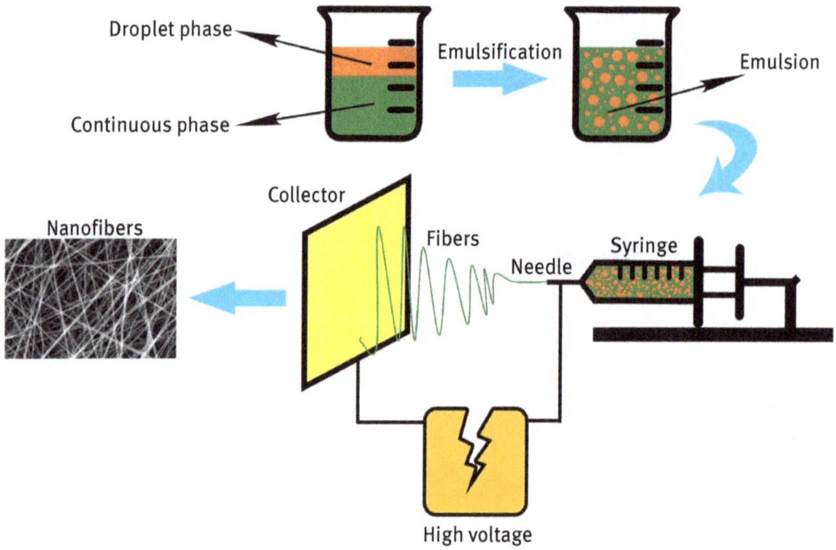

Figure 7.1: Representation of emulsion electrospinning.

Another potential favorable characteristic of emulsion electrospinning is providing nanofiber production from different polymer/solvent/drug systems. It is possible that hydrophilic or hydrophobic drugs can be incorporated into nanofibers using the emulsion electrospinning approach. There are some articles about encapsulating drugs into nanofibers with emulsion electrospinning in the literature. In these studies, natural polymers (gelatin, chitosan, etc.) were combined with synthetic polymers (PVA and PEO) and drugs. Kim et al. [23] produced scaffolds for biomedical applications. They used poly(lactide-*co*-glycolide) (PLGA) and block copolymer (PEG-b-PLA) as polymers, and mefoxin as a hydrophilic antibiotic drug. They investigated the fiber morphology, drug release behavior, and antimicrobial properties on *Staphylococcus aureus* cultures. They observed that fiber morphology depended on the drug concentration, drug release behavior at longer times, and good antibacterial activity against *S. aureus* in both a static (agar) and dynamic (liquid) environment. In another study, PLGA nanofibers were produced using hydrophilic and hydrophobic drugs – cefradine and 5-fluorouracil, respectively. Gelatin was also used as a bioactive protein to enhance the nanofibrous materials' surface properties for cell adhesion. It was determined that good mechanical properties, lower burst release, and lower toxicity provide and support fibroblast cells' attachment and growth. Authors have suggested that these electrospun nanofibers have a potential use in medical applications [19]. Gordon et al. [21] successfully incorporated lipophilic compounds into hydrophilic nanofibers. They used PVA as a hydrophilic polymer and celecoxib as a drug material. They determined that the average fiber diameter is between 21 and 93 nm. Differential scanning calorimetry and X-ray diffraction results showed that nanoparticles have an amorphous structure and better dissolution properties.

In addition, emulsion electrospinning is the most suitable method for combining nanofibers with essential oils. Essential oils are natural agents that are extracted from different parts of medicinal aromatic plants. It is well known that essential oils have different kinds of considerable therapeutic properties such as antiseptic, antifungal, antiviral, antibacterial, antioxidant, anti-inflammatory, antidiabetic, and antifungal. There are limited studies about emulsion electrospinning of nanofibers with essential oils in the literature. Kriegel et al. [24] produced PVA/eugenol nanofibers using an electrospinning method, and they investigated polymer solution properties, nanofiber morphology, and chemical structure of nanofibers. In another study, Mori et al. studied polylactic acid/candeia essential oil via electrospinning for multipurpose applications. They also determined the fiber morphology, structure, and thermal and microscopy analysis [25]. Rieger et al. [26] determined the correlation between solution rheology properties and chitosan/PEO/cinnamaldehyde nanofibers formation. Sadri et al. [27] developed a natural antimicrobial wound dressing nanofibrous material. They determined fiber morphology, chemical structure, and antibacterial activity. The results showed good antibacterial activity against *Staphylococcus aureus*, *Pseudomonas aeruginosa*, and *Escherichia coli*.

Lastly, Güler et al. [28] produced antibacterial PVP/cinnamon essential oil (CEO)–based nanofibers. According to the results, quite smooth, fine, and uniform nanofibers were obtained using emulsion electrospinning (Figure 7.2).

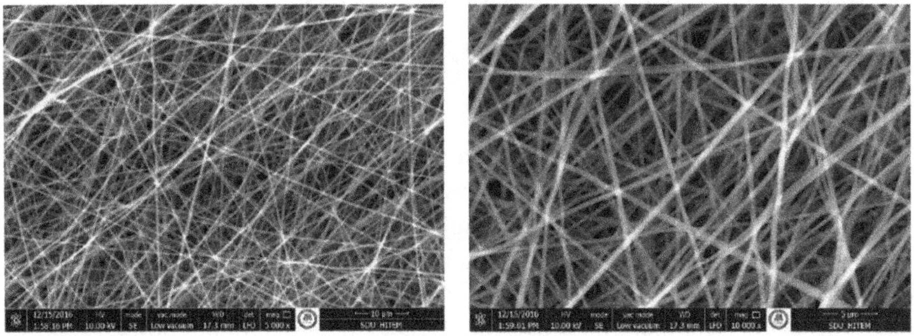

(A) (B)

Figure 7.2: Scanning electron microscopy images (5,000×–10,000×) of PVP (A) CEO (B) nanofiber samples. From Ref. [28].

In addition, according to the antibacterial test results, good antibacterial properties were obtained in terms of *Staphylococcus aureus* (ATCC 25923), *Escherichia coli* (ATCC 25922), and *Candida albicans* (ATCC 10231).

7.2.2 Suspension electrospinning

Blending of liquid and solid particles can be defined as a suspension. Solid particles disperse into the liquid homogeneously [17]. Suspension electrospinning is a new field, and a significant point for green electrospinning. A representation of suspension electrospinning is shown in Figure 7.3.

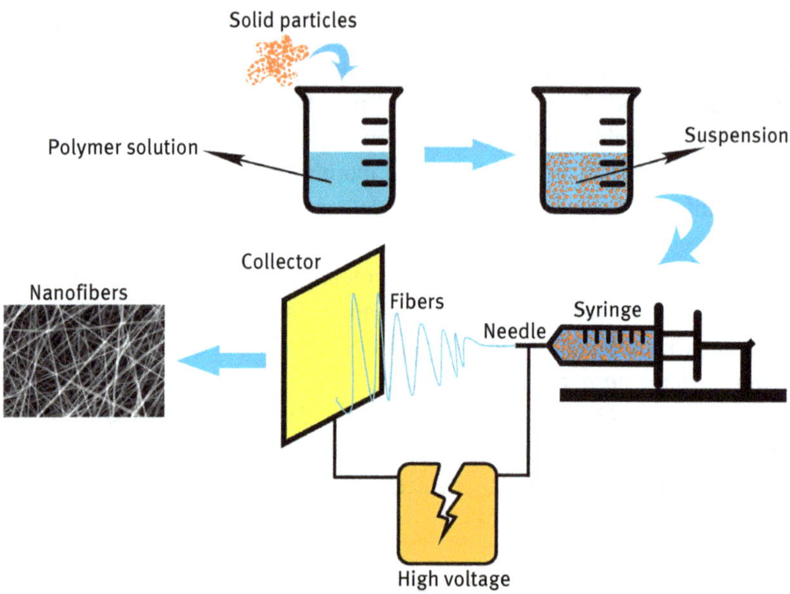

Figure 7.3: Representation of suspension electrospinning.

There are two approaches in the literature to suspension electrospinning. The first one is that solid particles are put into the polymer solution directly. Naveen et al. [29] studied the production of nanofibers with suspension electrospinning. Poly(3-hydroxybutyric acid) was used as a polymer because of its nontoxic properties. Kanamycin sulfate was used as a drug and it is a kind of hydrophilic antibiotic. They determined the fiber morphology, antimicrobial property, drug release profile, cell viability, and cell morphology. The results showed a good zone of inhibition against *Staphylococcus aureus*. When the polymer concentration increased, the surface tension and the viscosity also increased. Similarly, the average fiber diameter also increased with polymer concentration. Lastly, researchers analyzed that cells proliferated uniformly into the nanofiber structure.

The second approach involves suspension electrospinning using a template polymer. Sun et al. [16] produced nanofibers with suspension electrospinning using poly(hexamethylene adipate)-PEO block copolymers (PHA-b-PEO) as a polymer and

PEO as a template polymer. Once the nanofibers were produced and extracted with water by dialysis, PHA-b-PEO nanofibers were obtained. An important point of this study was the production of biodegradable polyesters using green solvents. Also, the authors suggest that nanofibrous surfaces can be used in medical, pharmaceutical, and agriculture application areas [16].

Han et al. [30] studied the water-soluble bioactive substance release behaviors with emulsion and suspension electrospinning. They used poly(L-lactide-*co*-glycolide) or poly(ethylene glycol)-b-poly(L-lactide-*co*-caprolactone) (PELCL) blends as a polymer, and chitosan hydrogel as a carrier. They investigated the structural morphology, mechanical properties, hydrophilicity, cytobiocompatibility, and release behaviors of nanofibers. The results of the study showed that the average fiber diameter was between 150 and 1,380 nm. All the nanofibrous surfaces had good hydrophilicity, mechanical properties, and good cytobiocompatibility. Lastly, the release behaviors were that EH-PELCL showed faster release than EH-PLGA and E-PLGA nanofibers.

7.3 Natural nanofibers

7.3.1 Natural polymer-based nanofibers

In recent years, electrospun natural nanofibers have become very attractive [31] because of their vital importance for some specific application areas (medical, food, etc.). In this section, natural polymer-based nanofibers, their properties, and previous studies in the literature are described and analyzed in detail. Natural nanofibers possess intrinsic biological and chemical properties than synthetic polymers, in terms of biocompatibility, hydrophilicity, nontoxicity, biodegradability, enhanced cell adhesion, proliferation, and reduced immune reaction [32–34]. Another attractive point is that they are renewable materials and come from a natural origin [35]. Natural raw materials are eco-friendly because these materials can be degraded via microorganisms [36].

Natural polymers consist of proteins like collagen, gelatin, and polysaccharides; for instance, cellulose and chitosan are given separately later. And natural nanofibers can be obtained from these proteins and polysaccharides via electrospinning.

Hyaluronic acid

Hyaluronic acid (HA) is a polyanionic polymer that is a naturally occurring linear polysaccharide. It consists of repeating disaccharide units β-1,4-D-glucuronic acid and β-1,3-*N*-acetyl-D-glucosamine. It is also a main component of extracellular matrices (ECMs). HA has ideal biocompatibility and biodegradability, because of its great viscoelastic and rheological properties. For this reason, it is remarkably attractive for biomedical and cosmetic applications; for instance, ophthalmology, drug delivery, dermatology, tissue engineering, medical implants, wound dressings, cell encapsulation, arthritis treatment, and surgery [34, 37–42].

It is well known that HA is also attractive for the electrospinning method. However, it is difficult to achieve electrospinning with a pure HA aqueous solution. A polymer solution of HA has fairly high viscosity and surface tension due to the high molecular weight of HA. High viscosity, high surface tension, and insufficient evaporation of solvent are major problems for the electrospinning process [34, 43]. To solve these problems, some researchers developed and used blowing-assisted electrospinning [34, 41, 44].

In the literature, there are also non-cross-linked [37, 41, 44] and cross-linked [42, 45, 46] forms of HA nanofibrous surfaces.

The cross-linking process is necessary to increase the nanofiber web stability, mechanical properties, and also to convert to a form nonsoluble to water. Electrospun nanofibrous HA surfaces can be cross-linked by chemical or heat treatment methods. However, chemical treatments may destroy the nanofiber structure [42].

Brenner et al. [37] produced HA-based nanofiber with different solvent systems, such as sodium hydroxide:dimethylformamide (NaOH:DMF) and ammonium hydroxide:dimethylformamide (NH_4OH:DMF). They determined that the average fiber diameter was 39 ± 12 nm for NH_4OH, and above 100 nm for NaOH:DMF. The solution conductivity increased with HA concentration for all solvent systems. It is known from the literature that DMF is quite suitable as a solvent for electrospinning. Because it can be evaporated quickly from polymer solutions and can dissolve several polymers. Also, DMF has high conductivity, and it is very important for spinnability performance. Cengiz and Jirsak [47] found that solution conductivity has a strong relation between spinning performance of electrospinning.

Chitin/chitosan

Chitosan is a natural polysaccharide obtained from chitin, and it comprises hydroxyl and amine groups [48]. Because chitosan has outstanding characteristics such as excellent biocompatibility, antimicrobial activity, nontoxicity, appropriate biodegradability, hydrophilicity, low cost, wound healing properties, antitumor effects, high mechanical strength, fast metal complexation, bioactivity, nonantigenicity, bacteriostatic, bactericidal, candidacidal activity, and excellent affinity to proteins [49–54].

These unique properties of chitosan will inspire the methodology for successful electrospinning of this natural material [55]. However, electrospinning of pure chitosan is quite difficult because of its low solubility and low stability. Therefore, chitosan can be blended with other polymers for electrospinning to utilize the advantages of chitosan, and also to enhance the mechanical properties of the nanofiber web [49]. Also, some additives (surfactant, salt, and cosolvent) can blend with chitosan solutions to decrease surface tension and also increase spinnability. There are many studies in the literature about blending different polymers with chitosan for various application areas, including chitosan/PVA [53], chitosan/poly(ethylene terephthalate) (PET) [56], chitosan/PEO [49, 57], and chitosan/collagen [48].

In the literature, Jung et al. [56] produced PET/chitosan, chitosan, and PET electrospun nanofibers for biomedical applications. They obtained fiber diameters between 500 and 800 nm. They investigated the nanofibers' antibacterial activity against *Staphylococcus aureus* and *Klebsiella pneumoniae* bacteria. According to their results, electrospun nanofibrous webs showed a high degree of growth inhibition, and PET/chitosan nanofibers had better cell adherence than PET nanofibers.

Chitosan can be used in medical, filtration, and food application areas such as pharmaceuticals, cosmetics, hemostatic, enzyme immobilization, biosensors, wound healing, neutralization of toxic metals, and absorption of heavy metal ions from water [48, 49, 51, 52, 54, 58]. The most common application area is filtration, because of the chemical structure of chitosan. It has several polar and ionizable groups [52]. It also contains high amounts of amino and hydroxyl groups in the chains of chitosan, so it has a high adsorption capacity. To enhance chemical and mechanical properties of nanofibrous material, chitosan is generally cross-linked with glutaraldehyde (GTA) for absorbent applications [59].

Alginate

Alginate is a member of the polysaccharides and is produced from marine brown-algae. It consists of (1,4)-β-D-mannuronic acid and (1,3)-α-L-guluronic acid in its chemical structure of linear chains. Alginate has many desired characteristics such as biodegradability, nontoxicity, biocompatibility, hydrophilicity, and relatively low cost. In addition, the main difference between alginate and other natural polymers is its high water-absorbing ability (200–300 times its weight) [60, 61]. This property provides effective absorption wound exudates and helps to maintain a moist wound environment. Alginate is a hydrophilic polymer; therefore, alginate-based materials provide uniform cell attachment and cell proliferation into the nanofibrous matrix [60, 62]. Nanofibrous materials produced from a low molecular weight of alginate can be suitable for in vivo medical applications (requiring degradation and removal from the body) [63].

However, there are some disadvantages of alginate, such as low solubility in general solvents and also difficult processability. In the literature, sodium alginate is generally used for the electrospinning of nanofibers because of easier processability of both solution properties and spinning performance [60–71].

Sodium alginate is a pH-sensitive and water-soluble polyelectrolyte. This is important for many application areas, so it has been studied extensively in tissue engineering, wound dressing, drug delivery, and regeneration of skin, cartilage, bone, and cardiac tissue due to its nontoxicity, compatibility, and biodegradability [63, 68, 72–77].

In the literature, there are various studies about the cross-linking of sodium alginate with polyvalent cations such as calcium chloride [60, 62, 63, 65]. The cross-linking process provides insolubility to the water and enhances the mechanical stability [60]. In addition to calcium chloride cross-linking, Leung et al. [66] suggested double cross-linking. With this method, in the structure of hydrophilic polymers, hydroxyl groups can be cross-linked using GTA, hexamethylene diisocyanate, and adipic acid hydrazide [66].

Disadvantages of natural polymers are low mechanical properties and processing difficulties compared to the synthetic polymers. It is common to blend natural polymers with synthetic polymers to reduce these disadvantages. Coşkun Üstündağ et al. [78] used PVA synthetic polymer with sodium alginate to increase spinnability and mechanical properties. A PVA/alginate nanofibrous wound dressing material was developed in this study. They were cross-linked nanofiber webs for improved water resistance. Lastly, they expressed that PVA/NaAlg nanofiber webs have a potential use as a wound dressing material [78].

Fibrin and fibrinogen

Fibrinogen is a water-soluble protein that forms a spherical and molecular mass of 340 kDa [79, 80], and is synthesized from liver. The fibrinogen molecules consist of six polypeptide chains – two Aα, two Bβ, and two γ [81]. It is found in the blood plasma about 80–90%, so it plays a vital role in hemostasis, wound healing, inflammation, angiogenesis, and neoplasia [81, 82]. Fibrin is obtained as a result of reactions between fibrinogen and thrombin. Fibrin and fibrinogen are used to produce scaffolds. Although they have appropriate properties for scaffolds, such as being easily degradable, nonimmunogenic, supporting cell proliferation, and promoting cell interactions, they have weak mechanical properties, thereby limiting their application in vivo [81–84]. Fibrinogen can be degraded easily, so fibrinogen scaffolds must cross-link to raise their lifetime both in vivo and in vitro. In the literature, transglutaminase and GTA are commonly used as cross-linking additives to enhance the mechanical properties of fibrinogen nanofibers [79, 82].

In the literature, there are some studies about electrospinning fibrinogen for the production of nanofiber scaffolds that mimic the structure of the natural ECM [81, 83]. Balasubramanian et al. [82] produced fibrinogen/gelatin electrospun nanofibers in two different weight ratios for natural polymeric scaffolds. They determined that the nanofiber webs were similar to native myocardial ECM in terms of their morphology, surface characteristics, tensile strength, and stiffness properties, and emphasized their compatibility with, and efficiency in proliferating, human cardiac cells. Fang et al. [83] developed scaffold from poly(L-lactide-*co*-ε-caprolactone) (PLCL)/fibrinogen polymer blends with different volume ratios. The nanofiber diameters were in the range of 224–450 nm. They determined that the hydrophilicity of nanofibers increased with the fibrinogen concentration. Lastly, cell viability results showed that human umbilical vein endothelial cells proliferate very well and had uniform morphology on the nanofibrous surfaces.

Collagen

Collagen is a natural polymer, which is the main component of an ECM. The main tissues, such as skin, bone, tendons, ligaments, and other connective tissues, include collagen in their structure. There are different kinds of collagens in tissue structures for all living organisms. The most common collagen types are type I, type II, and type III, and therefore there are various studies about nanofibers obtained from these types of collagens in the literature. According to these studies, type I collagen has an average diameter of about 50–500 nm, and a very uniform size for tissues [85, 86].

Collagen can be produced from different kinds of sources such as human placenta, rat tail tendon, bovine calf skin, and rabbit. Collagen has superior biocompatibility and nonimmunogenicity properties; however, its mechanical property is relatively low [86–89]. Collagen has good cell attachment and proliferation properties, and is easily degraded and resorbed in living organic structure [85, 89]. Therefore, collagen is quite an attractive natural polymer for tissue engineering and regenerative medicine applications [88, 90, 91].

It is possible to produce nanofibrous materials from pure collagen polymer solutions. Normally, collagen nanofibers exist in living ECMs with multifibrils as a three-dimensional network structure. Therefore, it is possible to obtain a structure such as a native ECM with electrospun collagen nanofibers [85, 86].

Collagen nanofibers can be chemically cross-linked by GTA, epoxy compounds, and 1-ethyl-3-(3-dimethyl-aminopropyl)-1-carbodiimide hydrochloride (EDC) [91–94]; physically cross-linked by heat, UV irradiation, and dehydrothermal treatment (DHT) [95, 96]; and photochemically cross-linked by laser irradiation as an attractive alternative [94].

There are some studies in the literature about collagen nanofiber blending with synthetic nanofibers. Chen et al. [87] produced nanofibers from thermoplastic polyurethane (TPU)/collagen polymers via coaxial electrospinning for tissue engineering and cross-linked with GTA. They found that the nanofibers' diameter was in the range of 700–800 nm, and the nanofibers' density was in the range of 0.30–0.40

g cm^{-3}. They observed that by increasing cross-linking time, the nanofibers became thicker, and porosity was reduced. The results showed that TPU/collagen electrospun nanofibrous material have potential for use in tissue engineering and functional biomaterials [87].

Gelatin

Gelatin is a natural polymer that has a protein-based structure, and can be produced from native collagen using a hydrolysis method. It can be easily extracted from living organisms such as skin, tendons, cartilage, bone, and muscle [97–106]. Gelatin has almost the same structure as collagen, and its biological properties are also similar. Gelatin has nonimmunogenicity, biodegradability, biocompatibility, and good cell adhesion and mucoadhesion properties, which are important for medical applications. In addition, gelatin does not cause antigenicity and shows a high hemostatic effect [107]. Because of these properties, gelatin has been widely used in tissue scaffolds, wound dressing, drug carriers, gene therapy, pharmaceutical, and cosmetic applications [97, 98, 104, 106, 108, 109]. When gelatin is used as a wound dressing, it can absorb wound discharge and provides a moist surrounding of the wound, so it can be healed speedily [104].

Gelatin polymers can be dissolved in water, but it is very difficult to obtain nanofibers from this aqueous solution via the electrospinning method. This is because the aqueous gelatin solution changes into a gel form in a short time at room temperature [97, 99, 100, 108]. In this situation, different solvents or solvent systems are used for electrospinning of gelatin polymers, including formic acid, trifluoroethanol (TFE), 1,1,1,3,3,3-hexafluoro-2-propanol, acetic acid, and acetic acid/water/ethyl acetate [99–101, 106, 109]. Gelatin nanofibers can be cross-linked using various methods to enhance water stability and mechanical properties to avoid the limitation of application areas [101, 106, 109]. There are physical cross-linking methods in the literature, such as ultraviolet and gamma irradiation, drying, chemical agents, heating, electron beam, DHT, and chemical methods. In general, GTA, tannic acid, formaldehyde, epoxy compounds, diisocyanates, carbodiimide and genipin are used as cross-linking agents [101, 109]. It is clear from the literature that gelatin nanofibers can be used as a promising biomaterial for the green electrospinning approach.

In the literature, there are some studies about gelation nanofibers blending with other polymers, such as poly(ε-caprolactone)/gelatin [98, 110], polyaniline/gelatin [111], poly(lactic-co-glycolic acid)/gelatin [103, 107], silk fibroin/gelatin [105], and chitosan/gelatin [107, 112]. There are also some studies about just gelatin nanofibers [99–101, 106, 109, 113–115]. Chen et al. produced gelatin nanofibers in various concentrations and used formic acid and ethanol solvents. They then cross-linked the nanofibers with GTA. They observed that bead formation increases as the concentration and viscosity decrease. In addition; fiber diameter increases with gelatin polymer concentration, and uniform gelatin nanofibers were obtained without beads [97].

Gums

Gum is another natural polymer; the structure of gum consists of anionic polysaccharides, which are obtained from trees and plants. Gum tragacanth (GT) is a common variety of this natural polymer and has attractive biological properties such as being renewable, biocompatible, biodegradable, antiallergenic, antibacterial, mutagenic, teratogenic, and carcinogenic with no adverse toxicological effects in nonallergic people [116]. Other advantages of GT are strong resistance to microbial attacks, cheap, and long shelf-life. Because of these properties, GT polymers can be used on their own, or blended with other synthetic polymers for applications in skin regeneration, drug carriers, peripheral nerve regeneration, and wound healing activity [117–119].

In the literature, there are few studies about electrospun gum nanofibers. Ranjbar-Mohammadi et al. [117] studied PVA/GT nanofiber production and cross-linked nanofibers obtained with GTA. According to the results of this study, smooth surface nanofibers showed good antimicrobial properties against gram-negative bacteria (*Pseudomonas aeruginosa*). They also observed good human fibroblast cell attachment and growth on scaffolds. It is well known from the literature that both molecular entanglement and solution viscosity are very important for nanofiber production. In this situation, there are some difficulties regarding insufficient chain entanglement of gum polymer because of solution rheological properties and polyanionic behavior [116]. In the near future, it is expected that natural gum-based nanofiber studies and application areas will increase.

Cellulose

Cellulose is the most common and strong natural biopolymer that can be obtained from various plants (wood pulp, algae, cotton lint, etc.) [120]. Cellulose raw material for nanofibers can be an attractive class of nanomaterials for obtaining lightweight and high-mechanical-property materials, especially for some application areas such as automotive, buildings, mobile computers, composites [121–123], filtration, biomedical [124], optically transparent functional materials [125–127], hygienic textiles, and protective textiles [128–131].

In the last few decades, studies about electrospun cellulose nanofibers are very popular because of the specific properties of both the cellulose raw material and nanofibers (high Young's modulus, high mechanical properties, lightweight, high surface-area-to-volume ratio, high porosity, etc.) [132–136]. However, there are some process limitations regarding cellulose solubility with common solvents [137].

In the literature, Kim et al. studied the electrospinning of cellulose solutions. They used lithium chloride/*N,N*-dimethyl acetamide and *N*-methylmorpholine oxide/water solvent systems for dissolving cellulose and produced nanofibers successfully. They determined that the average fiber diameter was in the range of 250–750 nm. They observed that the degradation time can be controlled with the average fiber diameter [137].

Others

Dextran is a natural polysaccharide polymer and has biocompatibility and biodegradability properties that are very important in biomedical applications (drug delivery and scaffolds) [138–141]. Cengiz-Çallıoğlu [142] produced dextran nanofibers with different surfactant concentrations via needleless electrospinning. The optimum surfactant concentration was chosen as a value of 2 wt% to produce a dextran nanofiber. The average fiber diameter was determined to be 162 nm, and the diameter uniformity coefficient was determined as 1.03. Generally, uniform, fine nanofibers and high fiber densities were obtained using the rod electrospinning method.

Silk is another natural polymer, which is a fibrous protein. It is produced by silkworms and spiders. The structure of silk includes two different polymers – fibroin and sericin. Silk fibroin has unique properties, for instance, good biocompatibility, biodegradability, good oxygen and water vapor permeability, and minimal inflammatory reaction. It can therefore be used in cosmetics, biomedical applications, and food additives [143, 144].

7.4 Application areas of natural nanofibers

7.4.1 Biomedical applications

Green electrospinning is very important for medical applications of both polymer and solvent systems when it comes to health. For this reason, the green electrospinning approach has become more and more popular over the last few decades. The biomedical field is an important and vital application area taking advantage of the electrospinning technique. Either solvents or polymers may have detrimental and toxic properties that can reflect on the final product. These detrimental properties play a vital role in some biomedical application areas. For this reason, most common biomedical application areas of nanofibers, such as wound dressing, tissue engineering, drug delivery, and controlled release are described in detail in this chapter.

Drug delivery and controlled release

Drug delivery technology is a new concept for the pharmaceutical and biotechnology industries [145]. Drug delivery is a more promising method than other potential applications, and many researchers have focused on improving drug delivery systems over the last decades. There are different kinds of nanoscale delivery systems, such as polymeric micelles, liposomes, complexes, and nanofibers. Among other nanoscale drug delivery systems, nanofiber drug delivery systems aim to improve the therapeutic effect, increase convenience, reduce the toxicity, and so on [146–151]. Systemic administration of drugs, especially antibiotics, is useful, but there are some drawbacks such as the need to increase oral doses to improve resistant bacterial strains [152].

Electrospun nanofibrous materials have unique properties for drug delivery, such as high porosity, high permeability, interconnecting porous structures, high surface area, small fiber diameter, high encapsulation efficiency, effective loading capacity, inexpensive, amorphous drug content, and high permeability [146, 148, 153, 154]. The high surface area contributes efficient drug release and mass transfer from electrospun nanofibers [155]. Also, the quantity of drugs released can be adjusted using the nanofiber diameter. Furthermore, the controlled release of drugs from nanofibers can be adjusted to environmental factors (pH and temperature), with different release behaviors such as immediate, constant, cyclic, and triggered [147, 149, 154, 156]. Generally, drug delivery systems are designed with a slow-release profile to increase drug presence time and drug action [152]. The electrospinning method presents a wide range of selection material for drug delivery system, including many drugs (antibiotics, anticancer drugs, etc.), hydrophobic and hydrophilic drugs, proteins, aptamer, and DNA can all be incorporated into the nanofibers via electrospinning [148, 152, 154]. Another advantage of the electrospinning method is that the process occurs in mild conditions, so the molecular structure and bioactivity of the encapsulated drugs or bioactive molecules can remain intact in the electrospun nanofibrous structure [23, 154]. Also, more than one drug can be incorporated into the nanofibers with electrospinning [147]. Moreover, it is possible to use various controlled drug release profiles as immediate, controlled, smooth, biphasic releases, and delayed [148].

In the literature, different kinds of polymers have been used as carrier materials, including natural, synthetic, and blends of both [148, 157]. Examples include Fenbufen/poly(D,L-lactide-*co*-glycolide)/gelatin [158], polycaprolactone/polyurethane/ketoprofen [159], PLGA/cefazolin, poly(L-lactide)/paclitaxel, doxorubicin hydrochloride, triethyl benzyl ammonium chloride [160], and PLGA/cefoxitin sodium [23]. Shen et al. produced Eudragit® L 100-55/diclofenac sodium (DS) using different solvents, such as methanol, ethanol, *N,N*-dimethylacetamide (DMAc), and methanol:DMAc (5:1, v/v), as well as with various drug concentrations via electrospinning. They investigated the effect of solvent and drug concentration on fiber spinning performance and morphology. They determined that the average fiber diameters were 1,046, 1,952, and 441 nm for ethanol, methanol, and methanol:DMAc, respectively. Also, scanning electron microscopy images showed that drug particles accumulated on the surface of nanoweb when the DS content was 33.3%. In vitro release tests showed that all the nanofibers had pH-sensitive drug release profiles, and it is possible to use these nanofibers as oral colon-targeted drug delivery systems [161].

Yang et al. investigated the release behavior of raspberry ketone (RK)-loaded gelatin/PVA nanofibers. They observed the effects of pH value, cross-linking time, different gelatin/PVA blending ratios, and drug concentrations on release behavior. They determined that generally, this was a large initial burst at short times. Increasing the cross-linking time, pH values, drug concentration, and gelation ratios caused a decrease in RK release [150].

Wound dressing

Skin is an important part of the human body. Many people have dermal wounds because of different reasons, such as fire, electricity, ultraviolet radiation, heat, infection, chemicals, nuclear energy, chronic diseases, split skin graft donor sites, postsurgical wounds, pressure ulcers, venous ulcers, and diabetic foot ulcers [162–166]. Wound dressing material should be chosen for the particular variety of wound. In recent years, researchers have started to improve and develop new regeneration wound dressing structures for wound healing [165]. Wound healing is a complex process and many contributing factors should be considered. For this reason, the ideal wound dressing must be chosen in terms of protecting the wound, removal of exudate, improved appearance, permeability of water, eliminating infection, limiting inflammation, wound cleansing, and oxygen, to supply an optimal healing environment [162–164, 167–169]. Another important property of wound dressing that can be present is higher porosity and a good barrier effect [168]. Ideally, wound dressing material should provide a native ECM structure to enhance cell attachment. Therefore, nanofibrous material structures can mimic to ECM, owing to the very small fiber diameter, porosity, and anisotropic fiber structure; hence, cells can be attached easily and healing can be quick [85]. In particular, diabetic patients have desired therapeutic effects for their diabetic wounds with expedition [165].

To provide these functions for wound healing, raw materials and additives are the key factors for the wound dressing material. In the literature, there are various studies about the applications of nanofibers as a wound dressing material for skin reconstruction. Generally, biocompatible, and in some cases, biodegradable polymers, and additives, such as chitosan/silver-nanoparticles/PVA [170], chitosan/silk fibroin [167], collagen/chitosan/PEO [171], PVA/poly(vinyl acetate) [162], polyurethane [168], polylactide–polyglycolide/collagen [85], and silk fibroin [143] can be used for this purpose.

In the literature, Jannesari et al. [162] fabricated PVA/poly(vinyl acetate)-based composite nanofibrous mats via electrospinning for wound dressing materials. They incorporated ciprofloxacin into nanofibrous materials and investigated the in vitro drug release. According to the results, the in vitro drug release analyses showed that polymer and drug additive properties (kind, concentration) have an important effect on drug release rate, swelling degree, weight loss, and initial burst. Moreover, the initial release and rate of drug release was significantly affected by the thickness of the nanofiber mats. Liu et al. [85] studied polylactide–polyglycolide (PLGA)/collagen nanofibrous membranes as a wound dressing material. They obtained an average fiber diameter of 250 nm, and the fiber diameter decreased with collagen concentration in the blend solutions. They investigated the effect of gauze, nanofibrous matrix, and commercial dressing for wound healing. They showed that electrospun nanofibrous surfaces were more efficient than gauze and commercial dressing.

It is well known in the literature that electrospun nanofibrous structures have unique properties, such as a porous structure, ultrafine pores, excellent oxygen permeability, huge surface area, hemostatic properties, potential for leaving no scar, and similarity to ECM, and they also have great potential as a skin regeneration material [85, 162, 168, 170]. Another important property of nanofibrous structures is that antibacterial and therapeutic agents can be impregnated into the nanofiber membranes easily [169, 172, 173].

Tissue engineering

Tissue engineering is a new scientific area in human health care that is open for improvement in terms of understanding the cellular biology and bioengineering applications [174, 175]. Fiber-based medical textile materials are very useful as a scaffold, which structurally mimics the native ECM [111].

In living systems, the ECM has a significant importance to control cell behavior [176]. Recently, electrospun nanofibrous materials have become very attractive as a way to mimic ECM structures for artificial tissue design, such as skin, bones, tendons, ligaments, cartilage, neurons, and blood vessels. Scaffold material design and architecture in terms of mechanical, biological, and kinetic properties are very important in tissue engineering, which can provide optimization in terms of polymer solution content and electrospinning process parameters [98, 177]. Nanofibrous web morphology (fiber diameter, uniformity, porosity, and anisotropic structure) and nanofiber deposition (orientation) are one of the significant features in designing an ideal tissue scaffold structure. This is because these properties significantly influence cell proliferation and other related functions in cells such as nerves and muscle cells [98, 175, 178, 179]. For instance, the high porosity and open pore structure of nanofibrous scaffolds for tissue engineering can allow the growth of a large number of cells, and provide uniform cell distribution and oxygen and nutrient diffusion [180, 181].

Various natural and synthetic biocompatible and biodegradable polymers can be used as raw materials to provide the architecture of nanofibrous artificial tissues [175]. Appropriate polymer selection is the key factor in producing scaffolds supporting cell attachment, adhesion, proliferation, function, and growth of cells, and provides the ECM function. In addition, another important factor is the adjustment of the artificial tissue degradation rate. For this aim, a polymer blending (synthetic and natural polymers) method can be used to increase cell adhesion and adjust the degradation rate [98, 182, 183]. There are many studies about polymers and polymer blends for tissue engineering in the literature, including poly(3-caprolactone)/gelatin (nerve tissue) [98], polycaprolactone/gelatin (vascular tissue) [184], polyaniline/gelatin (cardiac and neuronal tissue) [111], poly(ε-caprolactone) (bone tissue) [181], and hydroxyapatite/chitosan (bone tissue) [185].

An ideal tissue-engineered scaffold's mechanical properties and also stress–strain behavior should be compatible with the native tissue. The last aim of tissue engineering is the analysis of the artificial tissue scaffold material support with in vivo and in vitro studies.

7.4.2 Filtration

Filtration includes various kind of engineering fields (environmental, automotive, mechanical, biomedical, military, etc.) [186–191], and it is the most common application area for nanofibrous materials because of the superior properties of nanofibers, such as very low fiber diameter, high porosity, small pore diameter, interrelated pore structure, and a large specific surface area per unit volume [192–198]. Because of these attractive properties, nanofibrous filter materials are the unique commercialized product for various kinds of filtration applications (air, water, dye, oil, aerosol, etc.) [188, 199–201]. Nanospider technology is quite suitable for the production of commercial nanofibrous products. For this aim, Cengiz Çallıoğlu [202] studied the process parameters of this method. Another significant function of nanofibrous material is its sound absorption properties, which is also an important factor for noise insulation material. In the near future, it is expected that deficiencies about nanofibrous filter media will be corrected [199].

In the literature, there are many studies about nanofibers' applications as a filter medium. The nanofibrous filters' filtration performance and the general results obtained from these studies are that nanofibrous-material-covered filters are more effective than the standard glass fiber HEPA (high-efficiency particulate air) [192, 203–206]. However, because of the low mechanical properties of nanofibers, multilayer filter materials are generally designed, composed of spunbond or meltblown nonwovens and polymeric nanofibers to provide high-pressure drop and stability. To achieve this, a lamination method is generally used, and can be combined with nanofibrous webs and nonwoven surfaces using heat and pressure treatments. Yalcinkaya et al. invented a new nanofiltration membrane. This composite filtration material consists of polypropylene/polyethylene and nanofibrous material is laminated with nonwoven supporting material with this approach [198].

There are strong relations between filtration performance, and physical and chemical properties of nanofibrous filter material [199]. It is well known from the literature that nanofibrous material properties such as fiber diameter, diameter uniformity, web density, pore size, porosity, fiber deposition structure (randomly, oriented, etc.), and wettability are the most important key factors for filtering efficiency [188, 193, 194, 198, 199]. In addition, the raw materials of nanofibers, such as polymers (hydrophobic, hydrophilic, natural, synthetic, etc.), solvent type (blended, green, etc.), polymer solution properties (conductivity, surface tension, viscosity, etc.), and process parameters of electrospinning (voltage, solution feed rate, distance of the electrodes, etc.) are also indirectly important factors that affect filter performance [193, 195, 199]. For example, in general, a hydrophilic membrane surface is chosen for aqueous filtration applications [207]. Graham et al. [189] specified that filter efficiency (0.2–0.3 μm particle size) at the same pressure drop increases when the fiber diameter is less than 500 nm. Desai et al. [194] investigated whether the diameter of the nanofibers and nanofibrous polymer (chitosan) content are related to filtration efficiency.

7.4.3 Food industry

Nanotechnology has a significant potential to be useful in the food industry, and different kinds of nanostructures can be used, such as nanoliposomes, nanoemulsions, nanoparticles, nanofibers, and nanotubes [208, 209]. There are few studies in the literature about nanofibers for use in the food industry. It is thought that the polymer type for raw materials is generally synthetic, which is not suitable for the food industry. As advances in nanofiber production continue, using nanofibers in this area may increase with the green electrospinning approach. Electrospun nanofibers can be used for different purposes in the food industry, including food packaging material and imitation/artificial foods [210]. For example, electrospun nanofibrous membranes have been used for biosensor applications, because of their high specific surface area

(m^2 g^{-1}) [5]. Nanofibers are also very attractive for food packaging, and nanotubes encapsulated with nanofibers can have improved mechanical properties and provide antimicrobial effects for food packaging [209]. Cellulosic nanofibers can be used for the immobilization of enzymes [211], vitamins [212], and antimicrobials [213]; delivery and controlled release of materials; biosensors [5]; filtration [199]; and composites [214]. In addition, another potential use of nanofibers in the food industry is using nanofibrous filter material for juice clarification. Veleirinho and Lopes-da-Silva [215] produced PET nanofiber mats via electrospinning for the purifying of fruit juice. Ultrafiltration and conventional methods were used for clarification. They obtained a uniform, smooth surface, 420 nm average nanofiber diameter, 0.33 g cm^{-3} average density of the nanofiber mats, and 0.20 mm average fibrous material thickness. As a result, they determined that nanofibrous membranes for clarification process have high flux performance, simpler and more economical than the conventional processes [215]. It is thought that nanofibrous materials will be used much more in the near future in the food industry than today.

7.5 Conclusion

In this chapter, the necessity and importance of the green electrospinning approach, suitable polymer/solvent selection, and various electrospinning methods are explained. In addition, the natural polymers used to obtain natural nanofibers and application areas are given in detail. Electrospinning is a quickly growing method for obtaining polymeric nanofibrous materials, which has some disadvantages, especially because of harmful solvent usage. In recent years, to avoid problems related to hazardous polymer/solvent raw materials, researchers have focused on producing electrospun nanofibers using green production concepts (mild solvent, etc.). Green electrospinning, which is defined as nanofiber production using mild polymer/solvent systems, is a new and significant approach. Especially for some application areas such as medical, filtration, tissue engineering, and food, the green electrospinning concept is of vital importance. For that reason, polymer/solvent selection should be made very sensitively. Generally, emulsion and suspension electrospinning techniques are used for clean and safe electrospinning. Besides solvents, suitable polymer selection is also very important for this aim. Generally, water-soluble polymers, pure natural polymers, or synthetic/natural polymer blends have been used to obtain nanofibers as a raw material. In conclusion, it is a hoped that green electrospinning for clean and safe production will be much more common in the near future, and therefore more promising application areas of nanofibers will come to exist and be developed.

Abbreviations

CEO	Cinnamon essential oil
DHT	Dehydrothermal treatment
DMAc	*N,N*-Dimethylacetamide
DMF	Dimethylformamide
DNA	Deoxyribonucleic acid
DS	Diclofenac sodium
ECM	Extracellular matrix
EDC	1-Ethyl-3-(3-dimethyl-aminopropyl)-1-carbodiimide hydrochloride
GT	Gum tragacanth
GTA	Glutaraldehyde
h	Hour
HA	Hyaluronic acid
HEPA	High-efficiency particulate air
m^2	Square meter
NaAlg	Sodium alginate
NaOH:DMF	Sodium hydroxide:dimethylformamide
NH_4OH:DMF	Ammonium hydroxide:dimethylformamide
PAA	Poly(acrylic acid)
PEG-b-PLA	Poly(ethylene glycol)-b-poly(lactide)
PELCL	Poly(ethylene glycol)-b-poly (L-lactide-*co*-caprolactone)
PEO	Poly(ethylene oxide)
PET	Poly(ethylene terephthalate)
PHA-b-PEO	Poly(hexamethylene adipate)-PEO block copolymers
PLCL	Poly(L-lactide-*co*-ε-caprolactone)
PLGA	Poly(lactic-*co*-glycolic acid)
PVA	Poly(vinyl alcohol)
PVP	Polyvinylpyrrolidone
RK	Raspberry ketone
RNA	Ribonucleic acid
S. aureus	*Staphylococcus aureus*
TPU	Thermoplastic polyurethane
UV	Ultraviolet

References

[1] Jiang, S. et al. Polyimide nanofibers by "Green" electrospinning via aqueous solution for filtration applications, ACS Sustain. Chem. Eng., 2016, 4(9), 4797–4804.

[2] Padil, V.V.T., Wacławek, S., & Černík, M. Green synthesis: Nanoparticles and nanofibres based on tree gums for environmental applications, Ecol. Chem. Eng. S, 2016, 23, 4.

[3] Sridhar, R. et al. Green processing mediated novel polyelectrolyte nanofibers and their antimicrobial evaluation, Macromol. Mater. Eng., 2014, 299(3), 283–289.

[4] Wang, X. et al. Controlled release of protein from core-shell nanofibers prepared by emulsion electrospinning based on green chemical, J. Appl. Polym. Sci., 2015, 132(16), 1–9.

[5] Bhardwaj, N., & Kundu, S.C. Electrospinning: a fascinating fiber fabrication technique, Biotechnol. Adv., 2010, 28(3), 325–347.
[6] Haghi, A.K. Advances in Nanofibre Research, 2011, Smithers Rapra, Shawbury, Shrewsbury, Shropshire, United Kingdom.
[7] He, J.-H. et al. Electrospun nanofibres and their applications, 2008, ISmithers, Shawbury, UK.
[8] Salehi, R. et al. Interaction, controlled release, and antitumor activity of doxorubicin hydrochloride from pH-Sensitive P(NIPAAm-MAA-VP) nanofibrous scaffolds prepared by green electrospinning, Int. J. Polym. Mater. Polym. Biomater., 2014, 63(12), 609–619.
[9] Yang, X. et al. Green electrospun Manuka honey/silk fibroin fibrous matrices as potential wound dressing, Mater. Design, 2017, 119, 76–84.
[10] Chen, W. et al. Dexamethasone loaded core-shell SF/PEO nanofibers via green electrospinning reduced endothelial cells inflammatory damage, Colloids Surf. B Biointerfaces, 2015, 126, 561–568.
[11] Castilla-Casadiego, D.A. et al. "Green" electrospinning of a collagen/hydroxyapatite composite nanofibrous scaffold, MRS Commun., 2016, 6(04), 402–407.
[12] Bubel, K. et al. Tenside-free biodegradable polymer nanofiber nonwovens by "Green Electrospinning", Macromolecules, 2013, 46(17), 7034–7042.
[13] Bunyatova, U. et al. Fabrication and characterization of AgNPs incorporated PVA/ODA-MMT and PVP/ODA-MMT nanofiber structures by green electrospinning nanotechnology as excellent conducting and bioengineering nanomaterial, Acta Physica Polonica. A, 2016, 129 (4), 431–435.
[14] Fan, L. et al. Green electrospun pantothenic acid/silk fibroin composite nanofibers: fabrication, characterization and biological activity, Colloids. Surf. B. Biointerfaces, 2014, 117, 14–20.
[15] Lin, S. et al. Green electrospun grape seed extract-loaded silk fibroin nanofibrous mats with excellent cytocompatibility and antioxidant effect, Colloids. Surf. B. Biointerfaces, 2016, 139, 156–163.
[16] Sun, J. et al. Nanofibers by green electrospinning of aqueous suspensions of biodegradable block copolyesters for applications in medicine, pharmacy and agriculture, Macromol. Rapid. Commun., 2010, 31(23), 2077–2083.
[17] Agarwal, S., & Greiner, A. On the way to clean and safe electrospinning-green electrospinning: emulsion and suspension electrospinning, Polym. Adv. Technol., 2011, 22 (3), 372–378.
[18] Briggs, T., & Arinzeh, T.L. Examining the formulation of emulsion electrospinning for improving the release of bioactive proteins from electrospun fibers, J. Biomed. Mater. Res. A, 2014, 102(3), 674–684.
[19] Hu, J. et al. Preparation and characterization of electrospun PLGA/gelatin nanofibers as a drug delivery system by emulsion electrospinning, J. Biomater Sci. Polym. Ed., 2013, 24(8), 972–985.
[20] Sanders, E.H. et al. Two-phase electrospinning from a single electrified jet: microencapsulation of aqueous reservoirs in poly (ethylene-co-vinyl acetate) fibers, Macromolecules, 2003, 36(11), 3803–3805.
[21] Gordon, V., Marom, G., & Magdassi, S. Formation of hydrophilic nanofibers from nanoemulsions through electrospinning, Int. J. Pharm., 2015, 478(1), 172–179.
[22] Ajalloueian, F. et al. Emulsion electrospinning as an approach to fabricate PLGA/chitosan nanofibers for biomedical applications, Biomed. Res. Int., 2014, 2014, 475280.
[23] Kim, K. et al. Incorporation and controlled release of a hydrophilic antibiotic using poly (lactide-co-glycolide)-based electrospun nanofibrous scaffolds, J. Control. Release, 2004, 98 (1), 47–56.

[24] Kriegel, C. et al. Nanofibers as carrier systems for antimicrobial microemulsions, Part I: fabrication and characterization. Langmuir, 2009, 25(2), 1154–1161.

[25] Mori, C.L.S.d.O. et al. Nanostructured polylactic acid/candeia essential oil mats obtained by electrospinning, J. Nanomater., 2015, 2015, 1–9.

[26] Rieger, K.A., Birch, N.P., & Schiffman, J.D. Electrospinning chitosan/poly(ethylene oxide) solutions with essential oils: Correlating solution rheology to nanofiber formation, Carbohydr. Polym., 2016, 139, 131–138.

[27] Sadri, M. et al. New chitosan/Poly (ethylene oxide)/Thyme nanofiber prepared by electrospinning method for antimicrobial wound dressing, J. Nanostruct., 2016, 6(4), 322–328.

[28] Kesici Güler, H., Cengiz Çallıoğlu, F., & Sesli Çetin, E. , Emulsion Electrospinning of PVP/ Cinnamon Essential Oil Nanofibers, in XIV International Izmir Textile and Apparel Symposium, 2017, Izmir/Turkey., 243–249.

[29] Naveen, N. et al. Synthesis of nonwoven nanofibers by electrospinning - A promising biomaterial for tissue engineering and drug delivery, Adv. Eng. Mater., 2010, 12(8), B380–B387.

[30] Han, F. et al. In situ encapsulation of hydrogel in ultrafine fibers by suspension electrospinning, Polym. Eng. Sci., 2012, 52(12), 2695–2704.

[31] Haghi, A.K., & Akbari, M. Trends in electrospinning of natural nanofibers, Physica Status Solidi (a), 2007, 204(6), 1830–1834.

[32] Krishnan, R., Sundarrajan, S., & Ramakrishna, S. Green processing of nanofibers for regenerative medicine, Macromol. Mater. Eng., 2013, 298, 1034–1058.

[33] Bhattarai, N. et al. Alginate-based nanofibrous scaffolds: Structural, mechanical, and biological properties, Adv. Mater., 2006, 18(11), 1463–1467.

[34] Li, J. et al. Electrospinning of Hyaluronic Acid (HA) and HA/Gelatin Blends, Macromol. Rapid. Commun., 2006, 27(2), 114–120.

[35] Schiffman, J.D., & Schauer, C.L. A review: Electrospinning of biopolymer nanofibers and their applications, Polym. Rev., 2008, 48(2), 317–352.

[36] Šimkovic, I. What could be greener than composites made from polysaccharides?, Carbohydr. Polym., 2008, 74(4), 759–762.

[37] Brenner, E.K. et al. Electrospinning of hyaluronic acid nanofibers from aqueous ammonium solutions, Carbohydr. Polym., 2012, 87(1), 926–929.

[38] Collier, J.H. et al. Synthesis and characterization of polypyrrole–hyaluronic acid composite biomaterials for tissue engineering applications, J. Biomed. Mater. Res. Part A, 2000, 50(4), 574–584.

[39] Collins, M.N., & Birkinshaw, C. Hyaluronic acid-based scaffolds for tissue engineering--a review, Carbohydr. Polym., 2013, 92(2), 1262–1279.

[40] Fischer, R.L., McCoy, M.G., & Grant, S.A. Electrospinning collagen and hyaluronic acid nanofiber meshes, J. Mater. Sci. Mater. Med., 2012, 23(7), 1645–1654.

[41] Um, I.C. et al. Electro-spinning and electro-blowing of hyaluronic acid, Biomacromolecules, 2004, 5(4), 1428–1436.

[42] Ji, Y. et al. Dual-syringe reactive electrospinning of cross-linked hyaluronic acid hydrogel nanofibers for tissue engineering applications, Macromol. Biosci., 2006, 6(10), 811–817.

[43] Ji, Y. et al. Electrospun three-dimensional hyaluronic acid nanofibrous scaffolds, Biomaterials, 2006, 27(20), 3782–3792.

[44] Liu, Y. et al. Effects of solution properties and electric field on the electrospinning of hyaluronic acid, Carbohydr. Polym., 2011, 83(2), 1011–1015.

[45] Kim, T.G., Chung, H.J., & Park, T.G Macroporous and nanofibrous hyaluronic acid/collagen hybrid scaffold fabricated by concurrent electrospinning and deposition/leaching of salt particles, Acta. Biomater., 2008, 4(6), 1611–1619.

[46] Wang, X. et al. Formation of water-resistant hyaluronic acid nanofibers by blowing-assisted electro-spinning and non-toxic post treatments, Polymer, 2005, 46(13), 4853–4867.
[47] Cengiz, F., & Jirsak, O The effect of salt on the roller electrospinning of polyurethane nanofibers, Fiber. Polym., 2009, 10(2), 177–184.
[48] Chen, Z.G. et al. Electrospun collagen-chitosan nanofiber: a biomimetic extracellular matrix for endothelial cell and smooth muscle cell, Acta. Biomater., 2010, 6(2), 372–382.
[49] Aliabadi, M. et al. Electrospun nanofiber membrane of PEO/Chitosan for the adsorption of nickel, cadmium, lead and copper ions from aqueous solution, Chem. Eng. J., 2013, 220, 237–243.
[50] Geng, X., Kwon, O.H., & Jang, J. Electrospinning of chitosan dissolved in concentrated acetic acid solution, Biomaterials, 2005, 26(27), 5427–5432.
[51] Gomathi, P. et al. Fabrication of novel chitosan nanofiber/gold nanoparticles composite towards improved performance for a cholesterol sensor, Sens. Actuators B: Chem., 2011, 153 (1), 44–49.
[52] Haider, S., & Park, S.-Y. Preparation of the electrospun chitosan nanofibers and their applications to the adsorption of Cu(II) and Pb(II) ions from an aqueous solution, J. Membrane. Sci., 2009, 328(1–2), 90–96.
[53] Park, J.M. et al. Immobilization of lysozyme-CLEA onto electrospun chitosan nanofiber for effective antibacterial applications, Int. J. Biol. Macromol., 2013, 54, 37–43.
[54] Wang, W. et al. Effects of Schwann cell alignment along the oriented electrospun chitosan nanofibers on nerve regeneration, J. Biomed. Mater. Res. A, 2009, 91(4), 994–1005.
[55] Ohkawa, K. et al. Chitosan nanofiber, Biomacromolecules, 2006, 7(11), 3291–3294.
[56] Jung, K.-H. et al. Preparation and antibacterial activity of PET/chitosan nanofibrous mats using an electrospinning technique, J. Appl. Polym. Sci., 2007, 105(5), 2816–2823.
[57] Chu, X.H. et al. In vitro evaluation of a multi-layer radial-flow bioreactor based on galactosylated chitosan nanofiber scaffolds, Biomaterials, 2009, 30(27), 4533–4538.
[58] Gu, B.K et al. Fabrication of sonicated chitosan nanofiber mat with enlarged porosity for use as hemostatic materials, Carbohydr. Polym., 2013, 97(1), 65–73.
[59] Li, Y., Qiu, T., & Xu, X. Preparation of lead-ion imprinted crosslinked electro-spun chitosan nanofiber mats and application in lead ions removal from aqueous solutions, Eur. Polym. J., 2013, 49(6), 1487–1494.
[60] Park, S.A., Park, K.E., & Kim, W. Preparation of sodium alginate/poly(ethylene oxide) blend nanofibers with lecithin, Macromole. Res., 2010, 18(9), 891–896.
[61] Safi, S. et al. Study of electrospinning of sodium alginate, blended solutions of sodium alginate/poly(vinyl alcohol) and sodium alginate/poly(ethylene oxide), J. Appl. Polym. Sci., 2007, 104(5), 3245–3255.
[62] Bhattarai, N., & Zhang, M. Controlled synthesis and structural stability of alginate-based nanofibers, Nanotechnology, 2007, 18, 45.
[63] Bonino, C.A. et al. Electrospinning alginate-based nanofibers: From blends to crosslinked low molecular weight alginate-only systems, Carbohydr. Polym., 2011, 85(1), 111–119.
[64] Jeong, S.I. et al. Electrospun alginate nanofibers with controlled cell adhesion for tissue engineering, Macromol. Biosci., 2010, 10(8), 934–943.
[65] Kim, G. K.-e. Park, Alginate-nanofibers fabricated by an electrohydrodynamic process. Polym. Eng. Sci., 2009, 49(11), 2242–2248.
[66] Leung, V. et al. Postelectrospinning modifications for alginate nanofiber-based wound dressings, J. Biomed. Mater. Res. B Appl. Biomater., 2014, 102(3), 508–515.
[67] Li, Y. et al. Facile preparation and enhanced capacitance of the polyaniline/sodium alginate nanofiber network for supercapacitors, Langmuir, 2011, 27(10), 6458–6463.

[68] Ma, G. et al. Electrospun sodium alginate/poly(ethylene oxide) core–shell nanofibers scaffolds potential for tissue engineering applications, Carbohydr. Polym., 2012, 87(1), 737–743.
[69] Shalumon, K.T. et al. Sodium alginate/poly(vinyl alcohol)/nano ZnO composite nanofibers for antibacterial wound dressings, Int. J. Biol. Macromol., 2011, 49(3), 247–254.
[70] Yeom, C., & Lee, K.H. Characterization of sodium alginate membrane crosslinked with glutaraldehyde in pervaporation separation, J. Appl. Polym. Sci., 1998, 67(2), 209–219.
[71] Yu, Y. et al. Facile synthesis of polyaniline– sodium alginate nanofibers, Langmuir, 2006, 22 (8), 3899–3905.
[72] Augst, A.D., Kong, H.J., & Mooney, D.J. Alginate hydrogels as biomaterials, Macromole. Biosci., 2006, 6(8), 623–633.
[73] Dausse, Y. et al. Cartilage repair using new polysaccharidic biomaterials: macroscopic, histological and biochemical approaches in a rat model of cartilage defect, Osteoarthr. cartilage, 2003, 11(1), 16–28.
[74] Kanakis, J. et al. The crystallization of calcium carbonate on porcine and human cardiac valves and the antimineralization effect of sodium alginate, J.Cryst. Growth, 2001, 223(4), 557–564.
[75] Hashimoto, T. et al. Development of alginate wound dressings linked with hybrid peptides derived from laminin and elastin, Biomaterials, 2004, 25(7–8), 1407–1414.
[76] Thorvaldsson, A. et al. Electrospinning of highly porous scaffolds for cartilage regeneration, Biomacromolecules, 2008, 9(3), 1044–1049.
[77] Alsberg, E. et al. Engineering growing tissues, Proc. Nat. Acad. Sci., 2002, 99(19), 12025–12030.
[78] Coşkun Üstündağ, G. et al. In vivo evaluation of electrospun poly (Vinyl Alcohol)/sodium alginate nanofibrous mat as wound dressing, J. Textile Apparel/Tekstil ve Konfeksiyon, 2010, 20, 4.
[79] Sell, S.A. et al. Cross-linking methods of electrospun fibrinogen scaffolds for tissue engineering applications, Biomed. Mater., 2008, 3(4), 045001.
[80] Wnek, G.E. et al. Electrospinning of nanofiber fibrinogen structures, Nano letters, 2003, 3(2), 213–216.
[81] McManus, M.C. et al. Electrospun fibrinogen: feasibility as a tissue engineering scaffold in a rat cell culture model, J. Biomed. Mater. Res. A, 2007, 81(2), 299–309.
[82] Balasubramanian, P. et al. Human cardiomyocyte interaction with electrospun fibrinogen/gelatin nanofibers for myocardial regeneration, J. Biomater Sci. Polym. Ed., 2013, 24(14), 1660–1675.
[83] Fang, Z. et al. Preparation and biocompatibility of electrospun poly(l-lactide-co-ε-caprolactone)/fibrinogen blended nanofibrous scaffolds, Appl. Surf. Sci., 2011, 257(9), 4133–4138.
[84] Ravichandran, R. et al. Expression of cardiac proteins in neonatal cardiomyocytes on PGS/fibrinogen core/shell substrate for cardiac tissue engineering, Int. J. Cardiol., 2013, 167(4), 1461–1468.
[85] Liu, S.-J. et al. Electrospun PLGA/collagen nanofibrous membrane as early-stage wound dressing, J. Membr. Sci., 2010, 355(1–2), 53–59.
[86] Rho, K.S. et al. Electrospinning of collagen nanofibers: effects on the behavior of normal human keratinocytes and early-stage wound healing, Biomaterials, 2006, 27(8), 1452–1461.
[87] Chen, R. et al. Preparation and characterization of coaxial electrospun thermoplastic polyurethane/collagen compound nanofibers for tissue engineering applications, Colloids. Surf. B. Biointerfaces, 2010, 79(2), 315–325.
[88] Matthews, J.A. et al. Electrospinning of collagen nanofibers, Biomacromolecules, 2002, 3(2), 232–238.

[89] Venugopal, J.R., Zhang, Y., & Ramakrishna, S. In vitro culture of human dermal fibroblasts on electrospun polycaprolactone collagen nanofibrous membrane, Artif. Organs, 2006, 30(6), 440–446.
[90] Dong, B. et al. Electrospinning of collagen nanofiber scaffolds from benign solvents, Macromol. Rapid. Commun., 2009, 30(7), 539–542.
[91] Meng, L. et al. Electrospinning of in situ crosslinked collagen nanofibers, J. Mater. Chem., 2012, 22, 37.
[92] Charulatha, V., & Rajaram, A. Influence of different crosslinking treatments on the physical properties of collagen membranes, Biomaterials, 2003, 24(5), 759–767.
[93] Hartman, O. et al. Microfabricated electrospun collagen membranes for 3-D cancer models and drug screening applications, Biomacromolecules, 2009, 10(8), 2019–2032.
[94] Liu, T. et al. Photochemical crosslinked electrospun collagen nanofibers: synthesis, characterization and neural stem cell interactions, J. Biomed. Mater. Res. A, 2010, 95(1), 276–282.
[95] Itoh, S. et al. Evaluation of cross-linking procedures of collagen tubes used in peripheral nerve repair, Biomaterials, 2002, 23(23), 4475–4481.
[96] Kidoaki, S., Kwon, I.K., & Matsuda, T. Mesoscopic spatial designs of nano-and microfiber meshes for tissue-engineering matrix and scaffold based on newly devised multilayering and mixing electrospinning techniques, Biomaterials, 2005, 26(1), 37–46.
[97] Chen, H.-C., Jao, W.-C., & Yang, M.-C. Characterization of gelatin nanofibers electrospun using ethanol/formic acid/water as a solvent, Polym. Adv. Tech., 2009, 20(2), 98–103.
[98] Ghasemi-Mobarakeh, L. et al. Electrospun poly(epsilon-caprolactone)/gelatin nanofibrous scaffolds for nerve tissue engineering, Biomaterials, 2008, 29(34), 4532–4539.
[99] Huang, Z.-M. et al. Electrospinning and mechanical characterization of gelatin nanofibers, Polymer, 2004, 45(15), 5361–5368.
[100] Ki, C.S. et al. Characterization of gelatin nanofiber prepared from gelatin–formic acid solution, Polymer, 2005, 46(14), 5094–5102.
[101] Ko, J.H. et al. Characterization of cross-linked gelatin nanofibers through electrospinning, Macromol. Res., 2010, 18(2), 137–143.
[102] Meng, Z.X. et al. Fabrication of mineralized electrospun PLGA and PLGA/gelatin nanofibers and their potential in bone tissue engineering, Mater. Sci. Eng. C. Mater. Biol. Appl., 2013, 33 (2), 699–706.
[103] Meng, Z.X. et al. Electrospinning of PLGA/gelatin randomly-oriented and aligned nanofibers as potential scaffold in tissue engineering, Mater. Sci. Eng.: C, 2010, 30(8), 1204–1210.
[104] Xu, X., & Zhou, M. Antimicrobial gelatin nanofibers containing silver nanoparticles, Fiber. Polym., 2008, 9(6), 685–690.
[105] Yin, G. et al. Study on the properties of the electrospun silk fibroin/gelatin blend nanofibers for scaffolds, J. Appl. Polym. Sci., 2009, 111(3), 1471–1477.
[106] Zhang, Y.Z. et al. Crosslinking of the electrospun gelatin nanofibers, Polymer, 2006, 47(8), 2911–2917.
[107] Dhandayuthapani, B., Krishnan, U.M., & Sethuraman, S. Fabrication and characterization of chitosan-gelatin blend nanofibers for skin tissue engineering, J. Biomed. Mater. Res. B Appl. Biomater., 2010, 94(1), 264–272.
[108] Aduba, D.C., Jr. et al. Semi-interpenetrating network (sIPN) gelatin nanofiber scaffolds for oral mucosal drug delivery, Acta. Biomater, 2013, 9(5), 6576–6584.
[109] Panzavolta, S. et al. Electrospun gelatin nanofibers: optimization of genipin cross-linking to preserve fiber morphology after exposure to water, Acta. Biomater, 2011, 7(4), 1702–1709.
[110] Kim, M.S. et al. The development of genipin-crosslinked poly(caprolactone) (PCL)/gelatin nanofibers for tissue engineering applications, Macromol. Biosci., 2010, 10(1), 91–100.

[111] Li, M. et al. Electrospinning polyaniline-contained gelatin nanofibers for tissue engineering applications, Biomaterials, 2006, 27(13), 2705–2715.
[112] Zhuang, X. et al. Electrospun chitosan/gelatin nanofibers containing silver nanoparticles, Carbohydr. Polym., 2010, 82(2), 524–527.
[113] Okutan, N., Terzi, P., & Altay, F. Affecting parameters on electrospinning process and characterization of electrospun gelatin nanofibers, Food Hydrocolloid., 2014, 39, 19–26.
[114] Song, J.H., Kim, H.E., & Kim, H.W. Production of electrospun gelatin nanofiber by water-based co-solvent approach, J. Mater. Sci. Mater. Med., 2008, 19(1), 95–102.
[115] Zhang, S. et al. Gelatin nanofibrous membrane fabricated by electrospinning of aqueous gelatin solution for guided tissue regeneration, J. Biomed. Mater. Res. A, 2009, 90(3), 671–679.
[116] Ranjbar-Mohammadi, M., Bahrami, S.H., & Joghataei, M.T. Fabrication of novel nanofiber scaffolds from gum tragacanth/poly(vinyl alcohol) for wound dressing application: in vitro evaluation and antibacterial properties, Mater. Sci. Eng. C Mater. Biol. Appl., 2013, 33(8), 4935–4943.
[117] Ranjbar-Mohammadi, M., & Bahrami, S.H. Development of nanofibrous scaffolds containing gum tragacanth/poly (epsilon-caprolactone) for application as skin scaffolds, Mater. Sci. Eng. C Mater. Biol. Appl., 2015, 48, 71–79.
[118] Ranjbar-Mohammadi, M. et al. Fabrication of curcumin-loaded gum tragacanth/poly(vinyl alcohol) nanofibers with optimized electrospinning parameters, J. Ind. Text., 2017, 46(5), 1170–1192.
[119] Ranjbar-Mohammadi, M. et al. Antibacterial performance and in vivo diabetic wound healing of curcumin loaded gum tragacanth/poly(epsilon-caprolactone) electrospun nanofibers, Mater. Sci. Eng. C Mater. Biol. Appl., 2016, 69, 1183–1191.
[120] Abe, K., Iwamoto, S., & Yano, H. Obtaining cellulose nanofibers with a uniform width of 15 nm from wood, Biomacromolecules, 2007, 8(10), 3276–3278.
[121] Juntaro, J. et al. Creating hierarchical structures in renewable composites by attaching bacterial cellulose onto sisal fibers, Adv. mater., 2008, 20(16), 3122–3126.
[122] Liu, H. et al. Fabrication and properties of transparent polymethylmethacrylate/cellulose nanocrystals composites, Biores. Tech., 2010, 101(14), 5685–5692.
[123] Siqueira, G., Bras, J., & Dufresne, A Cellulose whiskers versus microfibrils: influence of the nature of the nanoparticle and its surface functionalization on the thermal and mechanical properties of nanocomposites, Biomacromolecules, 2008, 10(2), 425–432.
[124] Czaja, W.K. et al. The future prospects of microbial cellulose in biomedical applications, Biomacromolecules, 2007, 8(1), 1–12.
[125] Yano, H. et al. Optically transparent composites reinforced with networks of bacterial nanofibers, Adv. Mater., 2005, 17(2), 153–155.
[126] Nogi, M., & Yano, H. Transparent nanocomposites based on cellulose produced by bacteria offer potential innovation in the electronics device industry, Adv. mater., 2008, 20(10), 1849–1852.
[127] Nogi, M. et al. Optically transparent nanofiber paper, Adv. Mater., 2009, 21(16), 1595–1598.
[128] Iwatake, A., Nogi, M., & Yano, H. Cellulose nanofiber-reinforced polylactic acid, Comp. Sci. Tech., 2008, 68(9), 2103–2106.
[129] Azeredo, H.M. et al. Nanocomposite edible films from mango puree reinforced with cellulose nanofibers, J. Food. Sci., 2009, 74(5), N31–5.
[130] Helbert, W., Cavaille, J., & Dufresne, A. Thermoplastic nanocomposites filled with wheat straw cellulose whiskers, Part I: processing and mechanical behavior. Polym. Comp., 1996, 17(4), 604–611.

[131] Podsiadlo, P. et al. Molecularly engineered nanocomposites: layer-by-layer assembly of cellulose nanocrystals, Biomacromolecules, 2005, 6(6), 2914–2918.
[132] Azizi Samir, M.A.S., Alloin, F., & Dufresne, A. Review of recent research into cellulosic whiskers, their properties and their application in nanocomposite field, Biomacromolecules, 2005, 6(2), 612–626.
[133] Beecher, J.F. Organic materials: Wood, trees and nanotechnology, Nat. nanotech., 2007, 2(8), 466.
[134] Eichhorn, S.J. et al. Current international research into cellulose nanofibres and nanocomposites, J. mater. sci., 2010, 45, 1, 1.
[135] Habibi, Y., Lucia, L.A., & Rojas, O.J Cellulose nanocrystals: chemistry, self-assembly, and applications, Chem. Rev., 2010, 110(6), 3479–3500.
[136] Siró, I., & Plackett, D. Microfibrillated cellulose and new nanocomposite materials: a review, Cellulose, 2010, 17(3), 459–494.
[137] Kim, C.-W. et al. Structural studies of electrospun cellulose nanofibers, Polymer, 2006, 47 (14), 5097–5107.
[138] Hennink, W. et al. Biodegradable dextran hydrogels crosslinked by stereocomplex formation for the controlled release of pharmaceutical proteins, Int. j. pharm., 2004, 277(1–2), 99–104.
[139] Hovgaard, L., & Brøndsted, H. Dextran hydrogels for colon-specific drug delivery, J. Control. Release, 1995, 36(1–2), 159–166.
[140] Jiang, H et al. Optimization and characterization of dextran membranes prepared by electrospinning, Biomacromolecules, 2004, 5(2), 326–333.
[141] Lévesque, S.G., Lim, R.M., & Shoichet, M.S. Macroporous interconnected dextran scaffolds of controlled porosity for tissue-engineering applications, Biomaterials, 2005, 26(35), 7436–7446.
[142] Cengiz-Çallıoğlu, F. Dextran nanofiber production by needleless electrospinning process, e-Polymers, 2014, 14, 1.
[143] Min, B.-M. et al. Electrospinning of silk fibroin nanofibers and its effect on the adhesion and spreading of normal human keratinocytes and fibroblasts in vitro, Biomaterials, 2004, 25(7–8), 1289–1297.
[144] Zarkoob, S. et al. Structure and morphology of electrospun silk nanofibers, Polymer, 2004, 45(11), 3973–3977.
[145] Farokhzad, O.C., & Langer, R. Impact of nanotechnology on drug delivery, ACS Nano., 2009, 3 (1), 16–20.
[146] Balogh, A. et al. Alternating current electrospinning for preparation of fibrous drug delivery systems, Int. J. Pharm., 2015, 495(1), 75–80.
[147] Haider, A., Haider, S., & Kang, I.-K. A comprehensive review summarizing the effect of electrospinning parameters and potential applications of nanofibers in biomedical and biotechnology, Arab. J. Chem., 2015, 11(8), 1165–1188.
[148] Hu, X. et al. Electrospinning of polymeric nanofibers for drug delivery applications, J. Control Release, 2014, 185, 12–21.
[149] Okuda, T., Tominaga, K., & Kidoaki, S. Time-programmed dual release formulation by multilayered drug-loaded nanofiber meshes, J. Control Release, 2010, 143(2), 258–264.
[150] Yang, D., Li, Y., & Nie, J Preparation of gelatin/PVA nanofibers and their potential application in controlled release of drugs, Carbohydr. Polym., 2007, 69(3), 538–543.
[151] Yu, D.-G et al. Electrospun nanofiber-based drug delivery systems, Health, 2009, 01(02), 67–75.
[152] Zamani, M et al. Controlled release of metronidazole benzoate from poly epsilon-caprolactone electrospun nanofibers for periodontal diseases, Eur. J. Pharm. Biopharm., 2010, 75(2), 179–185.

[153] Agarwal, S, Wendorff, J.H, & Greiner, A Use of electrospinning technique for biomedical applications, Polymer, 2008, 49(26), 5603–5621.
[154] Cui, W., Zhou, Y., & Chang, J. Electrospun nanofibrous materials for tissue engineering and drug delivery, Sci. Technol. Adv. Mater., 2010, 11(1), 14108.
[155] Verreck, G et al. Incorporation of drugs in an amorphous state into electrospun nanofibers composed of a water-insoluble, nonbiodegradable polymer, J. Control. Release, 2003, 92(3), 349–360.
[156] Valarezo, E et al. Fabrication and sustained release properties of poly(ε-caprolactone) electrospun fibers loaded with layered double hydroxide nanoparticles intercalated with amoxicillin, Appl. Clay Sci., 2013, 72, 104–109.
[157] Sokolsky-Papkov, M et al. Polymer carriers for drug delivery in tissue engineering, Adv. drug deliver. rev., 2007, 59(4–5), 187–206.
[158] Meng, Z.X et al. Preparation and characterization of electrospun PLGA/gelatin nanofibers as a potential drug delivery system, Colloids. Surf. B. Biointerfaces, 2011, 84(1), 97–102.
[159] Kenawy, E.-R. et al. Processing of polymer nanofibers through electrospinning as drug delivery systems, Mater. Chem. Phys., 2009, 113(1), 296–302.
[160] Zeng, J. et al. Influence of the drug compatibility with polymer solution on the release kinetics of electrospun fiber formulation, J. Control Release, 2005, 105(1–2), 43–51.
[161] Shen, X. et al. Electrospun diclofenac sodium loaded Eudragit(R) L 100–55 nanofibers for colon-targeted drug delivery, Int. J. Pharm., 2011, 408(1–2), 200–207.
[162] Jannesari, M. et al. Composite poly(vinyl alcohol)/poly(vinyl acetate) electrospun nanofibrous mats as a novel wound dressing matrix for controlled release of drugs, Int. J. Nanomedicine, 2011, 6, 993–1003.
[163] Katti, D.S. et al. Bioresorbable nanofiber-based systems for wound healing and drug delivery: optimization of fabrication parameters, J. Biomed. Mater. Res. B Appl. Biomater, 2004, 70(2), 286–296.
[164] Uppal, R. et al. Hyaluronic acid nanofiber wound dressing--production, characterization, and in vivo behavior, J. Biomed. Mater. Res. B Appl. Biomater, 2011, 97(1), 20–29.
[165] Vargas, E.A. et al. Hyperbranched polyglycerol electrospun nanofibers for wound dressing applications, Acta. Biomater., 2010, 6(3), 1069–1078.
[166] Zhou, Y. et al. Electrospun water-soluble carboxyethyl chitosan/poly (vinyl alcohol) nanofibrous membrane as potential wound dressing for skin regeneration, Biomacromolecules, 2008, 9(1), 349–354.
[167] Cai, Z.X. et al. Fabrication of chitosan/silk fibroin composite nanofibers for wound-dressing applications, Int. J. Mol. Sci., 2010, 11(9), 3529–3539.
[168] Khil, M.S. et al. Electrospun nanofibrous polyurethane membrane as wound dressing, J. Biomed. Mater. Res. Part B: Appl. Biomater., 2003, 67(2), 675–679.
[169] Unnithan, A.R. et al. Wound-dressing materials with antibacterial activity from electrospun polyurethane-dextran nanofiber mats containing ciprofloxacin HCl, Carbohydr. Polym., 2012, 90(4), 1786–1793.
[170] Abdelgawad, A.M., Hudson, S.M., & Rojas, O.J. Antimicrobial wound dressing nanofiber mats from multicomponent (chitosan/silver-NPs/polyvinyl alcohol) systems, Carbohydr. Polym., 2014, 100, 166–178.
[171] Chen, J.-P., Chang, G.-Y., & Chen, J.-K. Electrospun collagen/chitosan nanofibrous membrane as wound dressing, Colloids Surf. A: Physicochem. Eng. Asp., 2008, 313–314, 183–188.
[172] Bölgen, N. et al. In vivo performance of antibiotic embedded electrospun PCL membranes for prevention of abdominal adhesions, J. Biomed. Mater. Res. Part B: Appl. Biomater., 2007, 81 (2), 530–543.

[173] Prabaharan, M., Jayakumar, R., & Nair, S. Electrospun nanofibrous scaffolds-current status and prospects in drug delivery, in Biomedical applications of polymeric nanofibers, Vol. 2011, Springer, 241–262.
[174] Li, W.J. et al. Electrospun nanofibrous structure: a novel scaffold for tissue engineering, J. Biomed. Mater. Res. Part A, 2002, 60(4), 613–621.
[175] Yang, F. et al. Electrospinning of nano/micro scale poly(L-lactic acid) aligned fibers and their potential in neural tissue engineering, Biomaterials, 2005, 26(15), 2603–2610.
[176] Xu, C. et al. Aligned biodegradable nanofibrous structure: a potential scaffold for blood vessel engineering, Biomaterials, 2004, 25(5), 877–886.
[177] Chong, E. et al. Evaluation of electrospun PCL/gelatin nanofibrous scaffold for wound healing and layered dermal reconstitution, Acta. Biomaterialia., 2007, 3(3), 321–330.
[178] Bini, T. et al. Poly (l-lactide-co-glycolide) biodegradable microfibers and electrospun nanofibers for nerve tissue engineering: an in vitro study, J. mater. sci., 2006, 41(19), 6453–6459.
[179] Murugan, R., & Ramakrishna, S. Design strategies of tissue engineering scaffolds with controlled fiber orientation, Tissue Eng., 2007, 13(8), 1845–1866.
[180] Ito, Y. et al. A composite of hydroxyapatite with electrospun biodegradable nanofibers as a tissue engineering material, J. Biosci. Bioeng., 2005, 100(1), 43–49.
[181] Yoshimoto, H. et al. A biodegradable nanofiber scaffold by electrospinning and its potential for bone tissue engineering, Biomaterials, 2003, 24(12), 2077–2082.
[182] Ciardelli, G et al. Blends of poly-(ε-caprolactone) and polysaccharides in tissue engineering applications, Biomacromolecules, 2005, 6(4), 1961–1976.
[183] Schnell, E. et al. Guidance of glial cell migration and axonal growth on electrospun nanofibers of poly-ε-caprolactone and a collagen/poly-ε-caprolactone blend, Biomaterials, 2007, 28(19), 3012–3025.
[184] Ku, S.H., & Park, C.B. Human endothelial cell growth on mussel-inspired nanofiber scaffold for vascular tissue engineering, Biomaterials, 2010, 31(36), 9431–9437.
[185] Zhang, Y et al. Electrospun biomimetic nanocomposite nanofibers of hydroxyapatite/chitosan for bone tissue engineering, Biomaterials, 2008, 29(32), 4314–4322.
[186] Dotti, F. et al. Electrospun porous mats for high efficiency filtration, J. Ind. Text., 2007, 37(2), 151–162.
[187] Gibson, P., Schreuder-Gibson, H., & Rivin, D. Transport properties of porous membranes based on electrospun nanofibers, Colloids Surf. A: Physicochem. Eng. Asp., 2001, 187, 469–481.
[188] Gopal, R. et al. Electrospun nanofibrous filtration membrane, J. Membr. Sci., 2006, 281(1–2), 581–586.
[189] Graham, K. et al. Polymeric nanofibers in air filtration applications. Fifteenth Annual Technical Conference & Expo of the American Filtration & Separations Society, Galveston, Texas, 2002.
[190] Qin, X.-H., & Wang, S.-Y. Electrospun nanofibers from crosslinked poly(vinyl alcohol) and its filtration efficiency, J. Appl. Polym. Sci., 2008, 109(2), 951–956.
[191] Sundarrajan, S. et al. Electrospun nanofibers for air filtration applications, Procedia Eng., 2014, 75, 159–163.
[192] Ahn, Y.C. et al. Development of high efficiency nanofilters made of nanofibers, Curr. Appl. Phys., 2006, 6(6), 1030–1035.
[193] Barhate, R., & Ramakrishna, S. Nanofibrous filtering media: Filtration problems and solutions from tiny materials, J. Membr. Sci., 2007, 296(1–2), 1–8.
[194] Desai, K. et al. Nanofibrous chitosan non-wovens for filtration applications, Polymer, 2009, 50(15), 3661–3669.

[195] Grafe, T., & Graham, K. Polymeric nanofibers and nanofiber webs: a new class of nonwovens, Nonwoven. Technol. Rev., 2002, 12, 51–55.
[196] Hassan, M.A. et al. Fabrication of nanofiber meltblown membranes and their filtration properties, J. Membr. Sci., 2013, 427, 336–344.
[197] Huang, L., Manickam, S.S., & McCutcheon, J.R. Increasing strength of electrospun nanofiber membranes for water filtration using solvent vapor, J. Membr. Sci., 2013, 436, 213–220.
[198] Yalcinkaya, B. et al. Thin Film Nanofibrous Composite Membrane for Dead-End Seawater Desalination, Hindawi Publishing Corporation/Journal of Nanomaterials, 2016, 1–12.
[199] Cho, D. et al. Further improvement of air filtration efficiency of cellulose filters coated with nanofibers via inclusion of electrostatically active nanoparticles, Polymer, 2013, 54(9), 2364–2372.
[200] Ramakrishna, S. et al. Science and engineering of electrospun nanofibers for advances in clean energy, water filtration, and regenerative medicine, J. Mater. Sci., 2010, 45(23), 6283–6312.
[201] Sambaer, W., Zatloukal, M., & Kimmer, D. 3D air filtration modeling for nanofiber-based filters in the ultrafine particle size range, Chem. Eng. Sci., 2012, 82, 299–311.
[202] Cengiz Çallıoğlu, F. The effect of glyoxal cross-linker and NaCl salt addition on the roller electrospinning of poly(Vinyl Alcohol) nanofibers, J. Text. Apparel/Tekstil ve Konfeksiyon, 2014, 24, 1.
[203] Podgórski, A., Bałazy, A., & Gradoń, L. Application of nanofibers to improve the filtration efficiency of the most penetrating aerosol particles in fibrous filters, Chem. Eng. Sci., 2006, 61(20), 6804–6815.
[204] Wang, J., Kim, S.C., & Pui, D.Y. Investigation of the figure of merit for filters with a single nanofiber layer on a substrate, J. Aero. Sci., 2008, 39(4), 323–334.
[205] Yun, K.M. et al. Nanoparticle filtration by electrospun polymer fibers, Chem. Eng. Sci., 2007, 62(17), 4751–4759.
[206] Yeom, B.Y., Shim, E., & Pourdeyhimi, B. Boehmite nanoparticles incorporated electrospun nylon-6 nanofiber web for new electret filter media, Macromole. res., 2010, 18(9), 884–890.
[207] Huang, L. et al. Improved mechanical properties and hydrophilicity of electrospun nanofiber membranes for filtration applications by dopamine modification, J. Membr. Sci., 2014, 460, 241–249.
[208] Ozimek, L., Pospiech, E., & Narine, S. Nanotechnologies in food and meat processing, ACTA Sci. Pol. Technol. Aliment., 2010, 9(4), 401–412.
[209] Sekhon, B.S. Food nanotechnology–an overview, Nanotech. Sci. Appl., 2010, 3, 1.
[210] Weiss, J., Takhistov, P., & McClements, D.J. Functional materials in food nanotechnology, J. Food Sci., 2006, 71(9), R107–R116.
[211] Lopez-Rubio, A., Gavara, R., & Lagaron, J.M Bioactive packaging: turning foods into healthier foods through biomaterials, Trends Food Sci. Technol., 2006, 17(10), 567–575.
[212] Taepaiboon, P., Rungsardthong, U., & Supaphol, P. Vitamin-loaded electrospun cellulose acetate nanofiber mats as transdermal and dermal therapeutic agents of vitamin A acid and vitamin E, Eur. J. Pharm. Biopharm., 2007, 67(2), 387–397.
[213] Tam, K. et al. Antibacterial activity of ZnO nanorods prepared by a hydrothermal method, Thin. Solid. films, 2008, 516(18), 6167–6174.
[214] Chauhan, V.S., & Chakrabarti, S.K. Use of nanotechnology for high performance cellulosic and papermaking products, Cell. Chem. Tech., 2012, 46(5), 389.
[215] Veleirinho, B., & Lopes-da-Silva, J.A. Application of electrospun poly(ethylene terephthalate) nanofiber mat to apple juice clarification, Process Biochem., 2009, 44(3), 353–356.

Julia L. Shamshina and Robin D. Rogers

8 Recent advances in the electrospinning of biopolymers

Abstract: Polymers constitute the largest group of chemicals that are currently used for manufacturing of nanofibers via electrospinning. Among those, use of synthetic polymers is nearly exclusive, mainly because of ease of their handling either through solution or melt processing. In recent years, there has been a growing shift toward the use of biopolymers including cellulose, chitin, and chitosan, because of many attractive properties of biopolymers (e.g., tissue biocompatibility and biodegradability), which inspired multiple big industry players in using them for the development of materials of the future. Yet, traditional electrospinning methods from common solvents (such as volatile organic solvents) are not suitable for biopolymers due to their insolubility in those solvents. This chapter focuses on recent advances in the electrospinning of biopolymers, which led to the development of an enabling technology based on a new class of materials known as ionic liquids (literally liquid salts) that allow electrospinning of virtually any biopolymer directly from its natural source. This innovative process technology leads to significant productivity enhancement, waste minimization, lower carbon footprint, and greener, more sustainable products and processes, as well as provides access to new greener products across a wide swath of industry.

Keywords: electrospinning, biopolymers, ionic liquids, single-needle syringe system, high-throughput base module, sustainable development

8.1 Can biopolymers solve the plastic waste accumulation problem?

The decline of fossil fuel resources combined with vast accumulation of post-consumer nonbiodegradable plastic waste in the landfills and natural habitats has become a definite threat for the environment. A very recent study by Geyer et al. [1] summarized data on manufacture, use, recycle, and utilization of all off-the-shelf synthetic plastics manufactured up to date. This study suggested that 8,300 million metric tons of plastics produced by date resulted in generation of approximately 6,300 million metric tons of plastic waste. Of this waste, only 9% has been recycled, 12% incinerated, and as much as 79% got amassed worldwide. The authors stated

Julia L. Shamshina, Richmond, USA
Robin D. Rogers, Tuscaloosa, USA

https://doi.org/10.1515/9783110581393-008

that if current plastics production proceeds at the current rate, and no new waste management solutions arise, the amount of plastic waste accumulated in the environment will nearly double by 2050. Other negative impacts of plastic production and usage include manufacture-related pollution, exposure of workers to toxic chemicals, and health risks to users. Transition from nonbiodegradable plastic materials to biodegradable biopolymeric products from renewable sources will be highly advantageous to society.

"Upfront" pollution reduction is possible through the development of environmentally friendly low-waste manufacturing processes from renewables. Abundant and renewable nature's polymers (such as agricultural or forest-based feedstocks) are created by nature and could also be completely broken down and returned to nature within a reasonably short period of time after customary disposal, can act as direct substitutes for the synthetic plastics, and have already attracted the attention of the community [2, 3]. The efficient and comprehensive utilization of biopolymers (e.g., cellulose, chitin, alginate, and silk; see Figure 8.1. For a few representative examples) in place of synthetic polymers is not only sustainable, but will also boost the development of high-end applications because of a wealth of useful properties of biopolymers (e.g., biocompatibility, biodegradability, and regenerative properties).

8.2 Ionic liquids to unlock nano- from nature?

Polymeric nanofibrous materials (or "nanomats") [4–6] are, perhaps, the most attractive architecture to be made of biopolymers. Three-dimensional nanomats are composed of nanometer-scale outer diameter fibers (<1,000 nm, generally 10–100 nm), with large aspect ratios (i.e., ratio between length and width) of above 50. Electrospun nanofibers, therefore, possess high surface area, high porosity, and low thermal expansion, are often lightweight, and thus represent a very attractive architecture for a variety of products and applications [6]. At the same time, one can broaden the scope of applications through the usage of biopolymers in place of synthetic polymers, by combining nanomaterial properties with properties inherent to the biopolymer itself (e.g., biocompatibility with living tissue and wound healing properties). These high-end applications include biomedical (e.g., tissue engineering [7], drug delivery, [8, 9], implant materials [10], wound dressings, and antimicrobial membranes), wastewater treatment (e.g., filtration systems [11]), energy, textiles, and cosmetic products. Electrospinning of biopolymers is also of special interest in the field of biotechnology, due to their potential to support controlled release of bioactive agents or the growth of cells [12].

The most common method to prepare nanomats is via melt or solution electrospinning, where the polymers are electrospun from their melts or volatile organic

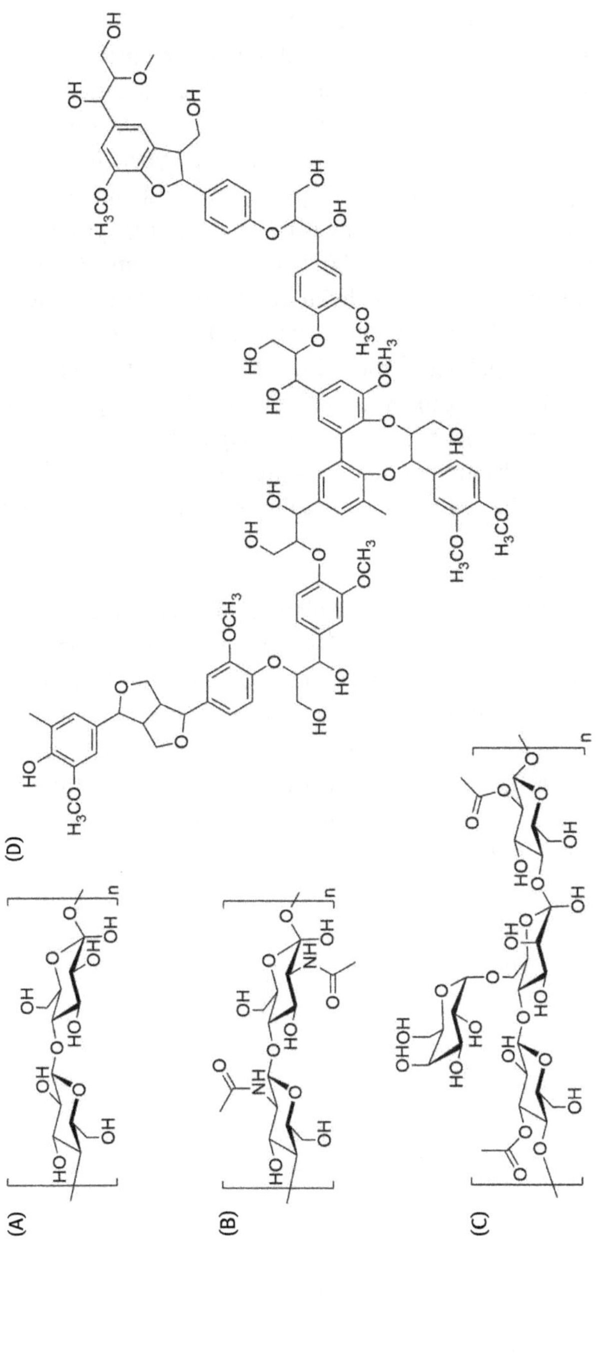

Figure 8.1: Biopolymeric structures: cellulose (A), chitin (B), hemicellulose (C), and lignin (D).

solvents (VOCs) [6, 13–15]. Such "solution processing" allows manipulation of plastics at low temperatures, simplifies incorporation of additives and fillers, and allows for quick changeovers of the constituent polymer(s). Such solution-based electrospinning has been shown to be effective for nanofiber production from a wide variety of synthetic polymers such as polyurethanes, polybenzimidazole, and polyacrylonitrile. For *biopolymers*, however, this process is challenging due to the fact that biopolymers are insoluble in commonly used solvents. This property is brought about by strong inter- and intramolecular hydrogen bonding. Complex systems are often used instead and include *N*-methylmorpholine-*N*-oxide, ethylenediamine/potassium thiocyanate, and *N,N*-dimethylacetamide (DMAc)/lithium chloride [16]. Another difficulty in biopolymer processing comes from the fact that biopolymers are obtained from living organisms, and therefore their properties depend on raw sources they have been obtained from. Thus, biopolymers vary in their molecular weight (MW), purity, distribution of charged groups (if present), crystallinity, and so on, thus requiring different loading, dissolution, and spinning conditions, even for the same polymer [17].

From this standpoint, ionic liquids (ILs, salts that melt below 100 °C [18]) have been shown to be an invaluable tool for both the isolation and dissolution of biopolymers (cellulose, chitin, amylose, pectin, silk, keratins, etc.) [19]. Among ILs, those possessing basic anions (e.g., 1-ethyl-3-methylimidazolium chloride ([C_2mim]Cl), bromide ([C_2mim]Br), and ([C_2mim][OAc])) are the most suitable for the purpose of biopolymer dissolution due to their ability to disrupt the strong hydrogen bonding network (Figure 8.2). These ILs have been successfully utilized for the preparation of fibers, films, hydrogels [20], and beads via solution processing of biopolymers dissolved in IL [16, 21–23].

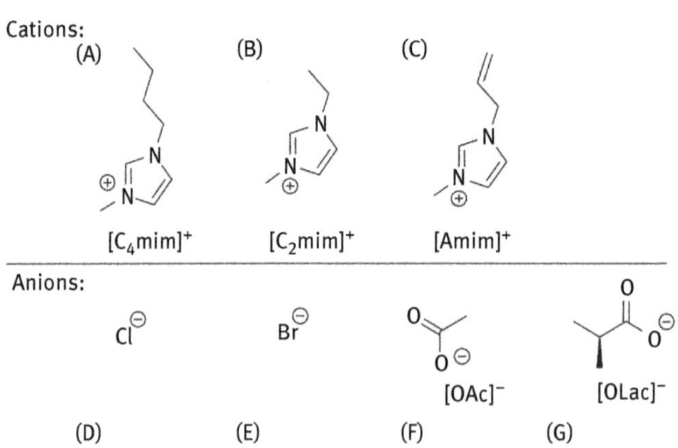

Figure 8.2: Representative cations and anions used in ILs proposed for biopolymer dissolution. Cations: 1-butyl-3-methylimidazolium (A), 1-ethyl-3-methylimidazolium (B), 1-allyl-3-methylimidazolium (C); and anions: chloride (D), bromide (E), acetate (F), and lactate (G).

8.3 "State-of-the-art" electrospinning of biopolymers from ionic liquids

While electrospinning of biopolymers from complex solvent systems is the subject of separate reviews [16, 24], we will focus on the electrospinning from ILs in this chapter. In this regard, the electrospinning from ILs is fundamentally different from common "state-of-the-art" technologies. The principle, however, is exactly the same – the process takes place through applying a high voltage to a liquid droplet coming out of the needle tip, charging the liquid [25]. Once the liquid becomes charged, the electrostatic repulsion counteracts the surface tension of the liquid and elongates the droplet which, at a critical point, ejects from the tip of the needle (i.e., the main principle of electrospinning is breaking the existing equilibrium between the surface tension of the liquid and the electrical force). When the voltage is increased, the charged jet is ejected in the form of a Taylor cone, forming a fiber. Fibers are collected on a collector (usually a drum), and VOC evaporates, leaving the nanofibrous mat on a collector plate or drum.

However, unlike the traditional wet – dry electrospinning (which involves dissolution of the desired polymer in a volatile solvent that evaporates in the electrospinning process leaving dry polymer fibers on a collector plate) – use of IL as a solvent requires total reconsideration of the process engineering. In general, electrospinning from IL is a wet–wet electrospinning process, where a polymer is dissolved in a nonvolatile IL as a primary solvent, and thus needs to be electrospun into a coagulation bath filled with an antisolvent (or "coagulant"). The coagulant is miscible with the IL, while the polymer is not. The nanomat, thus, is formed on the surface of the antisolvent and needs to be collected, washed, and dried after the process.

ILs that have been used for dissolution of biopolymers include butylimidazoliums: 1-butyl-3-methylimidazolium chloride ([C_4mim]Cl), bromide ([C_4mim]Br), and acetate ([C_4mim][OAc]), as well as ethylimidazoliums: 1-ethyl-3-methylimidazolium benzoate ([C_2mim][BA]), acetate ([C_2mim][OAc]), and lactate ([C_2mim][OLac], to name a few [19, 26]. Among those, [C_4mim]Cl and [C_2mim][OAc] are by far the most often used ILs. Overall, biopolymers that have been used for electrospinning include mostly celluloses such as pristine natural wood flour [27], pine wood [27], lignocellulosic biomass (hemp) [28, 29], cellulose pulp with DP of 1,100 [30–33], cotton linters with DP of 1,600 [34], biomimetic artificial cellulose (composed of 35–50% cellulose, 20–35% hemicellulose, and 5–30% lignin) [35], and purified celluloses of various MW (53,000 [36, 37], 194,400 [38], 5,800,000 [39, 40]). Other biopolymers include heparin [31, 32, 39, 40] (as composite with cellulose), carboxylated agarose (CA) with various extent of carboxylation (28%, 60%, and 93%) [41], silk fibroin extracted from cocoons of *Antheraea assamensis* [42], *Bombyx mori* silk [43], raw shrimp shell biomass, [44, 45], and chitin polymer regenerated from shrimp shell biomass by means of [C_2mim]-[OAc] IL [46, 47]. Electrospinning reports from ILs published to date are summarized in Table 8.1.

Table 8.1: Electrospinning reports of biopolymers from ILs published to date.

No.	Biopolymer(s)	IL	Biopolymer load (wt%)	Electrospinning conditions	Fiber diameter (Ø) (μm)	Reference
1	Cellulose (MW_{avg} = 5,800,000)	$[C_4mim]Cl$	10	*Setup*: Single-needle syringe/vertical assembly *Temperature*: NP[b] *Charge*: Grounded aluminum sheet beneath the collector *Voltage*: 15–20 kV *Nozzle-to-grounded-target distance*: 15 cm *Feed*: 0.030–0.050 mL min^{-1} *Collector*: Coagulation bath (ethanol)	<37 8.65[a]	[39, 40]
2	Cellulose (MW_{avg} = 5,800,000) - Heparin	$[C_4mim]Cl$ / $[C_2mim][BA]$ = 2.5 : 1 mol mol^{-1}	10 cellulose, 2 heparin		<41 14.45[a]	[39, 40]
3	Cellulose (MW_{avg} = 194,400)	$[C_4mim]Cl$	1.5, 2.4, 3, 4, and 5	*Setup*: Single-needle syringe/vertical assembly *Temperature*: 100 °C *Charge*: Grounded electrode (metal NP[b]), *Voltage*: 15 kV *Nozzle-to-coagulant surface distance*: 15 cm *Feed*: Free fall in electric field, *Collector*: Coagulation bath (water)	1.5 wt%: < 5 2.4 wt%: < 5 3 wt%: <2 4 wt%: < 1, 5 wt%: > 5	[38]
4	Raw cellulose fibers (MW_{avg} = 53,000)	$[C_2mim][OAc]$/ $[C_{10}mim]Cl$ = 9: 1 mol mol^{-1}	8	*Setup*: Single-needle syringe/horizontal position *Temperature*: 25 °C *Charge*: Grounded copper electrode *Voltage*: 20 kV *Nozzle-to-collector distance*: 12 cm *Feed*: 0.002 mL min^{-1} *Collector*: Coagulation bath (water), under agitation	0.10–0.15	[36, 37]

5	Pristine natural wood flour, and pine wood	[C_2mim][OAc]	5	*Setup*: Single-needle syringe/vertical assembly *Temperature*: Room temperature 25 °C *Charge*: Grounded electrode (metal NP[b]) *Voltage*: 30 kV *Nozzle-to-coagulant surface distance*: 7 cm *Solution flow rate*: Free fall in the electric field *Collector*: Coagulation bath (water)	5–20	[27]
6	Pristine natural wood flour, and pine wood	[C_2mim][OLac]	5		1–4	[27]
7	*Bombyx mori* silk	[C_4mim]Cl	5	*Setup*: Single-needle syringe/vertical assembly *Temperature*: 80 °C *Voltage*: 9 kV *Nozzle-to-coagulant surface distance*: 7 cm *Feed*: NP[b] *Collector*: Glass plate in coagulation bath (water)	50	[43]
8	Shrimp shell biomass, (containing 22 wt% chitin)	[C_2mim][OAc]	2	*Setup*: Single-needle syringe/vertical assembly *Charge*: Grounded aluminum plate *Voltage*: 28 kV *Nozzle-to-grounded-target distance*: 13 cm *Feed*: 0.500 mL min^{-1} *Collector*: Coagulation bath (water)	0.5–1	[44, 45]
9	Cellulose pulp (DP 1,100)	[C_2mim][OAc]	1.55	*Setup*: Single-needle spinneret (MECC)/vertical assembly *Charge*: Grounded aluminum foil (20 μm thick) *Voltage*: 28 kV *Nozzle-to-coagulant surface distance*: 10 cm *Feed*: 0.060 mL min^{-1} *Collector*: Coagulation bath (aq. suspension of magnesium hydroxide nanoparticles ($Mg(OH)_2$ NPs), 3–20 mg mL^{-1})	>3	[30]

(continued)

Table 8.1 (continued)

No.	Biopolymer(s)	IL	Biopolymer load (wt%)	Electrospinning conditions	Fiber diameter (Ø) (μm)	Reference
10	Silk fibroin from *Antheraea assamensis*	$[C_4mim][OAc]$	5, 7, and 10	*Setup*: Single-needle syringe/vertical assembly *Charge*: Grounded electrode (metal NP[b]) *Voltage*: 13, 18, and 20 kV *Nozzle-to-rotating drum distance*: 8 cm *Feed*: 0.0008 mL min^{-1} *Collector*: Rotating drum coated with aluminum foil (thickness of 0.5 ± 0.1 mm) slightly submerged into methanol coagulating bath, 1,000 rpm *Coagulation solvent*: Methanol	0.16–0.22	[42]
11	Biomimetic artificial cellulose (cellulose 35–50%, hemicellulose 20–35%, and lignin 5–30%)/polyoctylthiophene additive (P3OT)	$[C_2mim][OAc]$	NP[b]	NS[b]	<10	[35]
12	Cellulose pulp (DP 1,100)	$[C_2mim][OAc]$	1.5	*Setup*: Single-needle spinneret (MECC)/vertical assembly *Charge*: Grounded aluminum electrode, *Voltage*: 18 kV *Nozzle-to-coagulant surface distance*: 20 cm *Feed*: 0.040 mL min^{-1} *Collector*: Coagulation bath (water)	NP	[31, 32]
13	Cellulose (DP 1,100) $-Fe_3O_4-$ Heparin (monofilament)	$[C_2mim][OAc]$	0.8 cellulose 0.17 Fe_3O_4, 0.7 heparin		0.2–2 0.304(155)[a]	[31, 32]

14	Cellulose pulp (DP 1,100) + Fe_3O_4 (core–shell)	[C_2mim][OAc]	1.5 cellulose (shell), 2Fe_3O_4 (core)	*Setup*: The coaxial spinneret (MECC: core I.D. 1.27 mm; shell, I.D. 2.50 mm) fitted with blunt tip aluminum needles, fabricated using a coaxial–electrospinning technique *Voltage*: 15 kV *Electrode*: Grounded sheet of aluminum foil 7.5 × 7.5 cm^2 *Nozzle-to-coagulant surface distance*: 20 cm *Feed* (core, Fe_3O_4 solution): 0.030 mL min^{-1} *Feed* (shell, cellulose solution): 100 μL min^{-1} *Collector*: Coagulation bath (water)	NP [b]	[31, 32]
15	Cellulose (DP 1,100) –Fe_3O_4–heparin (core–shell)	[C_2mim][OAc]	1.5/1 cellulose/ heparin (shell), 2 Fe_3O_4 (core)	Same as above with the exception of feed rate: *Feed* (core, Fe_3O_4 solution): 0.030 mL min^{-1}, *Feed* (shell, cellulose–heparin solution): 0.230 mL min^{-1}	0.2–2.0 0.302(125)[a]	[31, 32]
16	Carboxylated agarose (CA), extent of carboxylation 28%, 60%, and 93%	[C_4mim]Cl	7.5	*Setup*: Single-needle syringe/vertical assembly *Temperature*: NP[b] *Charge*: Aluminum foil *Voltage*: 10.5 kV *Nozzle-to-coagulant surface distance*: NP[b] *Feed*: 0.125 mL min^{-1} *Collector*: Coagulation bath (ethanol)	CA28: 183(153),[a] CA60: 117(58),[a] CA93: 63(54)[a]	[41]

(continued)

Table 8.1 (continued)

No.	Biopolymer(s)	IL	Biopolymer load (wt%)	Electrospinning conditions	Fiber diameter (Ø) (µm)	Reference
17	Pulp	$[C_4mim]Cl$	2 and 3	*Setup*: Single-needle syringe/vertical assembly *Temperature*: 103 °C *Charge*: Grounded electrode (metal NP[b]) *Voltage*: 10 kV *Nozzle-to-coagulant surface distance*: 12 cm *Feed*: NP[b] *Collector*: Coagulation bath (water)	2%: <1, 3%: 1–20	[33]
18	Chitin regenerated from shrimp shells	$[C_2mim][OAc]$	0.3 and 0.5	*Setup*: Single-needle syringe/vertical assembly Temperature: 103 °C, *Charge*: Grounded electrode (metal NP[b]) *Voltage*: 10 kV *Nozzle-to-coagulant surface distance*: 12 cm *Feed*: NP[b] *Collector*: Coagulation bath (water)	0.020(10)[a]	[46, 47]

[a]Mean value; [b]NS, value/identity not stated.

The original setups for electrospinning from ILs were all single-needle syringe (SNS) systems (Figure 8.3) [27–32, 34–42, 44, 45, 48, 49]. These systems consist of four major components: a high-voltage power source connected to both a solution supplying polymer injection system (syringe with a capillary tip or blunt needle), and to a grounded electrode beneath the collector, and a collector itself (either coagulation bath or rotating drum). The assembly is usually vertical, although a single report demonstrated a horizontal assembly [36].

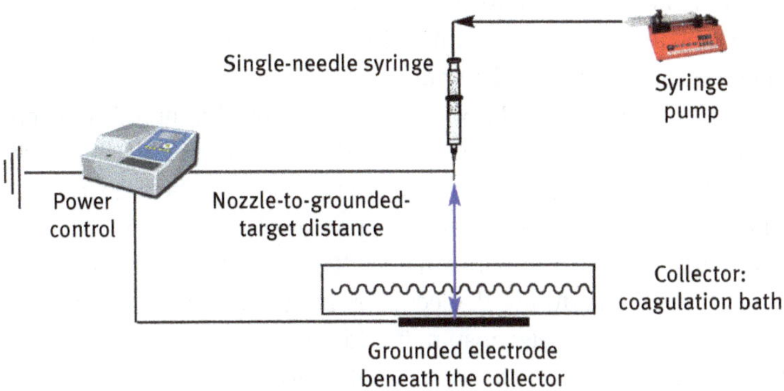

Figure 8.3: General single-needle syringe (SNS) system for electrospinning from ionic liquids.

A polymer solution is typically electrospun from a syringe under the pressure supplied by a syringe pump, although a few reports discuss the important advantage of decreasing flow rate using atmospheric pressure instead (free fall in the electric field) [27, 37], as it has been previously shown that fiber diameter decreases with reducing solution feed rate [50]. Typically, electrospinning is conducted at an accelerated potential of 15–50 kV. Distance between needle tips and collector is normally fixed; oftentimes, the tip-to-cosolvent surface distance was reported instead. Depending on the IL, experiments are carried out at either ambient or elevated temperature (above the melting point of the IL); elevated temperature is conditioned by an environmental constant temperature chamber or heating sleeve surrounding the syringe [38].

Nanomats are electrospun into alcoholic or aqueous coagulation baths, for removal of nonvolatile solvents from electrospun fibers. ILs can be recovered from the coagulation bath and can later be reused, reducing the waste generated and increasing the sustainability of the process, since the IL is one of the most expensive reagents used in this strategy. The collection of fibers and removal of the IL originally utilized a grounded ethanol coagulation bath [39]. However, while ethanol has been shown effective in removing [C_4mim]Cl from fibers, ethanol is flammable, and water became more popular coagulant system. A few reports utilize water–ethanol mixtures to reduce the fire hazard.

A few options other than coagulation baths have been developed for fiber collection and solvent removal, such as rotating collectors, usually in the form of a rotating drum coated with dacron mesh or aluminum foil slightly submerged into the coagulating bath [41, 51]. A few reports describe rotating drum collectors used in conjunction with high relative humidity chamber [32]. After electrospinning, fiber mats are collected from the coagulation bath for subsequent characterization.

Characterization of mats electrospun from ILs using SNS systems mainly revealed micron/submicron size fibers. The smallest fiber diameter of approximately 100–200 nm has been achieved by electrospinning of a solution of 8 wt% cellulose in [C_2mim][OAc] into a water coagulation bath, using a horizontal SNS system [36]. Slightly larger diameters, approximately 200 nm, have been reported for silk fibroin polymer [41], electrospun from 5, 7, and 10 wt% silk solution in [C_2mim][OAc], where nanomats were collected onto a rotating drum coated with aluminum foil slightly submerged into a methanol coagulating bath. Magnetic cellulose–heparin composites, both monofilament and core–shell, that were produced using an MECC spinneret system [52] were in the range of approximately 300 nm [30, 31]. Certainly, the fiber diameter and its distribution during IL electrospinning is very crucial for many uses, but to date the only scaled-up electrospinning (described in the following section [45]) was able to result in thin fibers of approximately 20 nm fiber diameter.

8.4 Start-up to scale-up

Practical application of electrospun fibers would require scale-up from a single- to a multineedle electrospinning setup. Multiple reports for the electrospinning of synthetic polymers presented the linear relationship between needle number and production rate. Therefore, multineedle spinnerets would decrease the overall cost ($/kg) of the material.

Our group presented efforts on developing a large-scale electrospinning setup and its application for electrospinning of chitin biopolymer from the IL [C_2mim][OAc] [45, 46]. The main differences between a SNS system and a scaled-up one was a substantially enlarged 300 mL feeding container, and a scaled spinneret head with seven-needle linear array, with pressure provided equally to each needle. The in-house built setup (Figure 8.4) contained a variable high-voltage power supply for applying electric potential to the system. The spinneret was placed into the specially designed frame stand that allowed the variability of nozzle-to-grounded-target distance.

Chitin–[C_2mim][OAc] solution was placed into a 300 mL glass feeding container, and the solution was fed into the spinneret using air pressure. Nanomats were electrospun into a water coagulation bath and collected from the coagulation bath using a conveyor belt for subsequent characterization. The conditions for electrospinning were: flow rates of 0–5 psi, voltages between 25 and 30 kV, and 19 cm

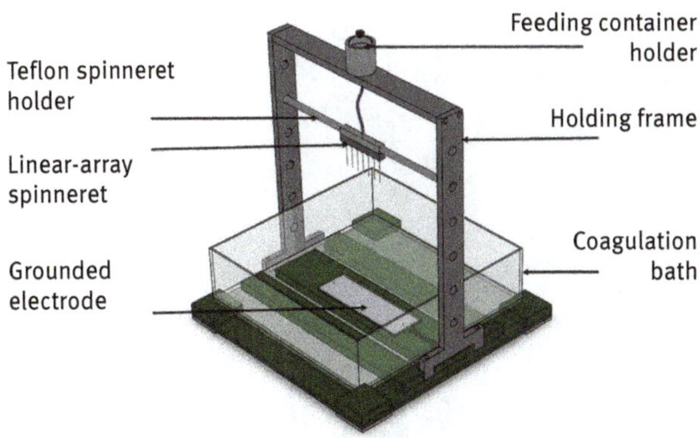

Figure 8.4: Large-scale electrospinning setup developed for electrospinning of chitin biopolymer from [C_2mim][OAc]. From Ref. [45, 46].

nozzle-to-grounded-target distance (9.5 cm nozzle-to-coagulant-surface distance), at ambient temperature. The collector system was an electrically driven horizontal conveyor, with inclined collector ramp for removal of the nanomat from the coagulation bath. (Please note that the motor controller was electrically isolated and operated well away from the effects of the high-intensity electrical field of the electrospinning system.)

The fiber morphologies of the electrospun mats were studied using scanning electron microscopy (SEM, Figure 8.5) and atomic force microscopy (AFM, Figure 8.6).

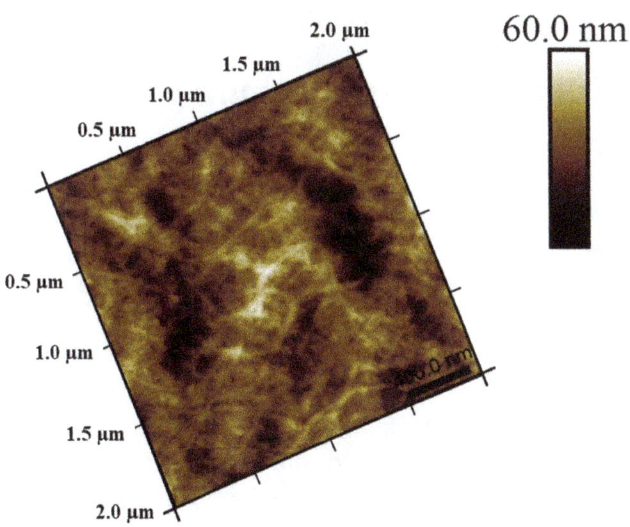

Figure 8.5: AFM image of an electrospun air-dried chitin mat (prepared from 0.23 wt% regenerated chitin–[C_2mim][OAc] solution): Scan size was 2 × 2 µm, *Z*-scale was 60 nm.

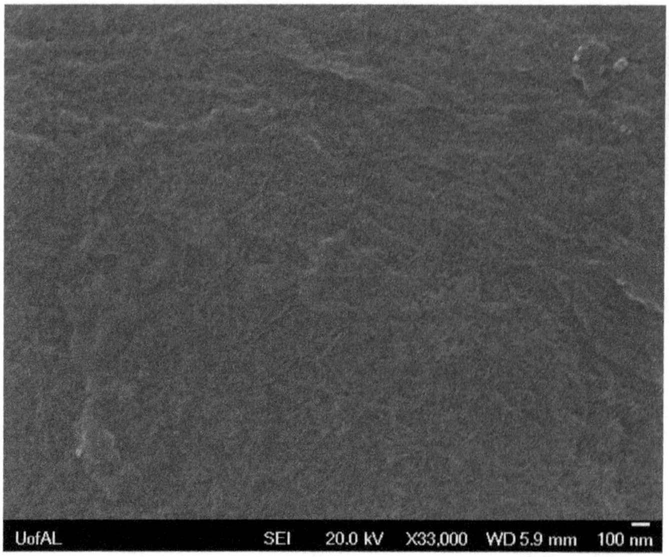

Figure 8.6: SEM image (20 kV accelerating voltage, gold coated) of the electrospun air-dried chitin mat at ×33,000 magnification.

To date, fiber diameters for the electrospinning from ILs are on the order of some hundred nanometers, and only the scaled-up electrospinning setup was able to result in nano-size fibers (ca. 20 nm fiber diameter).

8.5 Cosolvents with ILs for electrospinning of biopolymers

Biopolymer–IL solutions exhibit high conductivities because of the inherent ionicity of the ILs, yet ILs are more viscous than VOCs, more so with dissolved biopolymer. As a result, "beaded fibers" are often produced. One way to circumvent this limitation is to decrease the viscosity by decreasing the polymer load or increasing solution temperature. Alternatively, one can use a cosolvent to reduce the viscosity and surface tension of the solution. Dimethyl sulfoxide (DMSO) [26, 33, 50], dimethylformamide (DMF) [27, 28, 53], DMAc [50], and glycerol [53] were studied as cosolvents. As expected, a high conductivity combined with low surface tension and viscosity led to decreased fiber diameter. A summary of representative examples is included in Table 8.2.

Table 8.2: Representative reports of electrospinning of biopolymers from IL/cosolvent systems.

No.	Biopolymer	IL	Cosolvent	Ratio (w/w)	Biopolymer load (wt%)	Electrospinning conditions	Fiber diameter (Ø) (µm)	Reference
1	Cellulose	$[C_2mim]$-[OAc]	DMF	1:0.5–1:1	6.3–8.3	*Setup*: Single-needle syringe/vertical assembly *Temperature*: 25 °C *Charge*: Grounded electrode (metal NP[b]) *Voltage*: 30 kV *Nozzle-to-collector distance*: 15 cm *Flow rate*: NP[c] *Collector*: Rotating wired cylinder type collector; as-spun fiber was immersed in ethanol at 4 °C	0.39–0.45	[53]
2	Cellulose	$[C_2mim]$-[OAc]	DMAc	1:0.5–1:1	6.3–8.3	(as for no. 1)	0.61–1.10	[53]
3	Disintegrated pulp cellulose fractions	$[C_2mim]$-[OAc]	DMSO	1:1	1.5	*Setup*: Single-needle syringe/vertical assembly *Temperature*: 25 °C *Charge*: Grounded electrode (metal NP[b]) *Voltage*: 50 kV *Nozzle-to-grounded-target distance*: 10 cm *Solution flow rate*: 0.00083 mL min^{-1} *Collector*: Rotating drum (Ø 10 cm) coated with aluminum foil partly submerged into water bath	NPNS[b]	[51]
4	Cotton linters (DP 1,600)	[Amim]Cl	DMSO	1:2	3–5	*Setup*: Single-needle syringe/vertical assembly *Temperature*: 25 °C *Charge*: Grounded electrode (metal NP[b]) *Voltage*: 50 kV *Nozzle-to-grounded-target distance*: 12 cm *Solution flow rate*: 0.050 mL min^{-1} *Collector*: Rotating copper-wire drum coated with aluminum foil *Coagulation solvent*: None, solidifying the jet under high RH (>80% optimum)	0.10–0.80	[34]

(continued)

Table 8.2 (continued)

No.	Biopolymer	IL	Cosolvent	Ratio (w/w)	Biopolymer load (wt%)	Electrospinning conditions	Fiber diameter (Ø) (μm)	Reference
5	Lignocellulosic biomass (lignin and the hemicelluloses) and hemp	$[C_2mim]$-[OAc]	DMF	1: 1.75	14	*Setup*: Single-needle syringe/vertical assembly *Temperature*: 25 °C *Charge*: Grounded electrode (metal NP [b]), *Voltage*: 35 kV *Nozzle-to-grounded-target distance*: 15 cm *Solution flow rate*: 0.5 MPa[c] *Collection*: Rotating wired drum *Coagulation solvent*: None, mat immersed in ethanol after electrospinning	0.05–0.6	[28, 29]
6	Cellulose (MW_{avg} = 194,400)	$[C_4mim]Cl$	DMSO	94:6	4	*Setup*: Single-needle syringe/vertical assembly *Temperature*: 25 °C *Charge*: Grounded electrode (metal NP [b]) *Voltage*: 15 kV *Nozzle-to-grounded-target distance*: 15 cm *Solution flow rate*: Free fall in electric field without additional control *Collector*: Coagulation bath (water)	0.500[a]	[27]
7	Cellulose (MW_{avg} = 194,400)	$[C_4mim]Cl$	DMSO	93:7	5		0.800[a]	[27]
8	Cellulose (DP 270)	$[C_2mim]$-[OAc] + $AcOH_{cat}$	Glycerol	9:1	20	Not conducted. Directed to stability of cellulose in the IL	NS[b]	[54]

[a]Mean value; [b]NS, value not stated; [c]air pressure to pit solution out of the syringe.

8.6 Representative products for the biomedical industry

Nanomaterials prepared from biopolymers are suitable substrates for many applications in different fields, more so in biotechnology, due to a combination of high surface area and antimicrobial, biodegradability, and biocompatibility properties of many biopolymers. In addition, unlike many conventional synthetic polymers, biopolymers can be readily degraded. The ease of formation of composite materials coupled with possible surface modification for many biopolymers (e.g., chitin) or incorporation of organic and inorganic additives allows one to create novel types of functional materials. Considering all of the above, composite nanofibrous materials have become a subject of intensive academic research in recent years, primarily aiming at applications in the biotechnology sector. Below are a few examples of these composites.

8.6.1 Cellulose–heparin monofilament composites

Numerous researches have focused on embedding antibiotics, anticancer drugs, proteins and nucleic acids, cells, and so on into synthetic nanofibers [55]. At the same time, the area of electrospinning of biopolymers from ILs is at an early stage, and just a few reports demonstrated the advantage of synergetic usage of high surface-to-volume ratio of nanofibers with inherent properties of biopolymers in this drug delivery area. One such example is co-electrospinning of heparin and cellulose [38]. Heparin is a sulfated glycosaminoglycan with anticoagulant properties [56], which is used to treat and prevent thromboembolic complications, and is usually administered intravenously. The combination of heparin with cross-linked bovine collagen, coated onto knit fabric, was commercialized by MAQUET Cardiovascular, LLC [57], under the trade name HEMACAROTID Heparin knitted. To take advantage of the high surface-to-volume ratio, heparin was electrospun in combination with the synthetic plastic poly(L-lactide-*co*-caprolactone). [58, 59],

Linhardt et al. [38] first reported coelectrospinning a structurally sound cellulose biopolymer (MW_{avg} = 5,800,000) with heparin. Solutions of cellulose (10 wt% load) dissolved in [C_4mim]Cl and heparin (2 wt%) in [C_2mim][BA] were mixed in 2.5:1 ratio, and electrospun. The typical system for the electrospinning from ILs was used such as a high-voltage power source connected to both vertically positioned SNS, and to a grounded aluminum sheet beneath the ethanol-filled coagulation bath. The polymer solution was fed by a syringe pump with the feed rate of 0.03–0.05 mL min^{-1}. Electrospinning was conducted at an accelerated potential of 15–20 kV, with the distance between needle tip and grounded electrode of 15 cm. After electrospinning, the nanomats were freeze-dried and characterized, revealing micron- and submicron size fibers in a nanofibrous composite.

Biological testing of the cellulose–heparin fibers was performed by measuring the efficiency of human whole blood coagulation using thromboelastography, and determining the kinetics and strength of clot formation. It was found that heparin sustained its bioactivity even after high-voltage electrospinning. In brief, cellulose–heparin composite fibers resulted in extended clotting time in a concentration-dependent fashion, suggesting the presence of heparin in an electrospun fiber acted as an anticoagulant slowing clot formation without altering the final amount of clot formed.

8.6.2 Cellulose–heparin with magnetite monofilament composites

A logical continuation of the efforts mentioned earlier was to prepare magnetically responsive heparin-immobilized cellulose nanofiber composites by the same group [39]. The system for electrospinning of *monofilament* fibers from cellulose, heparin, and magnetite nanoparticles (Fe_3O_4 NPs) was very similar to that employed by the group earlier, with just a few modifications. First of all, the IL employed was [C_2mim][OAc], which was found to have better performance for cellulose dissolution than previously used [C_4mim]Cl. [C_2mim][OAc] is also a liquid at room temperature making dissolution and use easier. In this IL, 0.8 wt% cellulose and 0.7 wt% of heparin were codissolved, and 0.17 wt% Fe_3O_4 suspended prior to electrospinning. Such codissolution of two polymers in the same IL (as opposed to the previous mixing of two different ILs) resulted in a substantial increase of the amount of heparin to be impregnated into the composites. The cellulose-to-heparin ratio was approximately 1:1 w/w [39], compared to approximately 25:1 w/w in earlier work [38], at wt% of polymers in the IL as shown above.

With respect to the electrospinning setup, a spinneret (MECC, Japan) was used rather than a SNS, although the overall system looked pretty similar to the one described earlier. The feed rate was also similar to the previous study and water was used as a coagulant in place of ethanol. Morphological investigation using SEM on the nanocomposites revealed the presence of homogeneously dispersed magnetite within the cellulose matrix. This impregnation of magnetic NP enabled the manipulation of nonwoven fiber membranes in both dry and wet states by an external magnetic field. The fibers were substantially thinner than those produced previously with a mean diameter of approximately 300 nm. The anticoagulant activity of the immobilized heparin on the nanocomposite fiber surfaces was by Anti-Xa assay (heparin assay) and antifactor IIa (thrombin) designed to qualitatively assess the presence of heparin, and measure heparin levels through change in activity in samples containing heparin, low MW heparin, and related molecules with thrombin and/or factor Xa inhibitor activity. The tests confirmed immobilization of bioactive heparin into the monofilament fiber.

8.6.3 Cellulose–heparin with magnetite core–shell composites

In this work, core–shell fibers were manufactured through coaxial electrospinning by Linhardt and coworkers [39]. While Fe_3O_4 NPs are considered nontoxic to eliminate any cytotoxic effects and potential complications, they were included inside a continuous cellulose–heparin shell, through coaxial electrospinning. Here, the purpose of the preparation of coaxial fibers with magnetite NPs (Fe_3O_4) as the core component was the ability to manipulate the fibers in the presence of an applied magnetic field.

The coaxial electrospinning method is normally used when different characteristics from each component needed to be combined into single fiber. Moreover, components with low spinnability can be spun using this spinneret, when the polymer with good spinnability is used as core fiber, and the component with low spinnability or even nonspinnable material is made to as the sheath and immobilized on fibers. Coaxial electrospinning also allows shape (e.g., drop shape) inclusions inside a continuous shell, by controlling both the viscosity and feed rate of inner and outer solutions.

The procedure for the synthesis of coaxial fibers with magnetite NPs (Fe_3O_4) involved the initial preparation of the Fe_3O_4 NP core solution (2 wt%), through mechanic stirring of NPs in [C_2mim][OAc], at 80 °C until the NPs were homogeneously dispersed in the IL. A cellulose/heparin blended shell solution was prepared by co-dissolving 1.5 wt% cellulose and 1 wt% heparin in [C_2mim][OAc]. The electrospinning system consisted of a dual syringe system connected independently to ports on the coaxial spinneret using Teflon tubing. In turn, the coaxial spinneret was fitted with blunt tip aluminum needles of inner diameters Ø 0.94 mm for core and Ø 2.50 mm for shell, respectively. A high-voltage power supply was used to apply the necessary voltage (16–20 kV). The grounded collector was an aluminum sheet (7.5 cm × 7.5 cm) placed underneath a coagulation bath filled with water–ethanol mixture. The distance was optimized to be 15 cm from the tip of the needle to grounded collector. The core and shell solutions were fed into the coaxial spinneret at a "total" flow rate of 200 $\mu L\ min^{-1}$ (separate flow rates were not shown). The fibers were collected and freeze-dried.

Detailed structural analysis of the core–shell fibers using Fourier transform infrared (FT-IR) spectroscopy, SEM, transmission electron microscopy, powder X-ray diffraction, and thermal gravimetric analysis suggested that the core–shell fibers clearly exhibited an Fe_3O_4 NP core. Magnetoelectric characterization was performed visually with a 15 G neodymium block magnet, revealing significant magnetic character of the nanomats in both dry and wet states. Similarly to the previous study, routine quality control amydolytic essays, anti-Xa assay and antifactor IIa assay that measure the inhibition of the activity of factors Xa and IIa in the presence of the serine protease inhibitor antithrombin III, were conducted on nanomats; the results were compared with a blank and a positive control. Nanomats showed a significant decrease in absorbance at 405 nm, confirming the presence of bioactive heparin immobilized into the shell layer.

8.6.4 Silk fibroin nanofibrous mats for skin tissue engineering

Electrospinning is indispensable for fabrication of nanofibrous mats for skin tissue engineering applications [60, 61]. In this respect, another interesting study focused on fabrication of electrospun mats from silk fibroin [41], which possesses good biocompatibility, cell compatibility, and suitable mechanical properties, exhibits a unique combination of elasticity and strength, and can be used in regeneration of different tissues.

Electrospinning was conducted using a dope of 10% w/w biopolymer – [C_4mim][OAc], in the usual fashion, employing a vertically assembled SNS system. The spinning dope was supplied at a rate of 0.5 mL h^{-1}, and electrospinning conducted at 13 – 20 kV. The substantial difference of the setup with state-of-the-art SNS was the rotating drum collector coated with aluminum foil of 0.5(1) mm thickness, slightly submerged into a methanol coagulating bath. The needle tip was positioned 8 cm above the rotating drum.

The nanofibers were very thin with average fiber diameter of 160 nm. Mats were characterized through evaluation of mechanical, structural, physical, and thermal properties, prior to being subjected to evaluation of their bio-related properties. The cytocompatibility test was performed against the mouse fibroblast-like (L929) cell line. The silk materials showed good cell viability indicating that IL processing did not introduce toxicity to the final materials. In addition, the nanomats supported cell growth to a much larger extent compared to analogous cast film, suggesting a high surface area advantage. Finally, both nanomats and cast film were loaded with gentamycin sulfate, a drug used to treat skin conditions (e.g., eczema, burns, minor cuts, and wounds), to produce a nanofibrous patch. Again, both release rate and antimicrobial efficiency of the gentamycin sulfate-loaded nanomats were found to be superior compared to analogous cast films. The results indicated tremendous promise for the silk nanomaterials that could be used for commercial production of biomedical devices including scaffolds, wound healing gauges, and drug delivery devices, as well as being a promising material for skin tissue engineering.

8.6.5 Cellulose micro-/nanofiber matrices for transdermal drug release

In this work [32], bio-based cellulosic nanomats (composed of both micro- and nanofibers) were prepared for use as drug delivery devices. Cellulose pulp as raw material was chosen because of its low environmental impact and useful properties, such as lightweight, high elastic modulus, high thermal stability, and high oxygen barrier properties, all important in transdermal patches. The cellulose micro-/nanofiber matrices were prepared by electrospinning of pulp (using 2% and 3% w/v cellulose concentrations) in [C_4mim]Cl. Again, a vertically assembled SNS system was used, although optimization of electrospinning conditions was not a goal of this work. The syringe with IL–cellulose solution was heated to 103 °C, to keep the IL liquid, and electrospinning was conducted under relatively low voltage of 10 kV into a water bath, with nozzle-to-coagulant surface distance of 12 cm. The feed pressure was not provided.

Micro- and nanofibers were produced after coagulation, and the mats were dried (conditions not provided) and characterized. The fiber diameter was found to be in the range of approximately 1– 20 μm. These nanomat matrices were used for loading and release studies of a model compound, ibuprofen. The loading of ibuprofen was conducted by a simple immersing method and approximately 6% of ibuprofen was loaded (with respect to nanomat weight). Ibuprofen release was studied in an acetate buffer solution (pH 5.5) and the study revealed that 80% of the loaded ibuprofen was released during 300 min, but there was no further release. Burst release [62] was observed, where 60% of the drug was released within the first 60 min. This suggests that the fibers in the interior of the nanfibrous mats have high surface area, which allows for such quick release. While a relatively short time of release was observed (300 min), the nanomats were loaded by simple immersing.

Cell cytotoxicity assay of these materials demonstrated that the amount of viable cells on the loaded nanomats was comparable to the percentage of viable cells in the cells-only control, with a reduction in cell viability of only 16.8% (corresponding to a normalized cells viability of 83.2%). Commercial nonwoven cellulose, the second control, in turn showed a reduction in cell viability of 22.7% (normalized cells viability of 77.3%). (Please note that for use as drug delivery devices, the biocompatibility is assessed for the Medical Devices category [63], through observation of tissue cells in vitro for the cell growth, reproduction, and/or morphological effects caused by the device, in accordance with ISO 10993-5:2009 [64].) According to the United States Pharmacopeia standards [65], the tested nanomats showed toxicity of Grade I (normalized cells viability ≥80%).

The ability of the nanomats to be loaded and to release active compounds could be a promising attribute for their use as drug delivery devices. Of course, it should be noted that the ibuprofen was used only as model, and any commercialization of the nanomats would involve the loading and release of other actives, depending on the desired application.

8.7 Outlook

Most "state-of-the-art" electrospinning is still using synthetic plastics or biopolymer derivatives (e.g., cellulose acetate) rather than pure biopolymers, because the electrospinning of pure biopolymers in their native form has not been feasible due to poor solubility in conventional solvents. While the technological and industrial development of electrospinning of *biopolymers* is yet at an early stage, the technology of electrospinning from ILs is becoming routine. Nonetheless, the technology for electrospinning biopolymers continues to be based on custom in-house designs, or tailored solutions instead of building the universal machinery needed for any industrial-size effort. At the same time, the isolation of nanometer fibers, ability to functionalize the surface, possibility of using biopolymer blends, and the use of molecular additives/NPs could result in major new product platforms.

Another difficulty is that valorization of biomass at scale to produce biopolymer supply is needed before scaling up the biopolymer-based technologies, including electrospinning. Technoeconomic analyses conducted for biopolymer extractions found that not all biopolymer extracting technologies are yet at the stage of generating profit. However, these analyses focused on low-end products and examined the process as a substitute for a conventional one. As such, they fail to take into account the possibility of this technology leading to innovative products that are impossible with "state-of-the-art" methods, and the value of replacing the current supply chain with a renewable one.

Such combination of biopolymers supplied at scale with suitable electrospinning equipment from ILs will lead to large quantities of on-demand polymeric nanofibers, paving the way for utilizing their full potential. The technology, when scaled-up, can deliver dozens of innovative new products for the global nanofiber product market. (Global nanofiber product market constituted $383.7 million in 2015, projecting to grow to nearly $2.0 billion in 2020, and representing a compound annual growth rate of 38.6% between 2015 and 2020). Clearly this area is primed for innovation and growth.

Note

R. D. Rogers has partial ownership of 525 Solutions and and is a named inventor on related patent applications. J. L. Shamshina is an employee of Mari Signum Mid-Atlantic, LLC. and is a named inventor on related patent applications. The authors have no other relevant affiliations or financial involvement with any organization or entity with a financial interest in or financial conflict with the subject matter or materials discussed in the manuscript apart from those disclosed.

Abbreviations

Concentrations

mol/mol	Molar ratio
wt%	Weight percent
w/w	Weight ratio
w/v	Weight of solute to volume of solvent ratio

Chemicals

CA	Carboxylated agarose
DMAc	N,N-Dimethylacetamide
DMSO	Dimethyl sulfoxide
DMF	Dimethylformamide
DMAc/LiCl	*N,N*-Dimethylacetamide/lithium chloride

Ions

Cations

$[Amim]^+$	1-Allyl-3-methylimidazolium
$[C_4mim]^+$	1-Butyl-3-methylimidazolium
$[C_{10}mim]^+$	1-Decyl-3-methylimidazolium
$[C_2mim]^+$	1-Ethyl-3-methylimidazolium

Anions

$[BA]^-$	Benzoate
Br^-	Bromide
Cl^-	Chloride
$[OAc]^-$	Acetate
$[OLac]^-$	Lactate

General

Anti-Xa	Heparin assay (plasma heparin (UH and LMWH))
Antifactor IIa	Heparin assay (activities of low-molecular-weight heparins)
DP	Degree of polymerization
IL	Ionic liquid
I.D.	Inner diameter
Fe_3O_4	Magnetite
MW	Molecular weight
L929	Mouse fibroblast (cell line)
NPs	Nanoparticles
VOCs	Volatile organic solvents

Analytical methods

FTIR	Fourier transform infrared spectroscopy
SEM	Scanning electron microscopy

References

[1] Geyer, R., Jambeck, JR., & Lavender, K. Production, use, and fate of all plastics ever made, Sci. Adv., 2017, 3, e1700782. DOI: 10.1126/sciadv.1700782

[2] Mulhaupt, R. Green polymer chemistry and bio-based plastics: dreams and reality, Macromol. Chem. Phys., 2012, 214, 159–174. DOI: 10.1002/macp.201200439

[3] Duvall, M., & Carra, R. FTC backs off green guides biodegradability standard. Beveridge & Diamond, P.C., October 23, 2015 (Accessed January 17, 2018), at http://www.bdlaw.com/news-1799.html.

[4] Dzenis, Y., & Wen, Y. Continuous carbon nanofibers for nanofiber composites, MRS Proc., 2002, 702, 173–178. DOI: 10.1557/PROC-702-U5.4.1

[5] Shin, YM., Hohman, MM., Brenner, MP., & Rutledge, GC. Experimental characterization of electrospinning: the electrically forced jet and instabilities, Polymer, 2001, 42, 9955–9967. DOI: 10.1016/S0032-3861(01)00540-7

[6] Huang, ZM., Zhang, YZ., Kotaki, M., & Ramakrishna, S. A review on polymer nanofibers by electrospinning and their applications in nanocomposites, Compos. Sci. Technol., 2003, 63, 2223–2253. DOI: 10.1016/S0266-3538(03)00178-7

[7] Vasita, R., & Katti, DS. Nanofibers and their applications in tissue engineering, Int. J Nanomed., 2006, 1, 15–30. DOI: 10.2147/nano.2006.1.1.15

[8] Nagy, ZK., Nyul, K., Wagner, I., Molnar, K., & Marosi, G. Electrospun water soluble polymer mat for ultrafast release of Donepezil HCl, Exp. Polym. Lett., 2010, 12, 763–772. DOI: 10.3144/expresspolymlett.2010.92

[9] Khadka, DB., & Haynie, DT. Protein- and peptide-based electrospun nanofibers in medical biomaterials, Nanomedicine, 2012, 8, 1242–1262. DOI: 10.1016/j.nano.2012.02.013

[10] Nagy, ZK., Zsombor, K., Balogh, A., Vajna, B., Farkas, A., Patyi, G., Kramarics, A., & Marosi, G. Comparison of electrospun and extruded Soluplus®-based solid dosage forms of improved dissolution, J. Pharm. Sci., 2011, 101, 322–332. DOI: 10.1002/jps.22731

[11] Wang, N., Mao, X., Zhang, S., Yu, J., & Ding, B. Electrospun nanofibers for air filtration. B Ding & J. Yu. eds. Nanostructure science and technology: electrospun nanofibers for energy and environmental applications. Springer-Verlag, Berlin Heidelberg, Germany, 2014, 299–323. DOI: 10.1007/978-3-642-54160-5_12.

[12] Muzzarelli, RAA. Biomedical exploitation of chitin and chitosan via mechano-chemical disassembly, electrospinning, dissolution in imidazolium ionic liquids, and supercritical drying, Mar. Drugs, 2011, 9, 1510–1533. DOI: 10.3390/md9091510

[13] Reneker, DH., & Chun, I. Nanometre diameter fibres of polymer, produced by electrospinning, Nanotechnology, 1996, 7, 216–223. DOI: 10.1088/0957-4484/7/3/009

[14] Frenot, A., & Chronakis, IS. Polymer nanofibers assembled by electrospinning, Curr. Opin. Colloid. Interface Sci., 2003, 8, 64–75. DOI: 10.1016/S1359-0294(03)00004-9

[15] Li, D., & Xia, YN. Electrospinning of nanofibers: reinventing the wheel?, Adv. Mater., 2004, 16, 1151–1170. DOI: 10.1002/adma.200400719

[16] Frey, MW. Electrospinning cellulose and cellulose derivatives, Polym. Rev., 2008, 48, 378–391. DOI: 10.1080/15583720802022281

[17] Williams, CL., Westover, TL., Emerson, RM., Tumuluru, JS., & Li, C. Sources of biomass feedstock variability and the potential impact on biofuels production, Bioenerg. Res., 2016, 9, 1–14. DOI: 10.1007/s12155-015-9694-y

[18] Huddleston, JG., Willauer, HD., Swatloski, RP., Visser, AE., & Rogers, RD. Room temperature ionic liquids as novel media for "clean" liquid-liquid extraction, Chem. Commun., 1998, 1765–1766. DOI: 10.1039/A803999B

[19] Wang, H., Gurau, G., & Rogers, RD. Dissolution of Biomass Using Ionic Liquids. S Zhang, J Wang, X Lu, & Q Zhou. eds. Structures and Interactions of Ionic Liquids. Structure and Bonding, Berlin Heidelberg, Germany, Springer-Verlag, 2014, 79–105. DOI: 10.1007/978-3-642-38619-0_3

[20] Shen, X., Shamshina, JL., Berton, P., Bandomir, J., Wang, H., Gurau, G., & Rogers, RD. Comparison of hydrogels prepared with ionic-liquid-isolated vs. commercial chitin and cellulose, ACS Sustainable Chem. Eng., 2016, 4, 471–480. DOI: 10.1021/acssuschemeng.5b01400

[21] King, CA., Shamshina, JL., Zavgorodnya, O., Cutfield, T., Block, LE., & Rogers, RD. Porous chitin microbeads for more sustainable cosmetics, ACS Sustainable Chem. Eng., 2017, 5, 11660–11667. DOI: 10.1021/acssuschemeng.7b03053

[22] King, CA., Shamshina, JL., Gurau, G., Berton, P., Khan, NFAF., & Rogers, RD. A platform for more sustainable chitin films from an ionic liquid process, Green Chem., 2017, 19, 117–126. DOI: 10.1039/C6GC02201D

[23] Shamshina, JL., Gurau, G., Block, LE., Hansen, LK., Dingee, C., Walters, A., & Rogers, RD. Chitin–calcium alginate composite fibers for wound care dressings spun from ionic liquid solution, J. Mater. Chem. B, 2014, 2, 3924–3936. DOI: 10.1039/C4TB00329B

[24] Schiffman, JD., & Schauer, CL. A review: electrospinning of biopolymer nanofibers and their applications, Polym. Rev., 2008, 2, 317–352. DOI: 10.1080/15583720802022182

[25] Reneker, DH., & Yarin, AL. Electrospinning jets and polymer nanofibers, Polymer, 2008, 2387–2425. DOI: 10.1016/j.polymer.2008.02.002

[26] Wang, H., Gurau, G., & Rogers, RD. Ionic liquid processing of cellulose, Chem. Soc. Rev., 2012, 41, 1519–1537. DOI: 10.1039/c2cs15311d

[27] Polaskova, M., Cermak, R., Verney, V., Ponizil, P., Commereuc, S., Costa - Gomes, MF., Padua, AAH., Mokrejs, P., & Machovsky, M. Preparation of microfibers from wood/ionic liquid solutions, Carbohydr. Polym., 2013, 92, 214–217. DOI:10.1016/j.carbpol.2012.08.08

[28] Kang, Y., Ahn, Y., Lee, SH., Hong, JH., Ku, MK., & Kim, H. Lignocellulosic nanofiber prepared by alkali treatment and electrospinning using ionic liquid, Fiber. Polym., 2013, 14, 530–536. DOI: 10.1007/s12221-013-0530-8

[29] Ahn, Y., Lee, SH., Kim, HJ., Yang, YH., Hong, JH., Kim, YH., & Kim, H. Electrospinning of lignocellulosic biomass using ionic liquid, Carbohydr. Polym., 2012, 88, 395–398. DOI: 10.1016/j.carbpol.2011.12.016

[30] Zheng, Y., Miao, J., Maeda, N., Frey, D., Linhardt, RJ., & Simmons, TJ. Uniform nanoparticle coating of cellulose fibers during wet electrospinning, J. Mater. Chem. A, 2014, 2, 15029–15034. DOI: 10.1039/C4TA03221G

[31] Hou, L., Udangawa, WM., Ranodhi, N., Pochiraju, A., Dong, W., Zheng, Y., Linhardt, RJ., & Simmons, TJ. Synthesis of heparin-immobilized, magnetically addressable cellulose nanofibers for biomedical applications, ACS Biomater. Sci. Eng., 2016, 2, 1905–1913. DOI: 10.1021/acsbiomaterials.6b00273

[32] Hou, L., Udangawa, RN., Zhang, X., Onishi, A., Simmons, TJ., Linhardt, RJ., Dong, W., Zheng, Y., & Lin, L. Magnetically responsive anticoagulant heparin-cellulose composite nanofibers. Abstracts of Papers, 252nd, ACS National Meeting & Exposition, Philadelphia, PA, United States, August 21-25, 2016, 2016, BIOL-45.

[33] Liu, Y., Nguyen, A., Allen, A., Zoldan, J., Huang, Y., & Chen, JY. Regenerated cellulose micro-nano fiber matrices for transdermal drug release, Mater. Sci. Eng. C, 2017, 74, 485–492. DOI: 10.1016/j.msec.2016.12.048
[34] Xu, S., Zhang, J., He, A., Li, J., Zhang, H., & Han, CC. Electrospinning of native cellulose from nonvolatile solvent system, Polymer, 2008, 49, 2911–2917. DOI: 10.1016/j.polymer.2008.04.046
[35] Park, TJ., Jung, YJ., Park, H., Choi, SW., Kim, E., Lee, SH., & Kim, JH. Photoluminescent synthetic wood fibers from an ionic liquid via electrospinning, Macromol. Res., 2011, 19, 317–321. DOI: 10.1007/s13233-011-0416-9
[36] Freire, MG., Teles, ARR., Lopes-da-Silva, JA., & Coutinho, JAP. Electrospinning of cellulose in ionic liquids solutions. Abstracts of Papers, 239th, ACS National Meeting, San Francisco, CA, United States, March 21-25, 2010, 2010, CELL-81.
[37] Freire, MG., Teles, ARR., Ferreira, RAS., Carlos, LD., Lopes-da-Silva, JA., & Coutinho, JAP. Electrospun nanosized cellulose fibers using ionic liquids at room temperature, Green Chem., 2011, 13, 3173–3180. DOI: 10.1039/c1gc15930e
[38] Quan, SL., Kang, SG., & Chin, IJ. Characterization of cellulose fibers electrospun using ionic liquid, Cellulose (Dordrecht, Netherlands), 2010, 17, 223–230. DOI: 10.1007/s10570-009-9386-x
[39] Viswanathan, G., Murugesan, S., Pushparaj, V., Nalamasu, O., Ajayan, PM., & Linhardt, RJ. Preparation of biopolymer fibers by electrospinning from room temperature ionic liquids, Biomacromolecules, 2006, 7, 415–418. DOI: 10.1021/bm050837s
[40] Linhardt, RJ., & Murugesan, S. Cellulose composites prepared using room temperature ionic liquids. Abstracts of Papers, 233rd, ACS National Meeting, Chicago, IL, United States, March 25-29, 2007, 2007, CELL - 089.
[41] Forget, A., Arya, N., Randriantsilefisoa, R., Miessmer, F., Buck, M., Ahmadi, V., Jonas, D., Blencowe, A., & Shastri, VP. Nonwoven carboxylated agarose-based fiber meshes with antimicrobial properties, Biomacromolecules, 2016, 17, 4021–4026. DOI: 10.1021/acs.biomac.6b01401
[42] Srivastava, CM., & Purwar, R. Fabrication of robust Antheraea assama fibroin nanofibrous mat using ionic liquid for skin tissue engineering, Mater. Sci. Eng. C, 2016, 68, 276–290. DOI: 10.1016/j.msec.2016.05.020
[43] Fox, DM., Fylstra, P., Hanley, M., Henderson, W., Trulove, PC., Bellayer, S., Gilman, JW., & DeLong, HC. The preparation and characterization of Bombyx mori silk nanocomposites using ionic liquids, ECS Trans., 2007, 3, 11–20. DOI: 10.1149/1.2798642
[44] Barber, PS., Griggs, CS., Bonner, JR., & Rogers, RD. Electrospinning of chitin nanofibers directly from an ionic liquid extract of shrimp shells, Green Chem., 2013, 15, 601–607. DOI: 10.1039/c2gc36582k
[45] Swatloski, RP., Barber, PS., Opichka, T., Bonner, JR., Gurau, G., Griggs, CS., & Rogers, RD. Process for electrospinning chitin fibers from chitinous biomass solution and fibers and articles produced thereby, PCT Int. Appl., 2014, WO 2014018586, A1 20140130.
[46] Shamshina, JL., Zavgorodnya, O., Bonner, JR., Gurau, G., DiNardo, T., & Rogers, RD. Practical electrospinning of biopolymers in ionic liquids, ChemSusChem., 2017, 10, 106–111. DOI: 10.1002/cssc.201601372
[47] Zavgorodnya, O., Shamshina, JL., Bonner, JR., & Rogers, RD. Electrospinning biopolymers from ionic liquids requires control of different solution properties than volatile organic solvents, ACS Sustainable Chem. Eng., 2017. DOI: 10.1021/acssuschemeng.7b0086
[48] Meli, L., Miao, J., Dordick, JS., & Linhardt, RJ. Electrospinning from room temperature ionic liquids for biopolymer fiber formation, Green Chem., 2010, 12, 1883–1892. DOI: 10.1039/c0gc00283f

[49] Isik, M., Sardon, H., & Mecerreyes, D. Ionic liquids and cellulose: dissolution, chemical modification and preparation of new cellulosic materials, Int. J. Mole. Sci., 2014, 157, 11922–11940. 19: DOI: 10.3390/ijms150711922
[50] Sill, TJ., & Recum, HAV. Biomaterials, 2008, 29, 1989–2006.
[51] Haerdelin, L., Perzon, E., Hagstroem, B., Walkenstroem, P., & Gatenholm, P. Influence of molecular weight and rheological behavior on electrospinning cellulose nanofibers from ionic liquids, J. Appl. Polym. Sci., 2013, 130, 2303–2310. DOI: 10.1002/app.39449
[52] MECC, Co., Ltd. Nanofiber Division. Components for electrospinning: Spinneret series. (Accessed January 17, 2018), at http://www.mecc-nano.com/components01.html.
[53] Ahn, Y., Hu, DH., Hong, JH., Lee, SH., Kim, HJ., & Kim, H. Effect of co-solvent on the spinnability and properties of electrospun cellulose nanofiber, Carbohydr. Polymers, 2012, 89, 340–345. DOI: 10.1016/j.carbpol.2012.03.006
[54] Clough, MT., Griffith, JA., Kuzmina, O., & Welton, T. Enhancing the stability of ionic liquid media for cellulose processing: acetal protection or carbene suppression?, Green Chem., 2016, 18, 3758–3766. DOI: 10.1039/C6GC00027D
[55] Hu, X., Liu, S., Zhou, G., Huang, Y., Xie, Z., & Jing, X. Electrospinning of polymeric nanofibers for drug delivery applications, J. Control. Rel., 2014, 185, 12–21. DOI: 10.1016/j.jconrel.2014.04.018
[56] Essential medicines and health products. WHO Model list of essential medicines (20th List). World Health Organization. March 2017. (Accessed January 17, 2018), at http://www.who.int/medicines/publications/essentialmedicines/20th_EML2017_FINAL_amendedAug2017.pdf?ua=1.
[57] Maquet Getinge group. Hemacarotid patch heparin ultrathin. (Accessed January 17, 2018), at https://www.maquet.com/globalassets/downloads/products/hemacarotid-heparin-knitted-ultrathin-patches/mcv00004531-hemacarotidheparinpatchut_a4.pdf.
[58] Wang, J., An, Q., Li, D., Wu, T., Chen, W., Sun, B., Ei-Hamshary, H., Al-Deyab, SS., Zhu, W., & Mo, X. Heparin and vascular endothelial growth factor loaded poly(L-lactide-co-caprolactone) nanofiber covered stent-graft for aneurysm treatment, J. Biomed. Nanotechnol., 2015, 11, 1947–1960.DOI: 10.1166/jbn.2015.2138
[59] Su, Y., Li, X., Liu, Y., Su, Q., Qiang, MLW., & Mo, X. Encapsulation and controlled release of heparin from electrospun poly(L-lactide-co-ε-caprolactone) nanofibers, J. Biomat. Sci. – Polym. E, 2011, 22, 165–177. DOI: 10.1163/092050609X12583785588757
[60] Pham, QP., Sharma, U., & Mikos, AG. Electrospinning of polymeric nanofibers for tissue engineering applications: a review, Tissue Eng., 2006, 12, 1197–1211. DOI: 10.1089/ten.2006.12.1197
[61] Ngadiman, NHA., Noordin, MY., Idris, A., & Kurniawan, D. A review of evolution of electrospun tissue engineering scaffold: From two dimensions to three dimensions, J. Eng. Medicine, 2017, 7, 597–616. DOI: 10.1177/0954411917699021
[62] Huang, X., & Brazel, CS. On the importance and mechanisms of burst release in matrix-controlled drug delivery systems, J. Control. Rel., 2001, 73, 121–136. DOI: 10.1016/S0168-3659(01)00248-6
[63] Considerations for the biocompatibility evaluation of medical devices. (Accessed January 22, 2018), at https://www.mddionline.com/considerations-biocompatibility-evaluation-medical-devices.
[64] ISO 10993-5:2009: Biological evaluation of medical devices – Part 5: Tests for in vitro cytotoxicity. (Accessed January 22, 2018), at https://www.iso.org/standard/36406.html.
[65] USP Class Testing. (Accessed January 22, 2018), at https://www.namsa.com/toxicology/usp-class-testing/.

Seongjun Moon, Kyung Jin Lee

9 Needless and syringeless electrospinning for mass production

Abstract: It is well known that electrospinning is a convenient method of fabricating polymeric nanofiber with a high surface area. Despite the convenience, electrospinning has not been used in practical applications because it has a lower production rate than conventional spinning methods. The lower production rate is caused by the electrospinning solution supplier. Many fabrication methods have been invented to overcome this limitation. For instance, multineedle electrospinning is the easiest way to increase the electrospinning production rate. However, although multineedle electrospinning increases the production rate, the method causes the morphological destruction of electrospun nanofiber because of electrical interference between needles. Herein, we introduce a new fabrication method for mass production of nanofiber based on the electrospinning process. Needleless electrospinning and syringeless electrospinning are introduced in this section. Particularly, the syringeless electrospinning contributes to the green electrospinning because the water-based polymeric solution can be easily used for material. In addition, the basic mechanism and driving force of the processes is also introduced. Furthermore, a way to increase the electrospinning production rate is discussed.

Keywords: green electrospinning, mass production of electrospun fiber, mechanism of mass production, needleless electrospinning, syringeless electrospinning

9.1 Introduction

Among the production methods of nanofiber web using a polymer solution, electrospinning has recently received much attention because of several obvious advantages [1–5]. Since the first conceptual patents in 1902, electrospinning has been developed to generate from traditional nonwoven nanofiber web to a hierarchical nanofiber structure [6–10]. However, despite these evolutions, it is quite difficult to find practical examples of electrospinning systems in industrial fields except for some examples [11]. The limitation of industrial-level production is directly attributed to its lower production rate. In general, the typical setup for the electrospinning process is initiated from certain polymeric solutions (or melt), a feeder that can supply the polymer solution, an electrode to produce polymeric droplets of the supplied polymer solution,

---

Seongjun Moon, Kyung Jin Lee, Department of Chemical Engineering and Applied Chemistry, College of Engineering, Chungnam National University, Daejeon, South Korea

https://doi.org/10.1515/9783110581393-009

and a DC voltage supply to apply a high voltage. The formation of polymer nanofibers by electrospinning starts with polymeric droplets formed at the ends of electrodes, formation of a Taylor cone, and nanofiber jetting from the Taylor cone. Detailed parameters to be considered are described in other chapters or elsewhere [2, 12].

Therefore, it can be said that producing mass amounts of nanofibers depends entirely on the solution feed rate of the feeder supplying the polymer solution. Although the injection rate of the polymer solution injecting toward the nozzle varies depending on various parameters such as the type of polymer solution, the applied voltage, and the thickness of the nanofibers to be obtained, it is generally approximately 0.1–1.0 mL h^{-1} at typical lab scale. Considering that optimal solvent concentration is approximately 10–20 wt %, the production rate of the nanofibers produced by electrospinning is less than 0.1–0.2 g h^{-1}. Therefore, although the electrospinning method has a merit that various kinds of polymer materials can be produced in nanofibers form with a very thin and uniform diameter as compared to conventional methods, the production speed is much lower than that of conventional fiber production methods.

In addition, the mass production of electrospun nanofiber is important in the field of green electrospinning. The green electrospinning means eco-friendly system for electrospinning. There are many methods to contribute to the green electrospinning, such as melt spinning (no solvent system) and aqueous system. However, in general, most electrospinning system for polymeric nanofiber uses toxic solvents to prepare its polymeric solution, and it is inevitable that the system uses the toxic solution. Therefore, to reduce the effect of toxic solution, we must decrease the exposure time of toxic solvent in atmosphere. Thus, the mass production of electrospinning can contribute to the eco-friendly system to decrease the exposure of toxic solvent because the system dramatically reduces the exposure time of the toxic solvent through rapid processing time.

In this chapter, we will introduce several methods to increase the production rate of electrospinning systems, and then discuss how to increase the production rate for each case.

9.2 Requirement of mass production based on electrospinning

It is highly important to fabricate nanofibers in polymer engineering science because they can provide high specific surface area and high functionality. Among diverse pathways for nanofiber preparation, such as melt spinning, wet-laid spinning, centrifugal spinning, solution blow spinning, and electrospinning [13–19], electrospinning is the most attractive fabrication method in polymer engineering science because of such advantages as the following:

First, it is easy to reduce the thickness of the fiber to nanoscale, and thus nonwoven nanofiber webs with large surface area can be prepared. Unlike conventional spinning techniques where diameter control is possible mainly within the microscale, electrospinning makes it possible to produce very thin nanofibers because it accommodates electrical voltage during the procedure [20–22]. The applied high voltage not only changes the polymer droplet formed at the tip of the needle into a polymeric jet with very small thickness, but also pulls out the polymeric jet until solidification, so that very thin polymer nanofibers can be generated. Therefore, nonwoven webs from these thin nanofibers can maximize specific surface area, and thus the unique functional group of the material can be maximized for specific applications as well. As the specific area increases, the surface exposure of the functional groups of the structured nanofibers increases, as does the chance that the functional groups and reactants will come in contact. Thus, the functionality of nanofiber on the electrospun web can be fully adopted in certain reactions. Furthermore, recent studies have shown that functional groups of polymers were modified to specific functional groups [23–25]. For example, the surface of the nanofibers can be coated with zirconium using a surface hydroxyl group of a polymeric nanofiber, and a study on introducing a large amount of an amine functional group onto the surface of a polyacrylonitrile (PAN) nanofiber by treating the surface with an amine compound was reported [24].

The second advantage of electrospinning is its versatility in material selection. From the beginning of classical electrospinning, many researchers have been able to spin nanofiber using various polymers. For example, nanofibers could be fabricated using commercial polymers such as polyethylene, polypropylene, and polyvinylchloride, or step polymers such as thermoplastic urethane with high elasticity [23, 26–31]. In recent years, attempts have been reported to convert various types of engineering or electronic polymers into nanofibers. For example, polyvinylidene fluoride (PVDF) having high strength and crystallinity is used to form nanofiber using a mixture of acetone and dimethylformamide (DMF) as solvent [32].

Finally, electrospinning can be structured in nanounits, unlike other fiber production methods. Although conventional spinning methods are used to simply produce fibers to make textiles using yarning of the fibers, nanofibers from electrospinning can control their inner structures at the nanometer level. For example, it is possible to fabricate nanofibers with improved functionality through formation of nanoarchitectures such as core/shell nanofiber, hollow nanofiber, biphasic nanofiber, and highly aligned nanofibers [10, 24, 33–38]. Therefore, because the electrospinning technique has such various advantages, mass production of the electrospinning technique will contribute to maximizing these advantages.

However, the biggest problem is that mass production is almost impossible using conventional electrospinning systems. Generally, the electrospinning process for producing nanofibers consists of a syringe for supplying the solution, a needle or nozzle for applying electricity, a DC supply for supplying electricity, and a collector for collecting nanofibers. Then, the polymeric nanofibers are produced through droplet generation, a launching jet, whipping motion of the jet, and a solidification process [2, 39, 40]. At that time, because the production rate is entirely related to the mass transfer of the polymer solution from the syringe, it can be considered that the simplest way to increase the production rate of polymeric nanofiber is to increment the feed rate of polymeric solution using a single feeder. Although increasing the feed rate of polymeric solution can increase the production rate to a certain level, single-needle electrospinning cannot provide high-quality nanofibers, because the shape of the fibers will be collapsed because of the mismatch of the inlet and outlet mass balance. This phenomenon occurs because the formation of electrospun nanofibers is stabilized by the balance between the supplied polymer solution, which is controlled by the solution feeder, and the consumed polymer solution by applied DC voltage. In order to increase the amount of consumption at the Taylor cone, a higher level of electrical voltage is required, but it is obvious that there should be certain limitations because an excessive level of applied voltage will be problematic in terms of safety and energy. The increased feed rate will cause the accumulation of polymer solution in droplets and the destruction of a stable Taylor cone, resulting in the destruction of the morphology of the nanofibers.

Therefore, it is true that there is a limit to increasing the production rate of electrospun nanofiber using single-needle electrospinning, and other methods are required.

9.3 Difficulties of mass production based on electrospinning

If that is the case, the simplest way to increase the production rate is multijet electrospinning. Multijet electrospinning is a method for rapid production of electrospun nanofibers by increasing the number of polymeric jets launched from the nozzle or electrode. To increase the number of polymeric jets, the number of solution feeders is increased, or a specially designed multinozzle electrode can be adopted [41–44]. Both methods can simply increase the number of polymeric jets and the amount of nanofibers. In addition, because the number of polymeric jets can be increased by simply increasing the number of nozzles used in general electrospinning, it seems possible to easily increase the amount of nanofibers without replacing the existing equipment. However, in the multijet electrospinning method, although the production rate of polymeric nanofiber increases in proportion to the initial number of jets, the morphology of not only the nanofiber web but also the single fiber itself is not uniform (Figure 9.1) [41].

This morphological destruction of polymer nanofibers is affected by the formation of polymeric droplets and a Taylor cone. In the case of general electrospinning based on a single electrode (or nozzle), a polymer droplet is formed at the end of

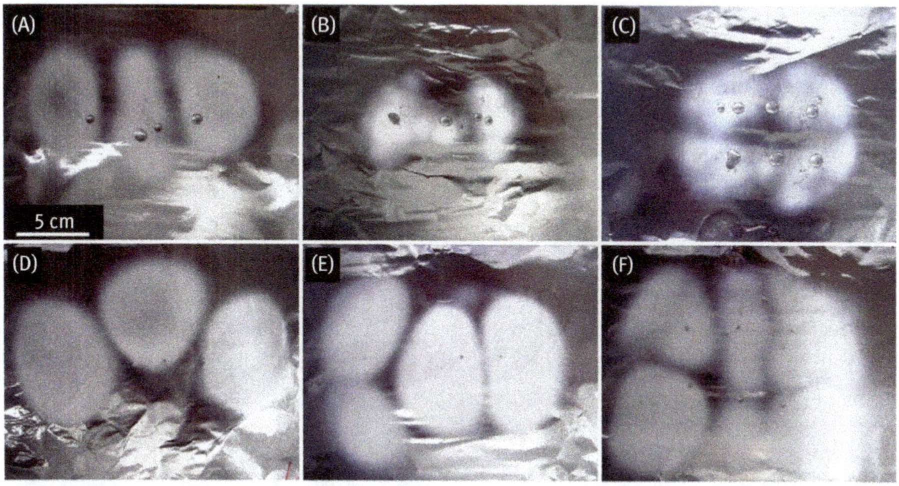

Figure 9.1: Divided nanofiber web due to electrical repulsion, (a) +30 kV charged collector at 20 cm working distance (3 needles), (b) +30 kV, 23 cm (3 needles), (c) +30 kV, 23 cm (6 needles), (d) −30 kV, 20 cm (3 needles), (e) −30 kV, 23 cm (3 needles), and (f) −30 kV, 23 cm (6 needles). from Ref. [41].

the electrode through a continuously supplied polymer solution, and a stable polymer jet is launched by high-voltage electrical energy. However, as the number of electrodes increases, the electrical fields are changed from that of general single-electrode electrospinning. In a typical single-needle-based electrospinning, the electric field starts from the end of a single electrode and moves to the surface of the collector. However, in the case of multineedle electrospinning, electrical interference occurs between each jet because high voltages are applied simultaneously to each nozzle. As a result, because of the formation of an unstable electric field and the effect of electric interference, the Taylor cone, which is essential for producing polymeric nanofibers, cannot be stably maintained, and the polymeric nanofibers cannot be produced continuously.

In addition, as shown in Figure 9.1, because each polymeric jet contains identical electrical charge, repulsive force between jets makes it impossible to obtain a homogeneous nanofiber nonwoven web [41]. This repulsive force can be generated between each polymeric jet, and the polymeric jet cannot direct to the collector. For example, as shown in Figure 9.1, in the case of single-needle electrospinning, the nanofibers are collected while forming a nonwoven web, but as the number of electrodes increases, the collected webs are separated and multiple fiber webs are obtained with unwanted solution droplets.

Furthermore, because of the repulsive force between the jets, the stretching process of the polymeric jet to produce nanosized fiber cannot be sufficiently performed. In a typical electrospinning process, the polymeric jet undergoes a stretching process through a complex whipping motion [45, 46]. This stretching process produces nanosized fibers. However, when the repulsive force is applied by the multiple jets, the space for the whipping motion is reduced. As a result, the polymeric jet cannot pass enough whipping motion, and relatively thick nanofibers are generated [47].

This phenomenon becomes more problematic when decreasing the gap distance between the electrodes and increasing the applied voltage (Figure 9.2) [42, 47–49]. Thus, as the number of electrodes increases in multijet electrospinning, the nanofiber quality is damaged. In order to decrease the electrical interference between the electrodes and to generate independent electrical fields, a certain distance between electrodes is crucial. However, increasing the gap distance between electrodes results in a decreasing number of electrodes per unit area, thus reducing productivity. Thus, controlling the gap between the needles is dilemmatic in multijet electrospinning in terms of nanofiber quality and production rate.

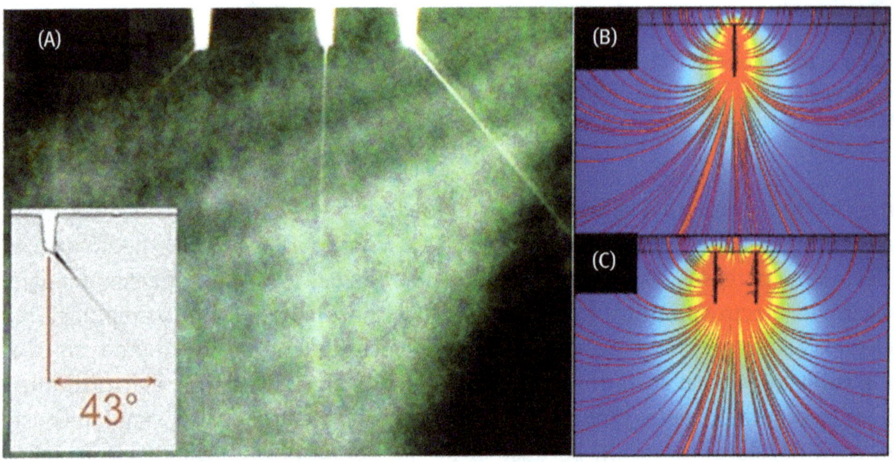

Figure 9.2: Electrical repulsion of multineedle electrospinning (A) realistic electrical repulsion [41]; simulation of electrical interference between needle (B) single needle, and (C) multineedle.

Some researchers have attempted to minimize the electrical interference by positioning additional same-charged electrodes around the multinozzle [41, 50]. As shown in Figure 9.3, the additional electrode located near the multinozzle effectively stabilizes the electrical interference between the nozzles and can also stabilize the shape of the electric field. As a result, the area of the nanofibers collected on the surface of the collector is reduced and a nanofiber web with higher density is collected compared to the case without the electrode. Furthermore, the production rate can be effectively increased by an effectively controlled electric field. However, unfortunately, it was confirmed that many fibers are agglomerated for two reasons. First, a large number of nanofibers are collected in a too small collecting area. Generally, the electrospun nanofibers should pass through the solidification of the solution jet during elongation

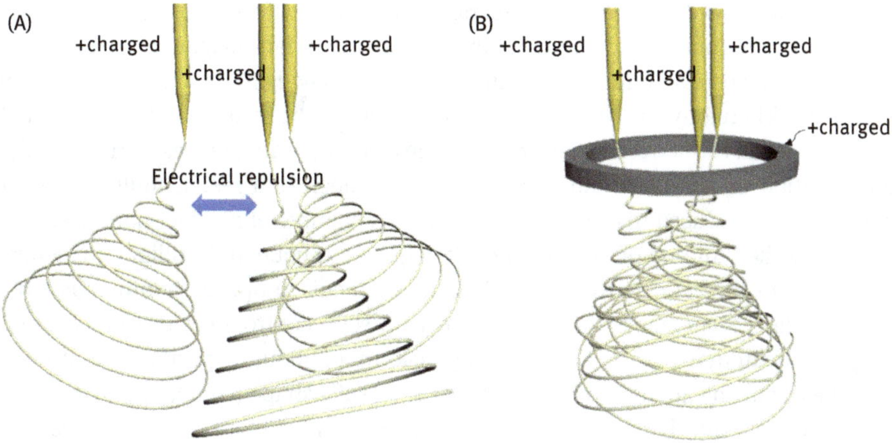

Figure 9.3: Multijet electrospinning a) without and b) with an auxiliary electrode.

of the polymeric jet, but because the additional electrode restricts the elongation, the solution jet cannot be sufficiently solidified before contacting other jet streams, resulting in nanofiber agglomeration. Second, the additional electrode can cause additional electrical interference onto the entire electrical fields. As mentioned previously, the addition of electrical interference can cause the generation of unstable electrical fields and make the Taylor cone unstable as well. Therefore, there is a limitation of multineedle electrospinning for the mass production of polymer nanofibers, and a new method to overcome the limit of multineedle electrospinning is required.

9.4 Needleless electrospinning

To overcome the limitations of multijet electrospinning, many researchers have attempted to devise new methods for mass production of electrospun nanofiber. Among these methods, the most effective and practically usable method is the so-called needleless electrospinning. These methods depend on different methods of supplying polymeric solution. Although the polymeric solutions are supplied using a nozzle and feeder such as a syringe in conventional single-nozzle electrospinning and multijet electrospinning, the new methods can supply polymeric solutions using various methods. These methods focus not on the generation of multiple nozzles but the generation of multiple droplets.

For example, polymer nanofibers could be produced using morphological changes of a ferromagnetic solution [51]. Yarin et al. located the ferromagnetic solution under the polymeric solution and applied high DC voltage into the solution. Since the ferromagnetic solution has higher density than the polymeric solution and it is not mixable with typical organic solvents, the ferromagnetic solution can be separated with several types of polymeric solution. Here, if the magnetic field is applied to the ferromagnetic solution, the shape of the ferromagnetic solution will change into a specific fluidal shape, such as a shape having many cones as shown in Figure 9.4. Therefore, the cone shape of the ferromagnetic solution acts as an end of the electrode in the conventional single-needle electrospinning process, and the polymer solution on the cone can become a Taylor cone. As a result, multiple Taylor cones can be produced in a short time, and a large number of polymer nanofibers are formed by the plurality of Taylor cones. It is quite a unique and brilliant challenge, but there are several disadvantages. First, it is difficult to operate continuously because of the system complexity. Simultaneous precise control of both electrical and magnetic hydrodynamics is a challenging task. In addition, a specific type of magnetic field must be applied to manufacture the specific shape (cone), and therefore it is difficult to control the properties of the polymer nanofibers uniformly. Despite these disadvantages, this method is significant in that it presents the direction of an experimental new method for mass production.

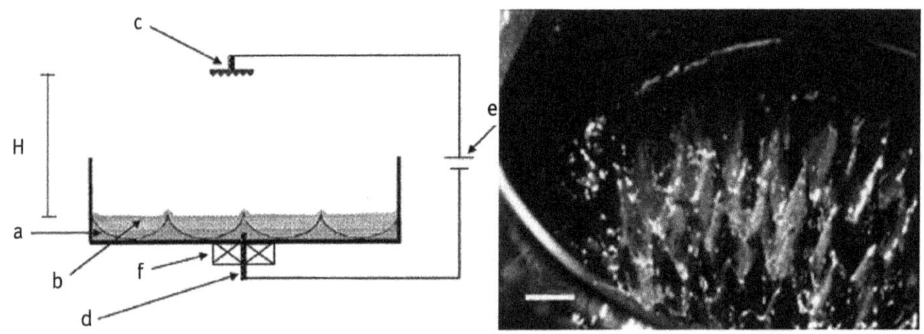

Figure 9.4: Needleless electrospinning system by ferromagnetic solution, a) magnetic liquid layer, b) polymeric solution, c) counter-electrode located at a distance H from the free surface of the polymer, d) electrode submerged into magnetic fluid, e) high voltage source, and f) strong permanent magnet or electromagnet. from Ref. [51].

Many researchers have attempted to produce webs with a large quantity of nanofibers by forming polymer droplets on the surfaces of solid objects, rather than forming polymeric droplets through needles [53]. These methods are called needleless electrospinning. For example, nanofiber webs can be produced using a cartridge-type solution feeder in large quantity [11, 52]. The solution cartridge, moving along the surface of a fixed conductive wire, can distribute a certain amount of the polymer solution on the surface of the conductive wire, and then it can form a large number of polymer droplets because of the surface energy difference between the dispensed polymer solution and surface of the conductive wire. The generated liquid droplets are launched toward the collector through the high-voltage electrical energy from the conductive wire, and the polymeric nanofibers can be generated in a large amount through solidification of

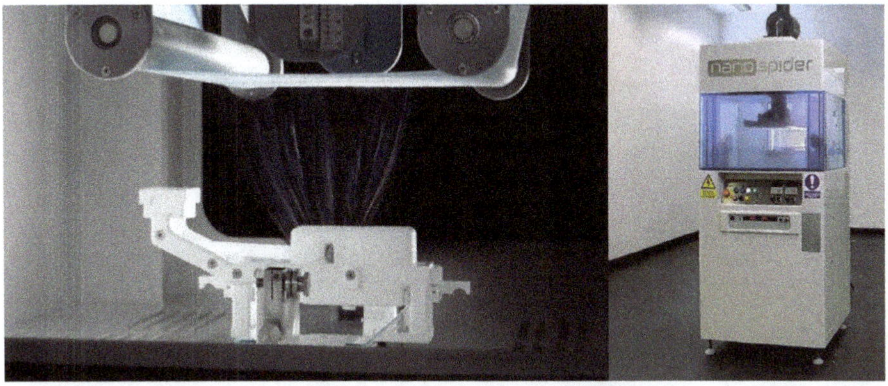

Figure 9.5: Cartridge-based electrospinning, Nanospider. From Ref. [52].

the launched polymer jets. This method is currently being developed by Elmarco Co., and it is being commercialized under the trade name of Nanospider^{TM} (Figure 9.5).

A method to produce nanofibers using centrifugal force was also reported (Figure 9.6) [18, 54]. In this method, a polymer solution is inserted into a cylindrical structure having many uniform pores, and the polymeric nanofibers can be obtained from small droplets that are generated by rotation of a cylindrical structure. A collector is arranged around the cylindrical structure in a circular shape, and high DC voltage is applied to the rotating cylinder to form nanofibers. This method has an advantage in that nanofibers can be formed at a relatively low voltage because the centrifugal force is added to the formation of the Taylor cone, compared to the conventional electrospinning method in which only electrical energy can be used for forming the Taylor cone. In addition, the position of the collector for collecting the nanofibers exists around the rotating barrel, so that it is possible to obtain a large amount of nanofibers having a very large area. However, although this method can obtain nanofibers in a large area, it is difficult to use for practical application because it requires the construction

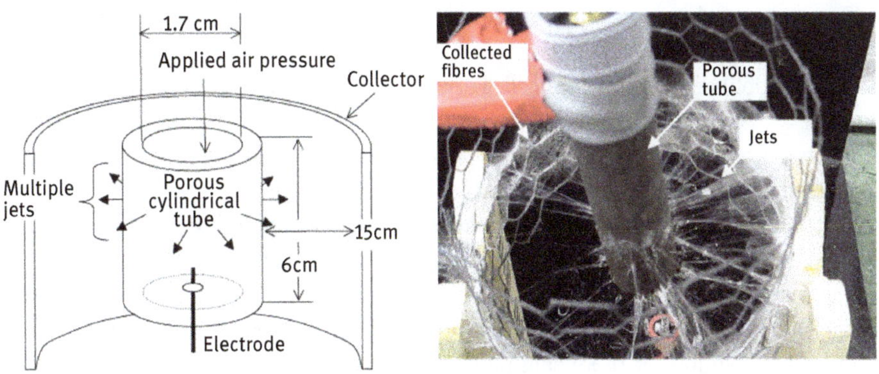

Figure 9.6: Electrospinning system using centrifugal force. From Ref. [54].

of a specific structure, and the morphologies of the obtained nanofiber show large variation.

Some researchers used a bubble as the initial structure of a polymeric jet as described in Figure 9.7 [55, 56]. The bubbles from the polymeric solution can be produced by direct injection of air or inert gas into a solution. Although this method shows the potential to produce large amounts of nanofibers in a short time, it is difficult to apply in practical mass production of nanofiber web because the size change of the nanofibers produced is not constant depending on the size of the bubble.

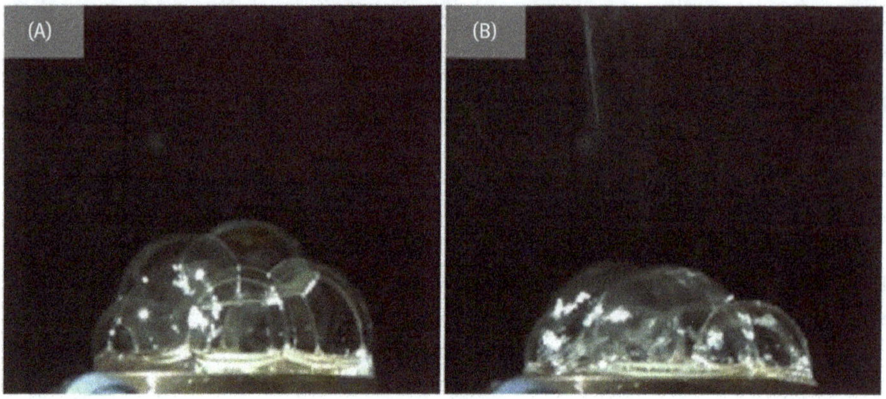

Figure 9.7: Bubble electrospinning for mass production of nanofiber a) digital image of liquid surface of the bubble and b) polymeric jet from the bubble. From Ref. [55].

Most needleless electrospinning is based on a system that produces polymeric droplets onto the surfaces of rotating objects (Figure 9.8). Here, half of the rotating body is immersed in the polymer solution, and the other half is exposed toward the collector. The polymer solution is supplied to the upper side of the structure by the slow rotation. With high-voltage electrical energy applied into the rotating body, the polymer solution shifts upward and can be changed into polymeric droplets and Taylor cones [53, 57]. Each polymeric droplet acts as an individual launcher, and thus a large amount of polymer nanofibers can be produced. For example, the shape of the rotating body in needleless electrospinning is designed in the form of a cylinder, disk, coil, or ball to produce nanofibers in large quantities [58–61]. These types of rotating bodies can be connected with high-voltage DC electrical energy through a conductive line in the polymeric solution, and then the nanofibers are collected from the top of the rotating body. Unlike conventional electrospinning, the needleless electrospinning method based on such a rotating body has an advantage in that it can be used as a continuous process because the polymer solution can be continuously supplied in the open system. Thus, the polymer solution for producing nanofibers can be continuously supplied without stopping the apparatus, and nanofibers can also be continuously produced.

Figure 9.8: Needleless electrospinning system using rotating objects: (A) rotating disk and (B) rotating cylinder.

The greatest difference between these needleless electrospinning and needle-based electrospinning (single-needle electrospinning and multineedle electrospinning) methods is related to the mechanism of droplet and Taylor cone formation. In the case of general single-needle electrospinning, the formation of the polymer droplet is derived from the surface tension of the polymer solution supplied through the nozzle, and then the formation of the Taylor cone is derived from the charge accumulated on the surface of the droplet only. By applying the high DC voltage onto the polymeric droplet, the droplet is changed into a Taylor cone to reduce the accumulated surficial charge, and the polymeric jet is launched from the end of the Taylor cone. However, in the case of needleless electrospinning based on a rotating body, the polymer solution is supplied through rotation of the rotating body immersed in the polymer solution, and the supplied polymer solution moves to the upper part of the rotating body to form droplets. At that time, unlike with needle-based electrospinning, the polymeric solution must be divided as small droplets and be self-shrunk on the surface of the rotating body. For this reason, the polymer solution for needleless electrospinning must have sufficient adhesive strength with the rotating body to move upward on the rotating body, and sufficient surface tension to be self-shrunk [62]. Therefore, the relationship between surface energy of the polymer solution and the surface of the rotating body is important.

The formation of droplets on the surface of a rotating body can directly affect the thickness and quality of the produced nanofiber (Figure 9.9) [63]. In the case of general needle-based electrospinning, the polymeric droplets maintain their diameter because the polymer solution is continuously fed by the feeder, and it is consumed at a constant rate by the applied high DC voltage, resulting in uniform-sized polymeric nanofiber. On the other hand, in the case of needleless electrospinning, the polymeric droplet is temporarily generated by the rotation of the rotating body, and the liquid droplet is instantaneously consumed to form the nanofiber. Thus, the size of the droplet is decreased by the consumption of polymeric solution in a single droplet, and the thickness of the produced fibers is also not constant. For example, it was confirmed that the fiber thickness decreases when needleless electrospinning using a cylinder-type rotating body is used [64, 65]. Thus, it is highly important to supply an accurately calculated quantity of polymeric solution to continuously generate the uniform polymeric nanofibers.

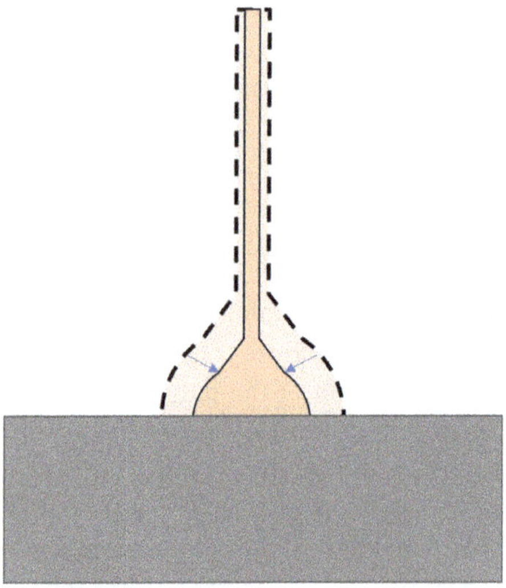

Figure 9.9: Schematic diagram of the mechanism for changing fiber diameters based on droplet consumption. From Ref. [63].

In addition, this droplet formation pathway can also affect the minimum driving voltage for producing nanofibers. Typical single-needle electrospinning requires a driving voltage in the range of approximately 10–20 kV, whereas typical needleless electrospinning requires a driving voltage of approximately 20–60 kV [53, 60, 66]. This increment in the driving voltage is related to the forces that generate the Taylor cone. The formation of the Taylor cone in typical electrospinning is related to the surface tension of the solution, gravity, and electrical elongation force. Thus, the applied high-voltage electrical energy is directly involved only in the formation of the Taylor cone in typical single-needle electrospinning. However, in the case of needleless electrospinning, the polymeric droplet on the surface of a rotating body should initially undergo a shrinkage process of the polymeric solution from the film-type polymeric solution to a spherical shape by the surface energy difference. Compared to syringe- and needle-based systems, it is difficult to form a polymeric droplet by surface tension only, and thus additional energy is required to form a polymer droplet, resulting in a requirement for high-voltage electrical energy. This is the key disadvantage of a needleless electrospinning system based on a flat cylindrical structure. The high-voltage operation is actually a serious problem in real processes. It is related not only to the energy saving (or cost) but also more seriously to the safety issue. When the operating voltage increases, there are more chances to generate sparking during the process. In this case, it is difficult to use common organic solvents with low flash points such as chloroform, acetone, and tetrahydrofuran (THF), which are powerful solvents for making polymeric solutions. Therefore, solvents with a low flash point or low vapor pressure are quite limited, and thus limited kinds of polymers can be adopted in the system.

Similarly, in the case of needleless electrospinning, only a very small number of polymer solutions can be used as raw materials because of the mechanism of generating droplets. As described previously, in needleless electrospinning, the formation of a polymer droplet is associated with the supply of the polymer solution and how well the supplied polymer solution shrinks on the surface. This feature is also related to the affinity between the polymeric solution and the surface of the rotating body [53, 61, 62, 65]. For example, in needleless electrospinning processes using a cylinder-type rotating body, if the affinity between the polymer solution and rotating body is very poor, the polymer solution cannot easily adhere on the surface of the rotating body; it is thus impossible to supply the solution continuously. On the other hand, if the affinity is very good, there is difficulty in solution to generate droplets on the surface. These points are another obstacle to extend this method to various types of polymers.

9.5 Syringeless electrospinning

In recent years, a syringeless electrospinning method using a helically probed rotating cylinder (HPRC) has been newly developed as a method for producing a large amount of nanofibers to overcome the limitations of needleless electrospinning [67]. In previous research, electrospinning using a probed rotating cylinder was investigated [11]. However, when a number of probes are added on a rotating cylinder in parallel to increase productivity, it is well known that the form of the product is collapsed for the reason described previously for multineedle electrospinning. Lee's group also described this phenomenon as shown in Figure 9.10 [67]. Simply incrementing the number of needles results in morphological deterioration of the nanofiber web because of the electrical interference between needles.

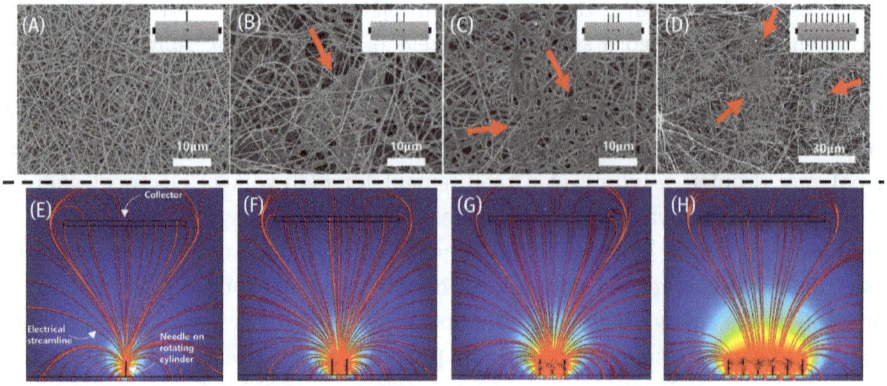

Figure 9.10: Electrical interference between probes according to the number of probes and obtained nanofiber morphologies. Insets were the type of the rotating cylinders with a) single needle, b) dual needles, c) triple needles, and d) multi-needles. e) ~h) represents electrical streamline simulation results according to number of needle.

Therefore, the crucial point is that the distance between needles should be secured and simultaneously the productivity should be increased, which is the quite dilemmatic task. Lee's groups showed that helically embedded probes can overcome this dilemma. In the syringeless electrospinning method, the probes were helically turned on the rotating cylinder to minimize the electrical interaction between the probes generated in the electrospinning process. Because the probes for generating nanofibers are helically arranged on the rotating cylinder, during the rotation of the rod, only the needle closest to the collector will generate an electric field, and thus no electrical interference will occur. Therefore, the distance between the next probes in parallel is increased enough to prevent electrical interference compared to the parallel arrangement maintaining continuous production of fibers, and as a result, nanofibers having a good morphology can be produced.

In addition, the electrode probes on the rotating rod can solve the problem of very low material diversity and the requirement of high voltage from needleless electrospinning. The probe located on the rotating rod can supply the polymer solution by simply drawing from the solution layer by rotation of the rotating body, and then the solution of the polymer can be simply changed to droplets because the end of the probe is sharp as in general single-needle electrospinning. Thus, the syringeless electrospinning system consumes lower driving voltage to produce nanofibers because the mechanism of the generation for droplets in the method does not comprise the shrinkage of polymeric solution to generate the droplet. In the case of syringeless electrospinning, the polymer solution could be spun using 15–20 kV of applied voltage (Figure 9.11).

In addition, the syringeless electrospinning technique can produce nanofibers through a simpler mechanism than the general single-needle electrospinning technique because syringeless electrospinning is based on a continuous-batch system. The general single-needle electrospinning technique can produce nanofibers through the

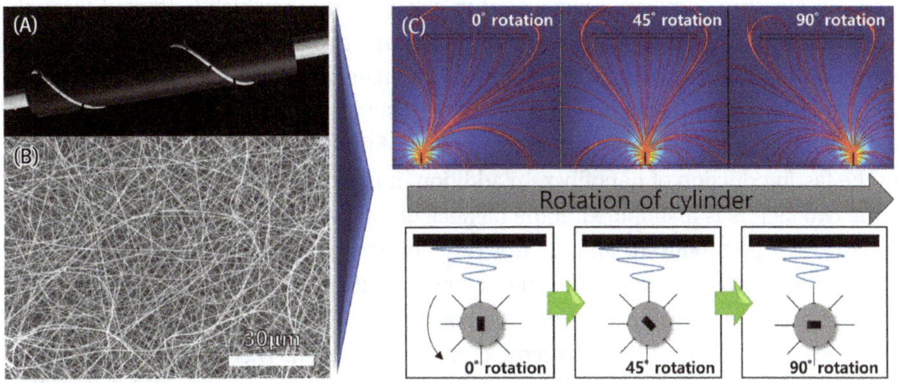

Figure 9.11: Syringeless electrospinning system using a helically probed rotating cylinder. a) schematic diagram of the helically probed rotating cylinder, b) SEM image of obtained polymeric nanofiber, and c) electrical streamline depending on rotating angle. from Ref. [67].

production of polymeric droplets by continuously supplying the polymer solution through the syringe, the formation of the polymer droplets through the syringe, and the consumption of the polymer solution by the applied electric field. For that reason, general single-needle electrospinning must control both the supply rate and consumption rate for polymeric solution. However, in the case of syringeless electrospinning, a single droplet acts like a batch system, and the consumption of the polymer solution by the electric field forms a continuous system because the initially generated polymer droplet was drawn from the solution. Nanofibers are generated using only the single droplets produced through the 360° rotation. Therefore, syringeless electrospinning considers only the consumption of polymeric solution for producing the nanofiber, and it has a simpler mechanism than general single-needle electrospinning.

As a result, syringeless electrospinning can provide various polymeric nanofibers using variety of number of polymers, such as organic-soluble polymers as well as water-soluble polymers. In addition, the system can provide not only nanofiber using general polymers but also nanofiber using semicrystalline polymeric solution and colloidal polymeric solution, which induce the nozzle blocking phenomena.

9.6 The way to increase the production rate

From the foregoing, the important factors for the mass production of electrospinning can be summarized as follows: first, the mass transfer in electrospinning must be rapid to maximize the amount of obtained nanofiber. The mass transfer of electrospinning is a measure of material movement in electrospinning, which is formed through the supply and consumption of a polymer solution. Thus, because the rapid supply of polymeric solution means that the supply of the material is fast, and the faster rate of consumption means that nanofibers produced using these materials are produced at high speed, both rapid supply and consumption of the polymeric solution are required for the fast production of nanofiber. In addition, it is highly important that the balance of the two factors is maintained. If the balance of the two factors is lost, uniform nanofibers cannot be produced, and disrupted nanofibers or beads on a string may form. For example, when the supply of the polymer solution is fast and the consumption speed of the polymer solution is low, the polymer droplet cannot be rapidly converted into the nanofiber although the electrical energy is applied. There is a possibility that a number of unspecific beads will be formed, related to the properties and the quality of nanofibers.

Second, the formed electric field should be constant and stable. In the case of typical single-needle electrospinning, the electric field begins at the end of the needle and is directed to the surface of the collector with a highly constant and stable shape. However, when the electric field is not stable, as in the case of multijet electrospinning because of electrical interference, the nanofibers cannot be precisely produced. The nanofibers formed by an electrically disturbed electrospinning system are not only unable to form uniform morphology, but also result in deserting the nanofiber from the collector because the formed electric field is not directed at the collector. Because this deserted nanofiber can cause a reduced amount of nanofibers to be collected, the interference can reduce the obtained amount of nanofiber. Therefore, stable and precise formation of these electric fields is essential for mass production of electrospinning.

Finally, to produce nanofibers in large quantities, it is necessary to use a mechanism capable of producing a large number of polymeric droplets at one time. This is an essential point to increase the production rate of the electrospinning system, and it is confirmed in recent studies. For example, in cylinder-type needleless electrospinning, the system can generate a large amount of nanofibers because a large number of droplets are simultaneously formed on the surface of the cylinder because of its surface tension. In addition, in the case of cartridge-type needleless electrospinning, using the process of dropping the polymer solution on the surface of the conductive wire as the cartridge passes through the wire, a small polymer droplet can be formed. Because the rate of dropping the polymer solution on the wire increases while increasing the speed of passing through the wire, the droplet can be rapidly formed on the wire; therefore, it is possible to increase the nanofiber production rate. In conclusion, in order to use electrospinning in industry, it is highly important to increase the electrospinning production rate by considering all the various factors described here.

9.7 Conclusion

Electrospinning can produce nanofiber using various kinds of polymeric materials, and it is easy to control the thickness of the fiber in nanoscale compared to other nanofiber spinning methods such as melt spinning, wet-laid spinning, and blow spinning. However, despite these advantages, electrospinning has been highly difficult to use in real industry because of its low production rate. The low production rate of electrospinning is strongly related to the mass transfer of the polymer solution, the electric field, and the mechanism of polymer droplet formation. In order to overcome the limitations of electrospinning, the multijet electrospinning technique was invented. However, although multijet electrospinning can provide nanofiber in large quantity, the obtained nanofiber is degraded because of the induced electrical interference. In addition, needleless electrospinning based on the fluidal solution and rotating bodies was developed. Among them, needleless electrospinning based on a rotating body was more useful and can be used in practical applications. And then, the system contributes to the green electrospinning because the system reduces the exposure time of toxic solvent. Furthermore, we introduced syringeless electrospinning using an HPRC, which was recently invented. Syringeless electrospinning can give us a number of nanofibers using various polymeric solutions such as PAN/DMF, TPU/THF, PMMA/THF, and polyvinyl alcohol/water. Furthermore, because of its special structure, the system can provide nanofiber using nozzle blockable polymeric solution such as PVDF and colloidal solution. However, until now, the system for producing such nanofibers in large quantities has remained at lab scale. Thus, as more research based on these methods proceeds, industrialization and commercialization of electrospinning will soon be possible.

Acknowledgments: This work was supported by the Korea Institute of Energy Technology Evaluation and Planning (KETEP) and the Ministry of Trade, Industry & Energy (MOTIE) of the Republic of Korea (20164010201070).

Abbreviations

DMF	Dimethylformamide
HPRC	Helically probed rotating cylinder
PAN	Polyacrylonitrile
PVDF	Polyvinylidene fluoride
THF	Tetrahydrofuran

References

[1] Z.-M. Huang, Y. Z. Zhang, M. Kotaki, S. Ramakrishna. A review on polymer nanofibers by electrospinning and their applications in nanocomposites, *Composites Science and Technology*. **63**, 2223–2253 (2003).

[2] A. L. Andrady. *Science and technology of polymer nanofibers*. John Wiley & Sons, New Jersey, United States (2008).

[3] G. Panthi, M. Park, H.-Y. Kim, S.-J. Park. Electrospun polymeric nanofibers encapsulated with nanostructured materials and their applications: a review, *Journal of Industrial and Engineering Chemistry*. **24**, 1–13 (2015).

[4] Q. P. Pham, U. Sharma, A. G. Mikos. Electrospinning of polymeric nanofibers for tissue engineering applications: a review, *Tissue engineering*. **12**, 1197–1211 (2006).

[5] I. S. Chronakis. Novel nanocomposites and nanoceramics based on polymer nanofibers using electrospinning process—a review, *Journal of Materials Processing Technology*. **167**, 283–293 (2005).

[6] J. F. Cooley. Apparatus for electrically dispersing fluids, US692631A (1902)

[7] D. H. Reneker, I. Chun. Nanometre diameter fibres of polymer, produced by electrospinning, *Nanotechnology*. **7**, 216 (1996).

[8] A. Abdal-hay, J. Lim, M. S. Hassan, J. K. Lim. Ultrathin conformal coating of apatite nanostructures onto electrospun nylon 6 nanofibers: mimicking the extracellular matrix, *Chemical engineering journal*. **228**, 708–716 (2013).

[9] K. J. Lee, T.-H. Park, S. Hwang, J. Yoon, J. Lahann. Janus-core and shell microfibers, *Langmuir*. **29**, 6181–6186 (2013).

[10] H. Yu, D. Y. Kim, K. J. Lee, J. H. Oh. Fabrication of one-dimensional organic nanomaterials and their optoelectronic applications, *Journal of nanoscience and nanotechnology*. **14**, 1282–1302 (2014).

[11] D. Petras, L. Mares, D. Stranska. Method and device for production of nanofibres from the polymeric solution through electrostatic spinning, EP1673493B1 (2008)

[12] J. M. Deitzel, J. Kleinmeyer, D. Harris, N. B. Tan. The effect of processing variables on the morphology of electrospun nanofibers and textiles, *Polymer*. **42**, 261–272 (2001).

[13] J. T. McCann, M. Marquez, Y. Xia. Melt coaxial electrospinning: a versatile method for the encapsulation of solid materials and fabrication of phase change nanofibers, *Nano letters*. **6**, 2868–2872 (2006).

[14] P. Pötschke, T. Andres, T. Villmow, S. Pegel, H. Brünig, K. Kobashi, D. Fischer, L. Häussler. Liquid sensing properties of fibres prepared by melt spinning from poly (lactic acid) containing multi-walled carbon nanotubes, *Composites Science and Technology*. **70**, 343–349 (2010).

[15] J. M. Samon, J. M. Schultz, J. Wu, B. Hsiao, F. Yeh, R. Kolb. Study of the structure development during the melt spinning of nylon 6 fiber by on-line wide-angle synchrotron X-ray scattering techniques, *Journal of Polymer Science Part B: Polymer Physics*. **37**, 1277–1287 (1999).

[16] C. Meier, M. E. Welland. Wet-spinning of amyloid protein nanofibers into multifunctional high-performance biofibers, *Biomacromolecules*. **12**, 3453–3459 (2011).

[17] E. S. Medeiros, G. M. Glenn, A. P. Klamczynski, W. J. Orts, L. H. Mattoso. Solution blow spinning: A new method to produce micro-and nanofibers from polymer solutions, *Journal of applied polymer science*. **113**, 2322–2330 (2009).

[18] X. Zhang, Y. Lu. Centrifugal spinning: an alternative approach to fabricate nanofibers at high speed and low cost, *Polymer Reviews*. **54**, 677–701 (2014).

[19] S. Iwamoto, A. Isogai, T. Iwata. Structure and Mechanical Properties of Wet-Spun Fibers Made from Natural Cellulose Nanofibers, *Biomacromolecules*. **12**, 831–836 (2011).
[20] R. Stepanyan, A. Subbotin, L. Cuperus, P. Boonen, M. Dorschu, F. Oosterlinck, M. Bulters. Fiber diameter control in electrospinning, *Applied Physics Letters*. **105**, 173105 (2014).
[21] S. V. Fridrikh, H. Y. Jian, M. P. Brenner, G. C. Rutledge. Controlling the fiber diameter during electrospinning, *Physical review letters*. **90**, 144502 (2003).
[22] C. Thompson, G. G. Chase, A. Yarin, D. Reneker. Effects of parameters on nanofiber diameter determined from electrospinning model, *Polymer*. **48**, 6913–6922 (2007).
[23] J. Choi, J. U. Jang, W. B. Yin, B. Lee, K. J. Lee. Synthesis of highly functionalized thermoplastic polyurethanes and their potential applications, *Polymer*. **116**, 287–294 (2017).
[24] S. Kim, W. B. Ying, H. Jung, S. G. Ryu, B. Lee, K. J. Lee. Zirconium Hydroxide-coated Nanofiber Mats for Nerve Agent Decontamination, *Chemistry–An Asian Journal*. **12**, 698–705 (2017).
[25] J. Choi, S. G. Ryu, B. Lee, W. B. Ying, K. J. Lee. N-chloro hydantoin functionalized polyurethane fibers toward protective cloth against chemical warfare agents, *Polymer*. **138**, 146–155 (2018).
[26] J. Ma, Q. Zhang, A. Mayo, Z. Ni, H. Yi, Y. Chen, R. Mu, L. M. Bellan, D. Li. Thermal conductivity of electrospun polyethylene nanofibers, *Nanoscale*. **7**, 16899–16908 (2015).
[27] K.-H. Lee, O. Ohsawa, K. Watanabe, I.-S. Kim, S. R. Givens, B. Chase, J. F. Rabolt. Electrospinning of syndiotactic polypropylene from a polymer solution at ambient temperatures, *Macromolecules*. **42**, 5215–5218 (2009).
[28] K. H. Lee, H. Y. Kim, Y. M. La, D. R. Lee, N. H. Sung. Influence of a mixing solvent with tetrahydrofuran and N, N-dimethylformamide on electrospun poly (vinyl chloride) nonwoven mats, *Journal of polymer science part B: polymer physics*. **40**, 2259–2268 (2002).
[29] K. Lee, B. Lee, C. Kim, H. Kim, K. Kim, C. Nah. Stress-strain behavior of the electrospun thermoplastic polyurethane elastomer fiber mats, *Macromolecular Research*. **13**, 441–445 (2005).
[30] A. Pedicini, R. J. Farris. Mechanical behavior of electrospun polyurethane, *Polymer*. **44**, 6857–6862 (2003).
[31] E. Berber, N. Horzum, B. Hazer, M. M. Demir. Solution electrospinning of polypropylene-based fibers and their application in catalysis, *Fibers and Polymers*. **17**, 760–768 (2016).
[32] W. A. Yee, M. Kotaki, Y. Liu, X. Lu. Morphology, polymorphism behavior and molecular orientation of electrospun poly(vinylidene fluoride) fibers, *Polymer*. **48**, 512–521 (2007).
[33] J. H. Jordahl, L. Solorio, H. Sun, S. Ramcharan, C. B. Teeple, H. R. Haley, K. J. Lee, T. W. Eyster, G. D. Luker, P. H. Krebsbach. 3D Jet Writing: Functional Microtissues Based on Tessellated Scaffold Architectures, *Advanced Materials*. (2018).
[34] J. H. Jordahl, S. Ramcharan, J. V. Gregory, J. Lahann. Needleless Electrohydrodynamic Cojetting of Bicompartmental Particles and Fibers from an Extended Fluid Interface, *Macromolecular rapid communications*. **38**, (2017).
[35] J. Lee, T.-H. Park, K. J. Lee, J. Lahann. Snail-like Particles from Compartmentalized Microfibers, *Macromolecular Rapid Communications*. **37**, 73–78 (2016).
[36] J. Yoon, T. W. Eyster, A. C. Misra, J. Lahann. Cardiomyocyte-driven actuation in biohybrid microcylinders, *Advanced Materials*. **27**, 4509–4515 (2015).
[37] K. J. Lee, J. Yoon, S. Rahmani, S. Hwang, S. Bhaskar, S. Mitragotri, J. Lahann. Spontaneous shape reconfigurations in multicompartmental microcylinders, *Proceedings of the National Academy of Sciences*. **109**, 16057–16062 (2012).

[38] T. Khudiyev, C. Hou, A. M. Stolyarov, Y. Fink. Sub-micrometer surface-patterned ribbon fibers and textiles, *Advanced Materials*. **29**, (2017).

[39] S. Ramakrishna. *An introduction to electrospinning and nanofibers*. World Scientific, Toh Tuck Link, Singapore (2005).

[40] J. H. Wendorff, S. Agarwal, A. Greiner. *Electrospinning: materials, processing, and applications*. John Wiley & Sons, Weinheim, Germany (2012).

[41] A. Varesano, R. A. Carletto, G. Mazzuchetti. Experimental investigations on the multi-jet electrospinning process, *Journal of Materials Processing Technology*. **209**, 5178–5185 (2009).

[42] Y. Zheng, R. H. Gong, Y. Zeng. Multijet motion and deviation in electrospinning, *RSC Advances*. **5**, 48533–48540 (2015).

[43] A. Varesano, F. Rombaldoni, G. Mazzuchetti, C. Tonin, R. Comotto. Multi-jet nozzle electrospinning on textile substrates: observations on process and nanofibre mat deposition, *Polymer International*. **59**, 1606–1615 (2010).

[44] H. Zhuo, J. Hu, S. Chen. Study of the thermal properties of shape memory polyurethane nanofibrous nonwoven, *Journal of Materials Science* **46**, 3464–3469 (2011).

[45] Y. Shin, M. Hohman, M. Brenner, G. Rutledge. Experimental characterization of electrospinning: the electrically forced jet and instabilities, *Polymer*. **42**, 09955–09967 (2001).

[46] Y. Sun, Y. Zeng, X. Wang. Three-dimensional model of whipping motion in the processing of microfibers, *Industrial & Engineering Chemistry Research*. **50**, 1099–1109 (2010).

[47] S. Xie, Y. Zeng. Effects of Electric Field on Multineedle Electrospinning: Experiment and Simulation Study, *Industrial & Engineering Chemistry Research*. **51**, 5336–5345 (2012).

[48] C. J. Angammana, S. H. Jayaram. The effects of electric field on the multijet electrospinning process and fiber morphology, *IEEE transactions on industry applications*. **47**, 1028–1035 (2011).

[49] L. Tian, C. Zhao, J. Li, Z. Pan. Multi-needle, electrospun, nanofiber filaments: effects of the needle arrangement on the nanofiber alignment degree and electrostatic field distribution, *Textile Research Journal*. **85**, 621–631 (2015).

[50] G. Kim, Y.-S. Cho, W. D. Kim. Stability analysis for multi-jets electrospinning process modified with a cylindrical electrode, *European Polymer Journal*. **42**, 2031–2038 (2006).

[51] A. L. Yarin, E. Zussman. Upward needleless electrospinning of multiple nanofibers, *Polymer*. **45**, 2977–2980 (2004).

[52] Elmarco. *Elmarco NS 1S500U*.

[53] H. Niu, X. Wang, T. Lin. *Needleless electrospinning:* developments and performances, in *Nanofibers-Production, Properties and Functional Applications*. InTech, Rijeka, Croatia (2011).

[54] J. S. Varabhas, G. G. Chase, D. H. Reneker. Electrospun nanofibers from a porous hollow tube, *Polymer*. **49**, (2008).

[55] Z. Shao, L. Yu, L. Xu, M. Wang. High-Throughput Fabrication of Quality Nanofibers Using a Modified Free Surface Electrospinning, *Nanoscale Research Letters*. **12**, 470 (2017).

[56] R. Yang, J. He, L. Xu, J. Yu. Bubble-electrospinning for fabricating nanofibers, *Polymer*. **50**, 5846–5850 (2009).

[57] H. Niu, X. Wang, T. Lin. Upward needleless electrospinning of nanofibers, *Journal of engineered fibers and fabrics*. **7**, 17–22 (2012).

[58] O. Jirsak, F. Sanetrnik, D. Lukas, V. Kotek, L. Martinova, J. Chaloupek. Method of nanofibres production from a polymer solution using electrostatic spinning and a device for carrying out the method, US7585437B2 (2009)

[59] H. Niu, W. Gao, T. Lin, W. X.G, L. Kong. Composite Yarns Fabricated From Continuous Needleless Electrospun Nanofibers, **54**, 1495–1502 (2014).

[60] X. Wang, H. T. Niu, X. G. Wang, T. Lin. Needleless Electrospinning of Uniform Nanofibers Using Spiral Coil Spinnerets, *Journal of Nanomaterials*. **2012**, 9–18 (2012).
[61] H. Niu, X. Wang, T. Lin. Needleless electrospinning: influences of fibre generator geometry, *Journal of the Textile Institute*. **103**, 787–794 (2012).
[62] X. Wang, T. Lin. *Needleless electrospinning of nanofibers: Technology and applications*. CRC Press, Rijeka, Croatia (2013).
[63] F. Yalcinkaya, B. Yalcinkaya, O. Jirsak. *Dependent and Independent Parameters of Needleless Electrospinning*, in *Electrospinning-Material, Techniques, and Biomedical Applications*. InTech, Rijeka, Croatia (2016).
[64] D. Wu, X. Huang, X. Lai, D. Sun, L. Lin. High throughput tip-less electrospinning via a circular cylindrical electrode, *Journal of Nanoscience and nanotechnology*. **10**, 4221–4226 (2010).
[65] H. Niu, T. Lin. Fiber generators in needleless electrospinning, *Journal of nanomaterials*. **2012**, 12 (2012).
[66] D. Lukas, A. Sarkar, P. Pokorny. Self-organization of jets in electrospinning from free liquid surface: A generalized approach, *Journal of Applied Physics*. **103**, 084309 (2008).
[67] S. Moon, M. Gil, K. J. Lee. Syringeless Electrospinning toward Versatile Fabrication of Nanofiber Web, *Scientific reports*. **7**, 41424 (2017).

Tuğba Isık, Nesrin Horzum, Mustafa M. Demir

10 A recycling route of plastics via electrospinning: from daily wastes to functional fibers

Abstract: Since large-scale plastic production has begun in the 1940s, plastics have been produced and used globally, bringing many advantages to modern life. The consumption of plastics has increased exponentially due to their low cost, chemical resistance, lightness, durability and ability to combine with other materials. However, plastic materials represent high tonnage in urban wastes, and it is known that these plastics discarded at the end of their useful life by filling the landfill sites. Electrospinning is a well-established and versatile technique for the fabrication of submicron fibers. In addition, it is a promising approach for the recycling of waste polymers without using complex methodologies. In this chapter, utilization of electrospinning approach for the recycling of daily wastes will be discussed. The literature about the daily wastes of both synthetic materials and natural/agricultural materials will be analyzed, and the applications of these materials will be given in detail.

Keywords: antibacterial, antimicrobial, battery, biomedical, electrospinning, filtration, photocatalytic, plastic waste, polymer waste, recycling

10.1 Recycling of daily wastes and challenges

Plastics that can be described as additive-included polymers are integral part of our daily life and current industry. They are getting more interest in manufacturing due to their benefits. Figure 10.1 shows the current applications of the polymeric products: packaging (39.9%), building and construction (19.7%), automotive (8.9%), electrical and electronic (5.8%), agriculture (3.3) and others (22.4%, consumer and household goods, furniture, sports, health and safety, etc.) [1].

The production of industrial plastics dates back to the end of 1930s when Carothers and DuPont patented Nylon [2]. The global plastic production has increased exponentially during the last 50 years. Although the first production was in the middle of twentieth century all around the world, in 2015, 322 million tons of

Tuğba Isık, Mustafa M. Demir, Department of Materials Science and Engineering, İzmir Institute of Technology, Urla, İzmir, Turkey
Nesrin Horzum, Engineering Sciences Department, İzmir Katip Çelebi University, Çiğli, İzmir, Turkey

https://doi.org/10.1515/9783110581393-010

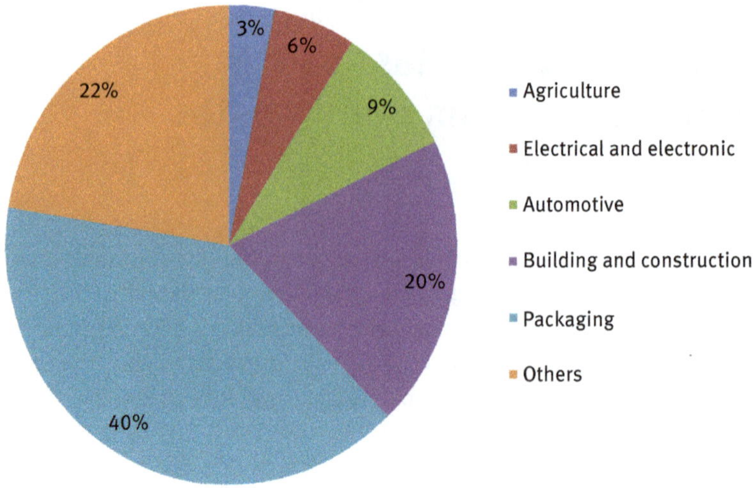

Figure 10.1: Distribution of plastic demand in Europe by 2015 (Source: Ref. [1]).

plastics were produced worldwide and this number reached to 49 million tons in Europe [1]. The synthetic plastics have replaced with glass, ceramic, wood and metal in many fields, and the commodity plastics have revolutionized the industry with the participation of polyethylene (PE), polypropylene (PP), polystyrene (PS), polyethylene terephthalate (PET) and polyvinylchloride (PVC) [3]. PE could be classified into two categories: low-density PE (LDPE) and high-density PE (HDPE). While toys, pipes and several types of bottles are produced from HDPE, the raw material of softer products such as reusable bags, food packaging films and agricultural films is LDPE. Another type of commodity plastics is PP, which is used for the production of several containers, automotive parts, single-used glasses, bank notes, food packaging and hinged caps. According to the Plastics Europe Report, the production of PP reaches up to 9.2 million tons per year in 2015, and it is the most demanded polymer with 19.1%. LDPE and HDPE demands follow PP with 17.3% and 12.1%, respectively. Besides PVC, PET and PS are other much-used plastics in industry for the production of plastic cups (PS), insulation [expanded PS (EPS)], bottles (PET) and window frame (PVC) [3].(Figure 10.2) However, most of the plastic materials are not biodegradable, and degradation of these polymers takes ages in normal environmental conditions [4].

In Europe, 25.8 million tons of plastics ended up in the waste stream and only 69.2% of them was recycled or used for energy recovery according to the European reports. Postconsumer wastes can be treated with various methods such as landfilling, energy recovery and recycling in accordance with the principle of waste hierarchy [5]. Landfilling is the first option in many European countries; however, some countries apply landfill ban to achieve high recycle rates. In addition, this method must be

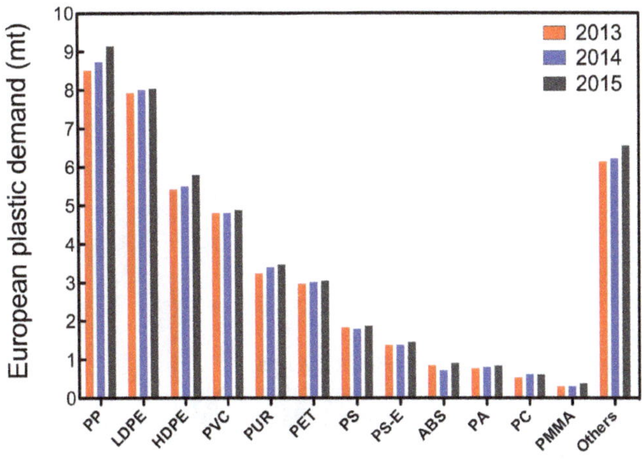

Figure 10.2: The plastic demand in Europe at 2015 including plastic materials (thermoplastics and polyurethanes) and other plastics (thermosets, adhesives, coatings and sealants). (Source: Ref. [1]).

the last choice due to requiring huge amount of space and long-term environmental problems. The use of recycled plastics is the most coherent approach among these methods. By means of recycle and reuse approach, plastics from the waste streams can be utilized in various applications instead of using virgin materials [4]. As another alternative approach to the recycling, energy recovery could be taken into account to stop the landfilling of plastics. They could be used as fuel for industrial processes, replacing fossil fuel. The energy value of plastics could be utilized for the production of electricity, heating and cooling for homes [1]. Figure 10.3 shows the change in the amounts of waste recovered, recycled and landfilled plastics in Europe (EU-28 + 2) between 2006 and 2014. The decrease in the amount of landfilled plastics from 12.9 to 8.0 million tons is the result of the increase in the recycling and energy recovery from 7.0 to 10.2 and from 4.7 to 7.7, respectively. Among all the types of wastes, packaging wastes have reached the highest recycling rate, which is around 40%, and this number represents 80% higher recycling rate than the other recycled wastes [1].

In Europe, some countries started to apply landfill ban to prompt people using another waste management methods such as energy recovery or recycling. Figure 10.4 shows the treatment percentages of postconsumer plastics in 2014. In 16 of 30 countries, the energy recovery rate is lower than the average value of 36.5%. Also, the average recycling rate was found as 27.1% and still 15 countries recycling rate is lower than this average value. Energy recovery rates are fairly high especially for the countries with landfill ban. However, the other remaining countries still prefer landfilling and even some countries do not have any energy recovery record [1].

The recycling/reprocessing approach of industrial wastes from the traditional materials has been used for many years. The similar procedure could be also

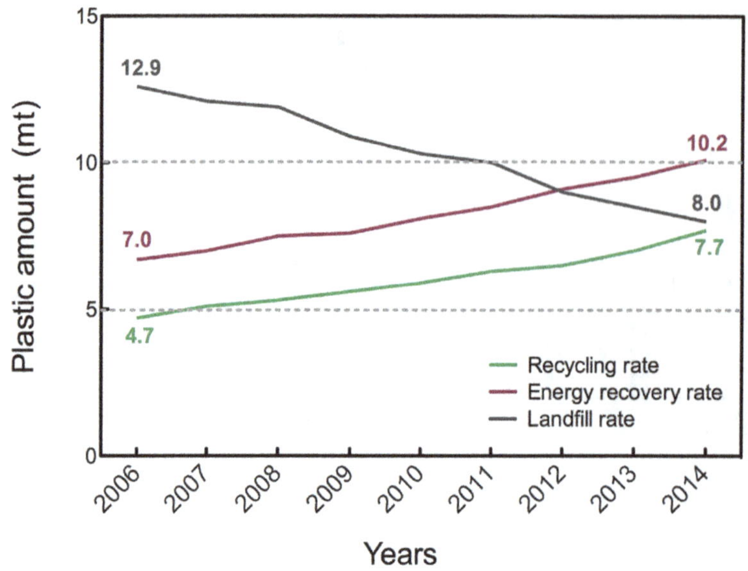

Figure 10.3: Plastic waste treatment evolution in EU-28 + 2 between 2006 and 2014 (Source: Ref. [1]).

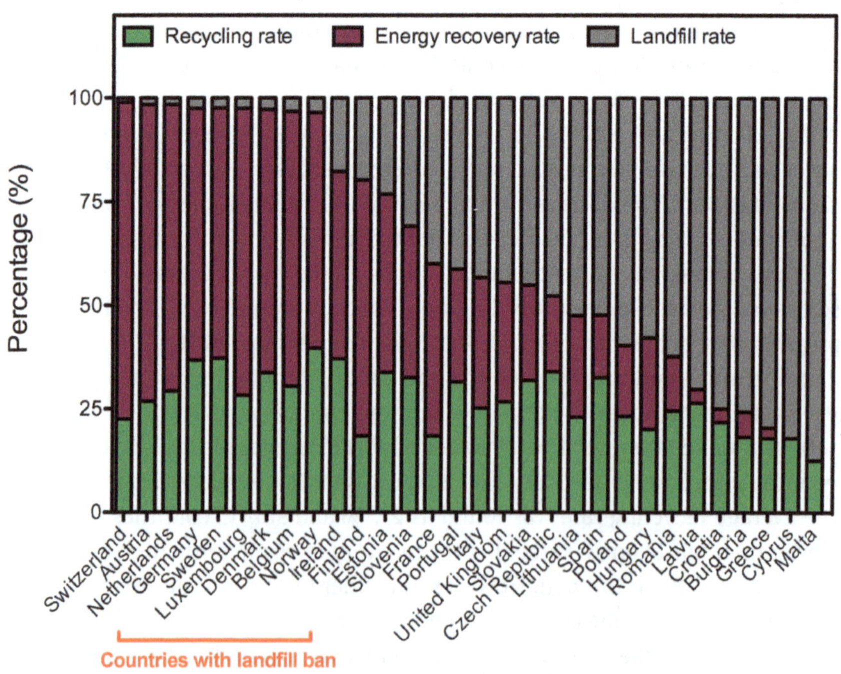

Figure 10.4: Treatment of postconsumer plastics by country in 2014 (Source: Ref. [1]).

applied to polymeric wastes to recover them for second use but there is an obstacle due to the loss of physical and mechanical properties of polymers. Because the traditional ones (i.e., glass and metal) are not organic materials, they could be recycled to secondary or tertiary products with the similar properties to primary materials. However, only the thermoplastic polymers could be recycled in this way and thermosets could not be recycled, remolded or reshaped [6]. Also, polymers lose their physical and mechanical properties after each process and final product does not have the similar properties with the raw one due to peroxidation [7–9]. Additionally, the recycling process itself uses energy and it is found that almost one third of the used energy for the production of PE is spent during its recycling. Waste plastics could be contaminated with biological residues or they could be the mixture of different materials (i.e., wood, metal or other plastics). When all the expenses (cleaning, purification, processing) for mechanical recycling are considered, the other alternatives in Scheme 10.1 could be more plausible because the ecological profits of mechanical recycling could be lost [10].

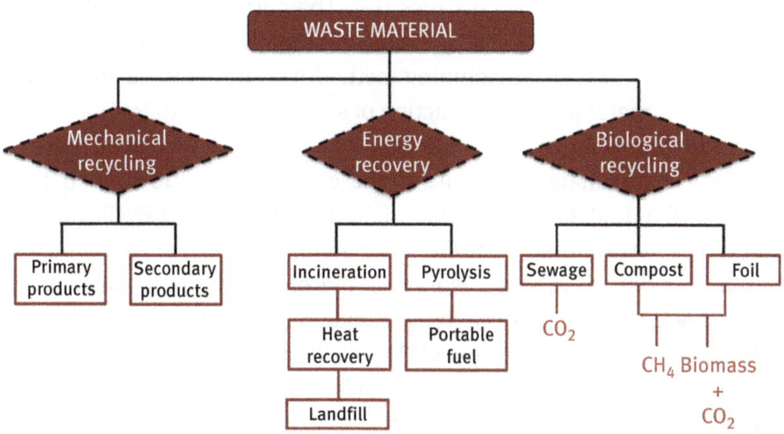

Scheme 10.1: Waste management routes (Source: Ref. [6]).

People approach with suspicion to incineration due to the emission of toxic substances to the environment. For example, polymers typically present the risk of releasing hazardous substances such as HCl from PVC, dioxins and furans from halogenated additive mixed plastics or polycyclic aromatic hydrocarbons (C_2H_2, C_2H_3, C_2H_4 and C_4H_3) from PS [11, 12]. However, incineration could be used with recovery of energy content in the plastic wastes. For local authorities, energy recovery by incineration is a preferred option because there is a financial gain by selling waste plastics as fuel due to the calorific value of some polymers is similar to fuel oil (especially for PE) [13]. Biodegradation also provides the return of nature's waste to carbon cycle, and biomass formation is an environmentally friendly approach.

However, the commodity polymers, which are currently used in industry, are not biodegradable because of the antioxidants and researches now focus on the development of polymeric materials that suit industrial requirements and return to biological cycle after their usage. Biological recycling of polymers occurs in two steps: first one is the biotic or abiotic hydrolysis and bioassimilation of biopolymer, and second one is peroxidation and bioassimilation. Because biodegradation is both an energetically and economically favorable method, it could be an alternative approach for agricultural plastics or modified synthetic plastics, which can be returned to biological cycle, to the current recycling methods [6].

10.2 Recyclable polymer wastes by electrospinning

There are visible economic and environmental consequences of plastic solid wastes due to the increasing levels of global plastic pollution both on land and in the oceans. If the plastic production continues in this direction, the current problems will be doubled in the following two decades [14]. The current task in public and private sectors is "to stay the plastics in economy and out of the oceans" [15]. Apart from the traditional methods, electrospinning method has started to be applied for the recycling of plastic materials and production of secondary materials. Thus, plastic nanofiber production from wastes could be a potentially value-added process. Synthetic plastics are dominant pollutants to the environment due to their long degradation time and abundant plastic wastes (i.e., PET, PS, PP and PVC) could be easily processed by electrospinning. On the other hand, there is a growing interest in the bio-based materials to produce economic and environmentally sustainable materials for large-scale uses. Also this type of materials represents a significant feature that they could be easily processed due to their renewability. The aim of this section is to focus on the applications of recyclable wastes from both synthetic and agricultural/natural sources in the literature.

10.2.1 Filtration and adsorption applications

Due to the growing and urbanization, water resources are rapidly contaminated with heavy metals, dyes, oil and other hazardous chemicals. Adsorption and filtration methods are widely applied processes for water and wastewater treatment, and this type of materials have been used in different forms such as beads, granules, powders, fibers and flakes [16]. There are several studies in the literature that process polymer wastes by electrospinning and fabricate recycled filtration membranes or adsorbents.

Water bottles are the major contributors to the waste streams and PET is the most common semicrystalline thermoplastic polymer, which is used in the form of different products such as water bottles and films. PET bottles have been discarded after their single usage and they are contaminated a lot less than the other type of municipal solid wastes. Thus, their recyclability and reusability have been gaining attention in the last years by considering the environmental pollution. When the applicability of PET in various applications was considered, recycled PET by electrospinning could be a cost-effective and practical approach [17, 18]. Strain et al. fabricated large quantities of PET electrospun fibers to filter polycyclic aromatic compounds, persistent organic pollutant and nanoparticles. The thin PET fibers were evaluated for the filtration of tobacco smoke, which contains several thousands of particulates and vapor phase components [19].

Because the effect of these pollutants is a current concern, this way of recycling is a convenient approach for the environment. Zander et al. reported the water filtration efficiency of bottle-grade PET fibers over 99% filtration efficiency for 1 μm particles. Also, the elastic modulus of recycled fibers is better than the commercial products due to increased interactions from the additives, which were not present in the commercial ones [20]. For the wastewater treatment, adsorption is a widely used technique and electrospun membranes are other forms of adsorbent materials due to their high adsorption capacity, better reusability, ease of scale-up and no need for posttreatment. After a basic activation process, PET fiber substrates could be modified with chitosan, which is a convenient adsorbent to metal ions due to its amino and hydroxyl groups. Khorram et al. proposed an adsorptive membrane for the removal of Cr (VI) ions from the wastewaters with an economical and efficient method. PET bottle waste was electrospun into the form of fibers and activated by cold plasma followed by chitosan functionalization. The maximum adsorption capacity and reusability of membranes were evaluated under optimum conditions (Figure 10.5) [21].

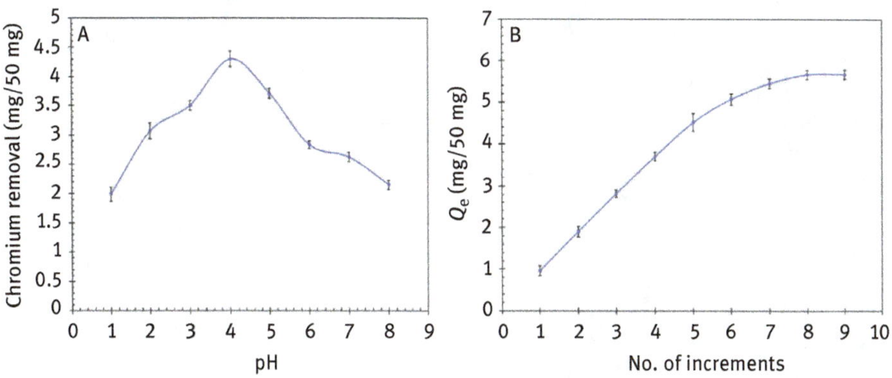

Figure 10.5: (A) The effect of pH on chromium removal and (B) cumulative amount of adsorbed chromium for filtration of 10 ppm chromium solution [Reprinted from Ref. [21]. Copyright (2017), with permission from Elsevier].

In another study, Zander et al. explored the usage of recycled PET electrospun fibers in water filtration for the removal of different size of particles ranging from 30 nm to 2.0 μm [22].

PS, especially EPS, is commonly used for insulation and packaging industries due to its versatility, dimensional stability, cleanliness and low cost. However, EPS wastes usually end up in landfill and occupy large spaces due to their low densities. Chemical and heating processed for EPS recycling is disadvantageous due to high-energy consumption. This obstacle can be overcome by using electrospinning due to its low energy requirement and fabrication of small size fibers. Shin et al. demonstrated the coalescence separation of water droplets in oil–water emulsions by using glass fiber media supplemented recycled EPS fibers. The effect of EPS fiber amount on the oil capture efficiency was evaluated, and 4% fiber amount is assumed optimal for pressure drop [23, 24].

One of the main products of chemical recycling for polycarbonate is bisphenol A [25], which is an endocrine disrupting compound. Although the conventional methods are convenient to recover waste PC materials, that is, compact disks, fabrication of fibrous materials by electrospinning of PC wastes is an appropriate method. Zander et al. carried out the formation of fibers from blends of two recycled polymers, which are PS and PC. The thermal and mechanical properties of these recycled blend fibers were investigated and the possible applications of these fibers were reported as ultra/microfiltration due to their improved elastic modulus (Figure 10.6) [20]. Isık et al. demonstrated the utilization of recycled high-impact PS (CD case) and EPS electrospun fibers for the removal of infectious clinical wastes due to their high protein adsorption capability. They also reported that the additives found in CD case,

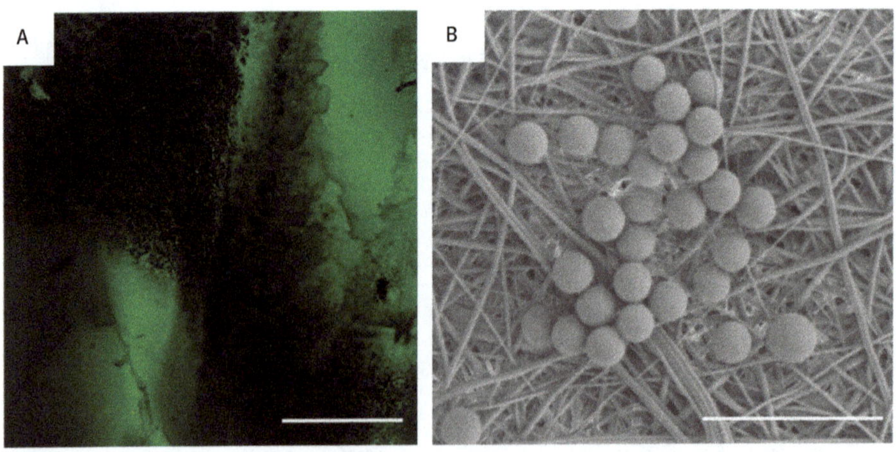

Figure 10.6: (A) Confocal laser scanning microscopy image of fluorescent ≈1 μm beads captured by the nanofiber mat and (B) SEM image of 1 μm beads on nanofiber mat. [Reprinted (adapted) with permission from Ref. [20]. Copyright (2015) American Chemical Society].

especially $CaCO_3$, enhanced protein adsorption capacity nearly eight times when CD case waste PS is compared with commercial PS [26]. Matulevicius et al. demonstrated the development of air filters from thermoplastic nylon and PS wastes by electrospinning. The filtration characteristics of electrospun filters have been evaluated by measuring the penetration of NaCl aerosol particles [27].

There is a considerable effort to create economic and environmentally sustainable polymeric membranes by using renewable sources rather than synthetic polymer membranes. Biopolymers are the recent and favorite resources as an alternative to synthetic polymers due to their biodegradability [28, 29] (Figure 10.7). By taking into consideration that keratin is one of the most abundant proteins, it could be evaluated in filtration membranes for the removal of hazardous substances. Moreover, keratin is an important renewable material and distinguished from the other proteins due to its high stability, which comes from its intramolecular structure. However, there is only one disadvantage that it has very low molecular weight that results in low physical properties, for instance, low mechanical properties. Thus, sometimes there is a need to blend this biopolymer with another suitable polymer for further applications. Aluigi et al. studied the air and water cleaning ability of keratin/polyamide 6 blend nanofibers by using chromium ions and airborne formaldehyde. The authors prepared different portions of blend solutions and fabricated filters. However, the keratin becomes soluble when its ratio in blend exceeds over half of the polyamide 6. Even at low keratin amounts, the membranes demonstrated a significant adsorption capacity to chromium ion and they are also good formaldehyde absorbers by reducing the airborne formaldehyde amount up to 70% [29]. The same group of authors also fabricated wool-derived keratin nanofibers by electrospinning for the selective removal of Cu(II) cations in the presence of Ni(II) and Co(II) ions. The results demonstrate that the adsorption of Cu(II) ions occurs by ion-exchange reactions, and the selectivity of nanofibers was reported as toward Cu(II) > Ni(II) > Co(II) [30].

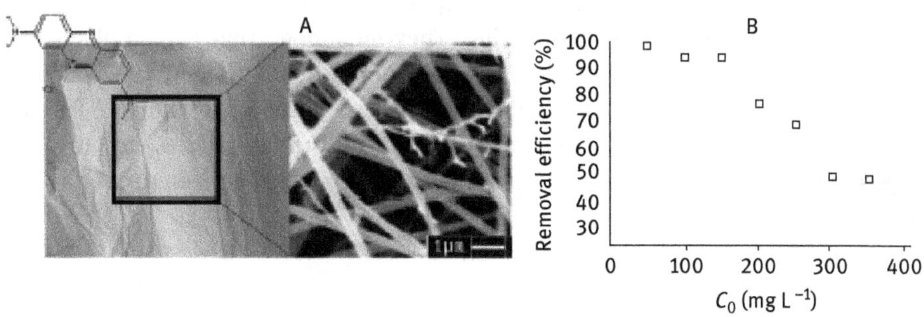

Figure 10.7: (A) Keratin nanofibrous membrane and related SEM micrograph, and (B) the effect of initial dye concentration on the removal efficiency of keratin membranes. [Reprinted from Ref. [28]. Copyright (2014), with permission from Elsevier].

Rice starch-based functional nanofibers have also interested by both scientific and industrial committees due to their biodegradability, biocompatibility and ease of formability. Woranuch et al. fabricated bio-based functional electrospun fibers by blending starch with poly(vinyl alcohol) for solution stability. They studied the solution parameters systematically to optimize the most plausible condition for good processability, well-defined morphology and uniform diameter. Also, the effect of β-cyclodextrin addition was investigated on the volatile organic compound adsorption capacity [31].

10.2.2 Photocatalytic applications

Fly ash is a by-product from coal combustion and it is a complex anthropogenic material that creates serious environmental and health concerns. Due to the increasing landfill cost, the recycling of fly ash gains an increasing attention, and the applications of waste fly ash take place in construction industry, ceramic industry, catalysis, environmental protection or valuable metal recovery [32]. By taking into consideration of this wide range of applications, fabrication of composite materials by electrospinning technique is a potential candidate as a recycling approach for fly ash. Also, fly ash has several favorable properties such as sphericity, porosity, non-toxicity, lightweight and high strength property, and being cost effective. Therefore, Saud et al. investigated the methylene blue adsorption/degradation of fly ash-doped TiO_2 nanofibers by electrospinning followed by calcination. Incorporation of fly ash into TiO_2 nanofibers strongly enhanced the photocatalysis and this reusable material can be used effectively for the purification of water [33] (Figure 10.8).

EPS wastes are also convenient candidates for the fabrication of composite materials for several applications. Rajak et al. investigated the photocatalytic activity of EPS by fabricating TiO_2/EPS composite fibers. The as-fabricated fibers showed a significant decrease in the concentration of a blue textile dye with 69% degradation [34]. In another study, Aluigi et al. prepared freestanding and flexible keratin nanofiber membranes by electrospinning. They mentioned that the problems are caused by industrial dye pollution and studied on a model system, which is prepared with methylene blue. The results show that proposed biodegradable keratin fibrous membranes are promising candidates for adsorption studies due to their large surface area and high porosity, and they are claimed to be regenerate followed by reusing due to their dynamic conditions [28].

10.2.3 Antimicrobial applications

Biomass pyrolysis and gasification result in the fabrication of an industrial by-product, tar, which is a potential carbon-based feedstock. However, the disposal of this product

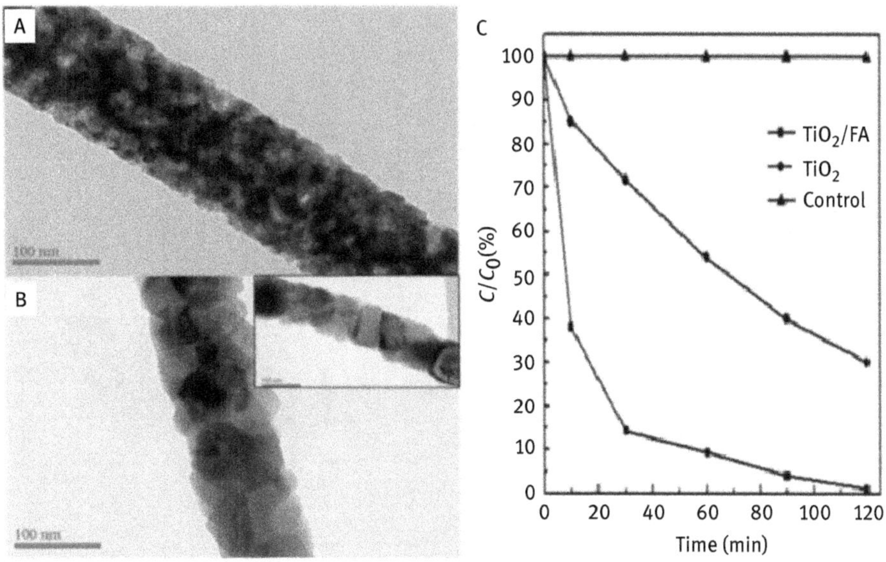

Figure 10.8: (A) and (B) TEM images of TiO_2 and fly ash/TiO_2 composite nanofibers and (C) comparison of methylene blue photodegradation by two specimens under UV radiation. [Reprinted from Ref. [33]. Copyright (2014), with permission from Elsevier].

is excessively difficult due to its complex chemistry and resistance to degradation. Due to its high carbon content, tar can be a low-cost precursor for the fabrication of carbon nanofibers (CNFs), which have promising applications in energy conversion, storage, gas adsorption and biomedical fields [35, 36]. Song et al. reported a novel route for the fabrication of porous CNFs with antimicrobial capability by electrospinning recycled tar and polyacrylonitrile (PAN) (Figure 10.9). As-prepared fibers demonstrated enhanced average fiber diameter distribution, surface roughness and antimicrobial capacity by the increment in tar content [37]. The same group of authors also investigated the effect of silver nanoparticle addition into the tar/PAN composite fibers on their antimicrobial activities. The resulting fibers have mesoporous characteristics with micropores, which come from the thermal decomposition of tar and phase separation of tar–PAN system [38]. Also, Saud et al. investigated the antimicrobial activity of fly ash doped TiO_2 nanofibers by electrospinning followed by calcination. Incorporation of fly ash into TiO_2 nanofibers strongly enhanced antibacterial resistance [33].

Regarding the sustainable material fabrication studies, natural or bio-based materials have been gaining attention in the recent years. These agricultural wastes or industrial by-products have several advantages such as their low cost and availability in large amounts. Woranuch et al. demonstrated silver nanoparticle added starch/PVA blend electrospun fibers and investigated the antimicrobial activity of these fibers. They reported that these fibers have superior bacterial inhibition against both

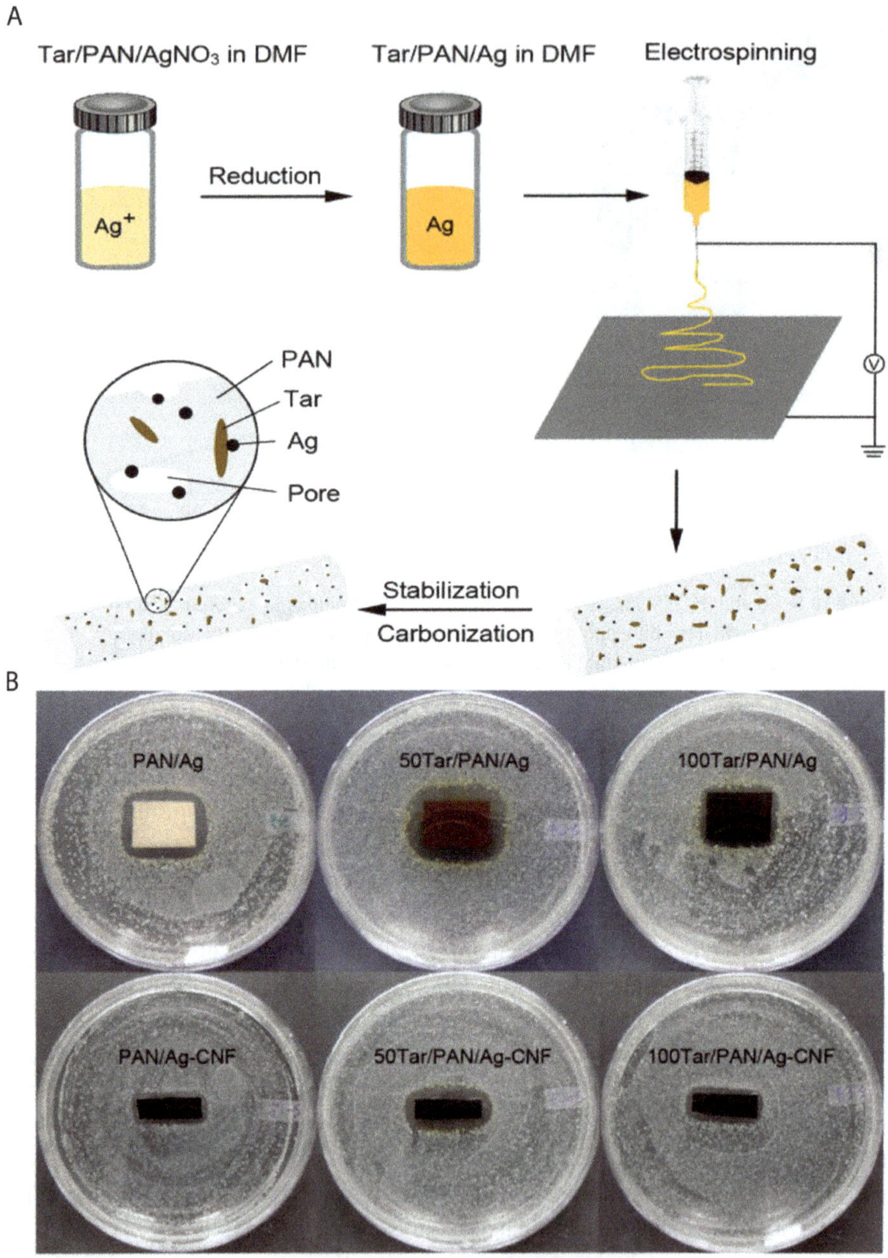

Figure 10.9: (A) Schematic illustrating the fabrication of the porous tar-derived CNFs through electrospinning followed by stabilization and carbonization processes and (B) antimicrobial activities of the as-spun nanofibers (upper row) and CNFs (lower row) with different tar contents against *S. aureus*. [Reprinted (adapted) with permission from Ref. [38]. Copyright (2015) American Chemical Society].

E. coli and *S. aureus* bacteria cultures [31]. Antibacterial activity of cellulose electrospun fibers was reported by Kampeerapappun et al. that they treated the cotton waste cellulose electrospun nanofibers with 3-(trimethoxysilyl) propyldimethyl octadecyl ammonium chloride and the results showed that proposed nanofibers exhibited excellent antibacterial activity against *E. coli* and *S. aureus* [39].

10.2.4 Battery and capacitor applications

Lithium-ion batteries (LIBs) are the commonly used power sources in recent years due to their stable cycling performances and they are suitable energy storage devices. However, there is a need to develop alternative anode materials for LIBs because the energy and power density are insufficient for large rechargeable devices [40]. Jiang et al. successfully synthesized silicon nanorods from industrial fly ash waste by electrospinning for the fabrication of LIB electrodes. Compared with the other literature results, this method has several advantages such as no need to delicate equipment and easy to scale up, high and sustainable silica derivation, and the integrity due to porous nanostructure of the material. When considering the cost-effectiveness and scalability of process, fly ash-derived silicon nanorods are effective materials for LIB anodes [41].

Apart from the aforementioned synthetic wastes, there are other less abundant wastes such as poly(vinyl butyral) (PVB). PVB is a commonly used adhesive in safety glass production due to its transmittance and superior adhesion. However, waste PVB materials are generally incinerated in landfill sites and cannot be recycled due to the costly pre-separation process. Park et al. synthesized CNFs from urea containing PAN/w-PVB (waste poly(vinyl butyral)) composite fibers (N-CNFO) via cost-effective electrospinning followed by carbonization processes. The unique open channels were created by the vaporization of w-PVB and provide numerous active sites on the surface of CNFs. The resulting material exhibits a high reversible capacity, superior rate capability and excellent cycling stability, and promising candidates for their utilization in the anodes of LIBs (Figure 10.10) [40].

The recent developments on the renewable polymers have been attracted attention on the LIB applications, too. The cellulosic fibers are another type of renewable raw materials due to their easy availability abundance, remarkable physical–mechanical properties and biodegradability. The utilization of cellulose acetate extract from cigarette filter tips and poly(vinylidene fluoride-*co*-hexafluoropropylene) (PVDF-HFP) blend solutions was processed by coaxial electrospinning, and Huang et al. fabricated a promising separator for high-performance LIBs. This study presents an environmentally friendly material, cellulose acetate, from waste cigarette filters and exhibits a separator with high storage capacity, better cycling feature, flame retardant property, excellent thermal

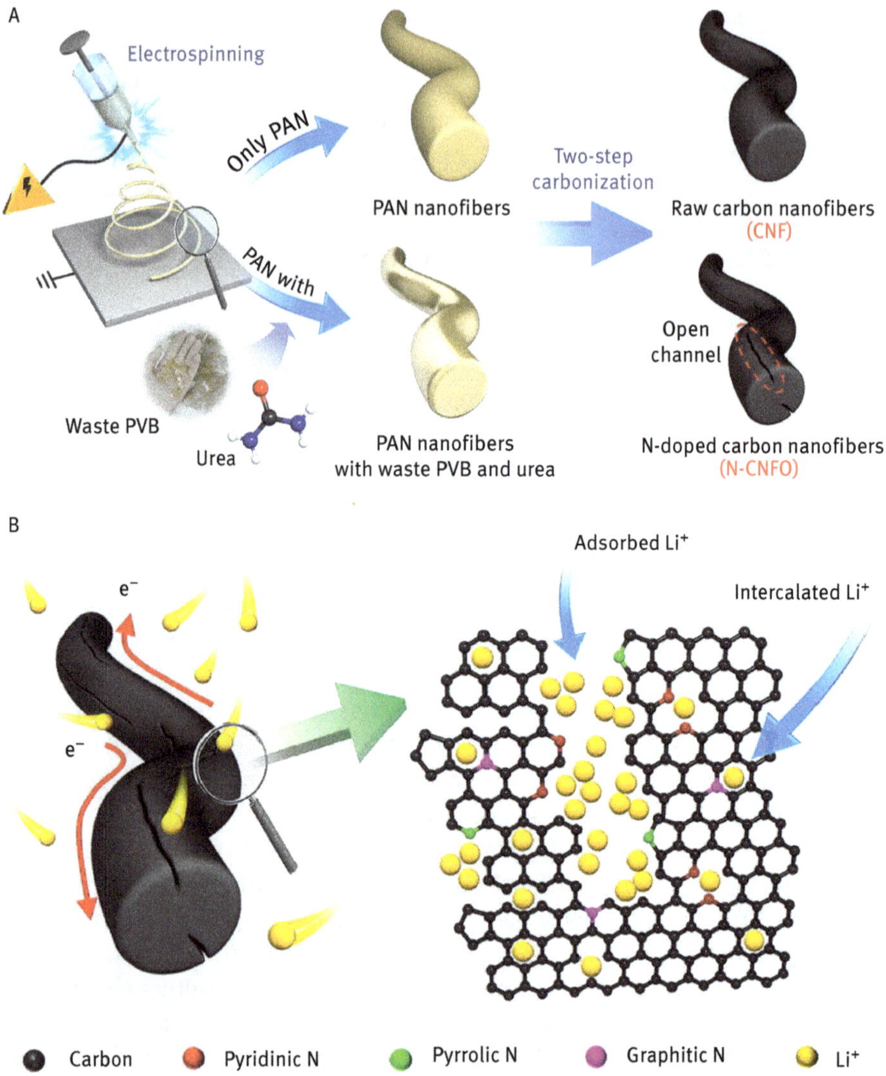

Figure 10.10: (A) Schematic illustration of the preparation of CNF and N-CNFO and (B) lithium-ion storage mechanism in N-CNFO. [Reprinted from Ref. [40]. Copyright (2017), with permission from Elsevier].

stability and enhanced rate performance [42]. Yang et al. developed a green method for the fabrication blend nanofibers from wheat straw and PAN precursors. These nanofibers are promising electrode materials with high specific capacitance and current density (Figure 10.11) [43].

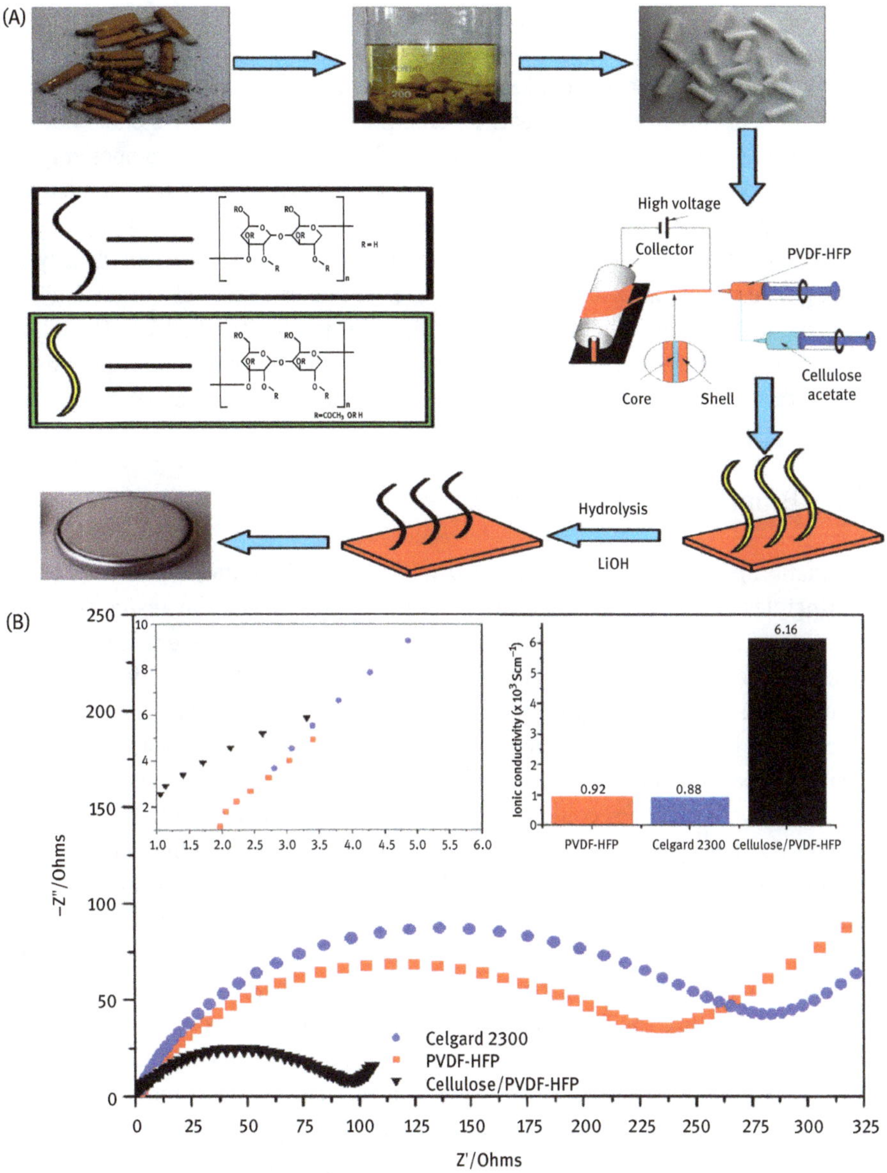

Figure 10.11: (A) Schematic illustration of the preparation process of cellulose-based coaxial nanofiber separators for lithium-ion battery and (B) Nyquist plots of Li/electrolyte-soaked separator/Li cells at 20 °C; insets are magnified Nyquist plot and ionic conductivity of separators. [Reprinted (adapted) with permission from Ref. [42]. Copyright (2015) American Chemical Society].

Lignin is also a green and economically viable precursor for the fabrication of electrode materials due to its obvious low cost and environmental benefits. It is the second most abundant polymer after cellulose. On the other hand, sodium- ion batteries have been gaining attention in recent years as promising electrical energy storage devices due to the abundance of sodium resources through the world. Jin et al. proposed PAN/lignin nanofibrous materials, which are mechanically flexible and have 3D interconnected structure. This composite electrode material exhibited a reversible capacity of about 292.6 mA h g^{-1} and high initial efficiency of 70.5% at a current density of 20 mA g^{-1}. Also, the proposed material has a good rate capability of 80 mA h g^{-1} at 1 A g^{-1} and high cycling stability over 200 cycles as 247 mA h g^{-1} at 0.1 A g^{-1}. These low-cost and freestanding nanofibers have outstanding electrochemical performances in SIBs and their green precursor source makes them a sustainable electrode material [44].

10.2.5 Biomedical applications

Biomimetic approaches have been used in tissue engineering, which combines the principles and methods of life sciences. The control of cellular responses, manipulation of healing environment to control the structure of regenerated tissue, production of cells and tissues for transportation into body, and development a quantitative understanding of biological processes are important issues in biomedical field [45]. Wound healing is one of these biological applications that is a complex process and requires collaboration of many cell strains and their products [46]. The nonwoven tissues could be used for wound healing applications to protect the tissue damage against any infections. Morganti et al. developed lignin-doped chitin nonwoven tissues to represent a skin-friendly and 100% biodegradable tool for wound repair. During wound healing, matrix reorganization process is assured by matrix metalloproteinase (MMPs), which has an important role in pathological tissue repair, fibrotic repair and scarring remedies [47]. The proposed material could modulate MMPs and reduce the dependence from the fossil fuel resources by encouraging the application of these biomaterials as a biomedical tool [48]. Sericin has excessive properties that it is biocompatible, biodegradable, antibacterial, UV and moisture absorbent, antioxidant and wound healing material [49]. In tissue engineering, the desirable properties of extracellular matrix are met by sericin. Eslah et al. fabricated PVA–sericin nanofibrous materials by electrospinning for the utilization of sericin in potential tissue engineering applications [50].

Plant extracts have extraordinary influences on various diseases and nearly 25% of drugs are derived from medical plants. *Indigofera aspalathoides* (*I. aspalathoides*), *Azadirachta indica* (*A. indica*), *Memecylon edule* (*M. edule*) and *Myristica andamanica* (*M. andamanica*) have been used for wound healing since the ancient times, and their application in this process was demonstrated several times in literature. However, there is a disadvantage that plant extracts are in oily or hydrogel forms and it is hard to shape them during their utilization. Jin et al. developed a suitable approach for the fabrication of plant extract containing wound-healing materials by electrospinning. The individual plant extracts were blended with polycaprolactone, and the cell biocompatibilities of these composite fibers are evaluated. The results of F-actin and collagen staining confirmed that PCL/*M. edule* composite fibers are the most suitable substrates for tissue engineering applications [51]. The antioxidant property of phenolic compounds has been used by Vashisth et al. The authors used agrowaste-derived ferulic acid, which is a food additive to prevent peroxidation of lipid, but limited availability. Ferulic acid was encapsulated by poly(D, L-lactide-*co*-glycolide)/polyethylene oxide (PEO) polymer blend and nanofibrous membranes were fabricated via electrospinning. The resulting composite fibers showed cytotoxic and antioxidant activities against HepG2 cells [52]. Peršin et al. also used plant extracts for the development of antioxidant and antimicrobial materials. The olive leaf extract was blended with carboxymethyl cellulose, alginate and PEO for the fabrication of electrospun fibers, which were then examined in wound healing application. The oil leaf extract increased the solution conductivity and decreased the surface tension by making easier the electrospinning process. Also the phenolic components in electrospun fibers are responsible for the antioxidant property and the proposed fibers show antibacterial property beside the Korsmeyer–Peppas model release mechanism [53]. Guzman-Puyol et al. also showed the utilization of fibroin and cellulose (from waste parsley) blends for the fabrication of biocompatible nanofibers, which can have adhesion and proliferation properties of fibroblasts. Also the addition of cellulose from parsley enhances the stiffness and elastic modulus of fibrous material. The proposed parsley/fibroin mats show excellent biocompatibility and could extend the utilization of these naturally derived polymers in sustainable management of food waste (Figure 10.12) [54].

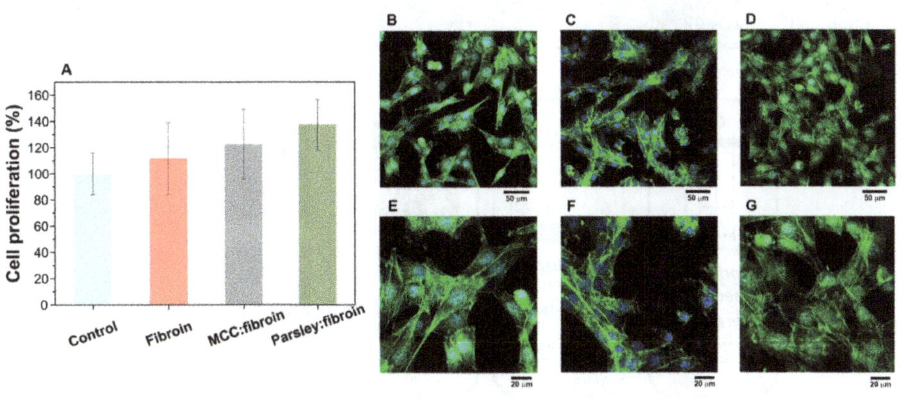

Figure 10.12: (A) Viability of fibroblast cells (MTT ('3-(4,5-Dimethylthiazol-2-yl)-2,5-diphenyltetrazolium bromidefor') assay) on the different types of nanofibers, and (B)–(G) fluorescence micrographs of fibroblast cells cultured on parsley/fibroin nanofibers. [Reprinted (adapted) with permission from Ref. [54]. Copyright (2016) American Chemical Society].

Ramphul et al. reported the fabrication of composite nanofibrous membranes from cellulose blended poly(L-lactide) or polydioxanone. The cellulose was extracted from bagasse, which is the waste of sugarcane with approximately 30% cellulose content. For the enhanced processability of solutions by electrospinning, cellulose was converted to cellulose acetate and blended with biocompatible polymers. The as-prepared fibers show biocompatibility proved by in vitro fibroblast cell studies, and MTT assay indicates their higher cell densities than the control samples [55]. Another recycling approach for sugarcane wastes was reported by Correia et al. such that polyhydroxybutyrate was extracted from sugarcane wastes and processed by electrospinning. The proposed electrospun fibers demonstrated suitable materials for biomedical applications due to their cell adhesion ability [56].

Polyhydroxyalkanoates are biopolyesters produced by bacteria as carbon storage materials. However, their high operational cost is the main problem for their use in packaging materials, although they could be replaced with PET in industry. Thus, Martínez-Sanz et al. reported the synthesis of polyhydroxybutyrate-*co*-hydroxyvalerate using food industry waste feedstocks, which are cheese whey and olive oil wastewaters, as substrates for the fabrication of bacterial cellulose reinforced nanocomposites by melt compounding. The proposed materials are suitable for food packaging applications [57]. In another study, PHA production was achieved by valorization of crude glycerol waste from biodiesel plant. The electrospun PHA membranes were tested in terms of cytotoxicity, hMSC adherence and proliferation in vitro, and it is claimed that they are potential candidates in regenerative medicine and tissue engineering applications (Figure 10.13) [58].

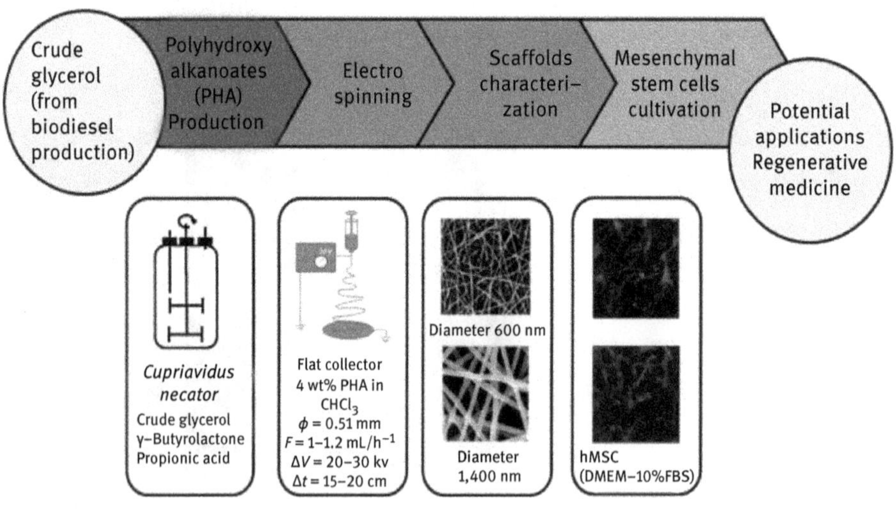

Figure 10.13: Schematic of integrated approach for waste glycerol bio-valorization into fabrication of electrospun scaffolds for stem cells. [Reprinted from Ref. [58], Copyright (2014), with permission from Elsevier].

There is a wide range of agricultural wastes that could be recycled and processed for their utilization in biomedical field. Corncob and wheat straw waste materials have also found an application in delivery systems for bioactive and functional ingredients. Kuan et al. treated the above-mentioned agricultural wastes and processed by electrospinning. The results showed that the materials are potential candidates for bioactive and nano-sized delivery agents due to their mineral binding capability (calcium, copper, iron and zinc), DPPH (2,2-diphenyl-1-picrylhydrazyl) radical scavenging ability and water holding capability [52]. Living functional cell containing composite fibers were also reported by Letnik et al. that the living cell, yeast, was isolated from olive mill wastewaters. The immobilized yeast was encapsulated by coaxial electrospinning, PVDF-HFP shell and yeast/PVP core. The resulting coaxial fibers have a unique structure and could be used in phenol degradation and ethanol fermentation applications (Figure 10.14) [59].

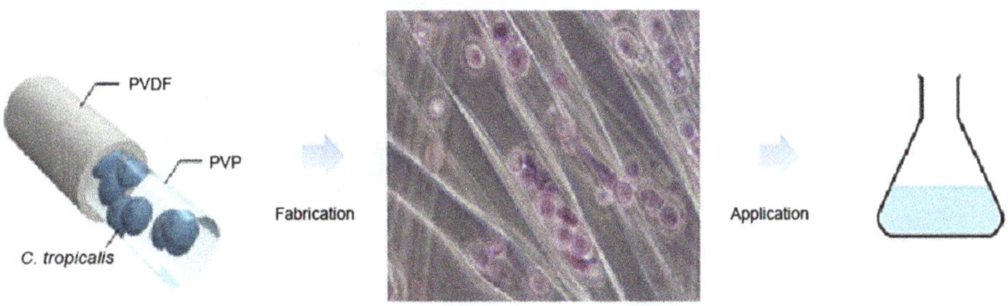

Figure 10.14: Schematic representation of PVDF-HFP@PVP/yeast composite fibers. [Reprinted (adapted) with permission from Ref. [59]. Copyright (2015) American Chemical Society].

10.2.6 Other applications

Dyeability of PET fibers fabricated from waste bottles was reported by Mahar et al. for advanced apparel applications. Their low energy consumption, less processing time, low material cost, good color yield and high mechanical strength make these recycled fibers favorable than the previously reported ones [60]. The encapsulation of magnetic or conductive materials was also studied by using recycled PS cups (EPS waste) for the determination of thermal, electrical, magnetic and surface properties of composite fibers. Khan et al. converted recycled EPS into fibrous form by electrospinning with the addition of multiwall carbon nanotubes and NiZn ferrite nanoparticles. The results showed that ferrite-based recycled EPS fibers show superparamagnetic behavior in addition to the enhancement of superhydrophobic, thermal conduction and dielectric constant properties. These composite fibers are asserted to be promising materials for diverse industrial applications [61].

Because of the low surface energy and roughness of materials fabricated from PET, they possess highly hydrophobic character, which represents a less fouling resistance than the hydrophilic ones. Thus, the improvement of PET fibrous membranes' hydrophilicity is an important issue in biological systems [62]. Lignocellulosic sisal fiber reinforced PET fibers were reported by Santos et al. that the contact angle of fibers was claimed to vary from 134° (hydrophobic) to 0° (superhydrophilic) (Figure 10.15), thereby both stiffness and hydrophilicity of electrospun fibers could be controlled with these composite fibers [63].

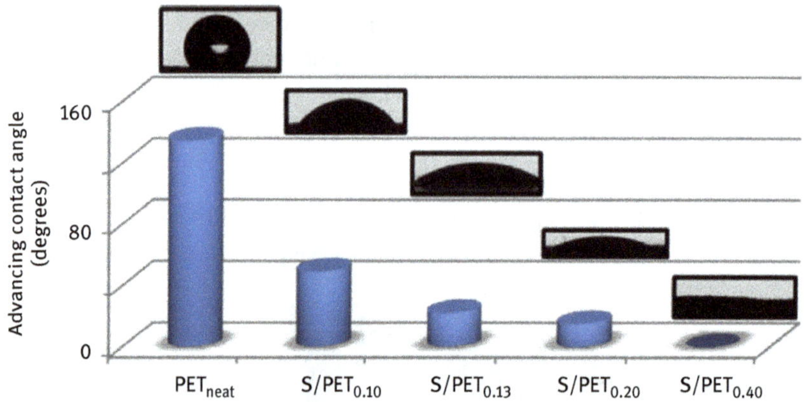

Figure 10.15: Advancing contact angle for PET$_{neat}$ and S/PET electrospun mats and respective snapshots taken after the first/second of contact between a water droplet and its surfaces. [Reprinted from Ref. [63]. Copyright (2015), with permission from Elsevier].

10.3 Conclusion and future perspective

Electrospinning has been started to use for recycling purpose due to being cost effective and becoming diversified easily. This recent recycling approach minimizes the utilization of primary raw sources and provides the manufacturing of the secondary materials by using the plastic waste, which create considerable environmental problems. These resulting secondary materials take place in several applications such as biomedical field, photocatalysis and battery production. However, there is an adverse affect of this method from the environmental point of view that solvent-based approach in electrospinning could be harmful. Due to the evaporation of organic solvents during the electrospinning process, aerosol formation possibility goes up dramatically and the atmosphere as well as people could be affected in a harmful way. To overcome this adverse affect of this approach, melt electrospinning method could be applied for the recycling of plastic wastes without any requirement of organic solvents. The plastic waste materials could be utilized by melting with an extruder and the resulting melt could be electrospun under high potential difference in a combined extruder–electrospinning system. Thus, dangerous wastes could be remediated by recycling of plastic waste without any adverse affect.

Abbreviations

AFD	Average fiber diameter
A. indica	*Azadirachta indica*
CMC	Carboxymethyl cellulose
CNFs	Carbon nanofibers
DPPH radical	2,2-Diphenyl-1-picrylhydrazyl
E. coli	*Escherichia coli*
EPS	Expanded polystyrene
FESEM	Field emission scanning electron microscopy
I. aspalathoides	*Indigofera aspalathoides*
HDPE	High-density polyethylene
LIBs	Lithium-ion batteries
LDPE	Low-density polyethylene
MMP	Matrix metalloproteinase
MTT	3-(4,5-Dimethylthiazol-2-yl)-2,5-Diphenyltetrazolium bromide
M. edule	*Memecylon edule*
M. andamanica	*Myristica andamanica*
POPs	Persistent organic pollutant
PAN	Polyacrylonitrile
PCL	Polycaprolactone
PC	Polycarbonate
PAHs	Polycyclic aromatic compounds
PDX	Polydioxanone
PE	Polyethylene
PEO	Polyethylene oxide
PET	Polyethylene terephthalate
PHAs	Polyhydroxyalkanoates
PHB	Polyhydroxybutyrate
PHBV	Polyhydroxybutyrate-*co*-hydroxyvalerate
PLA	Poly(lactic acid)
PLLA	Poly(L-lactide)
PP	Polypropylene
PS	Polystyrene
PVA	Poly(vinyl alcohol)
PVB	Poly (vinyl butyral)
PVC	Polyvinyl chloride
PVDF-HFP	Poly (vinylidene fluoride-*co*-hexafluoropropylene)
PVP	Polyvinyl pyrrolidone

SEM	Scanning electron microscope
SIBs	Sodium-ion batteries
S. aureus	*Staphylococcus aureus*
TEM	Transmission electron microscopy
UV	Ultraviolet
VOC	Volatile organic compound
w-PVB	Waste poly(vinyl butyral)

References

[1] Plastics Europe. An analysis of European latest plastics production, demand and waste data, plastics – the facts 2016 2016.
[2] Foundations of polymer science: Wallace Carothers and the development of nylon. American Chemical Society National Historic Chemical Landmarks. Accessed 28.10.2017,2017, at http://www.acs.org/content/acs/en/education/whatischemistry/landmarks/carotherspolymers.html.
[3] Mohanty, AK., Misra, M., & Hinrichsen, G. Biofibres, biodegradable polymers and biocomposites: An overview, Macromol. Mater. Eng., 2000, 276, 1–24.
[4] Gu, L., & Ozbakkaloglu, T. Use of recycled plastics in concrete: A critical review, Waste Manage., 2016, 51, 19–42.
[5] Gertakis, J., & Lewis, H. Sustainability and the waste management hierarchy2003.
[6] Scott, G. 'Green' polymers, Polym. Degrad. Stab., 2000, 68, 1–7.
[7] Sadrmohaghegh, C., & Scott, G. Effect of reprocessing on polymers 2. Polypropylene. Polym. Degrad. Stab., 1981, 3, 333–340.
[8] Sadrmohaghegh, C., Scott, G., & Setoudeh, E. Effects of reprocessing on polymers 3, photooxidation of polyethylene-polypropylene blends. Polym. Degrad. Stab., 1981, 3, 469–476.
[9] Sadrmohaghegh, C., Scott, G., & Setudeh, E. Recycling of mixed plastics, Polym.-Plastics Technol. Eng., 1985, 24, 149–185.
[10] Dealy, JM. Handbook of industrial energy analysis, I. Boustead & G. F. Hancock, John Wiley & Sons, New York, 1979, 422. pages.$69.50. 1980; 26:- 174.
[11] Hopewell, J., Dvorak, R., & Kosior, E. Plastics recycling: challenges and opportunities, Philos. Trans. Royal. Soc. B-Biol. Sci., 2009, 364, 2115–2126.
[12] You, JH., Chiang, PC., Chang, SC., & Wang Wuu, S. Polycyclic aromatic hydrocarbons (PAHs) and mutagenicity in air emissions from the two-stage incineration of polystyrene with various metallic salt additives, J. Hazard. Mater., 1996, 48, 69–82.
[13] Scott, G. Polymers and the environment: Royal society of chemistry; 1999.
[14] Garcia, JM., & Robertson, ML. The future of plastics recycling Chemical advances are increasing the proportion of polymer waste that can be recycled, Science, 2017, 358, 870–872.
[15] MacArthur, DE. Beyond plastic waste, Science, 2017, 358, 843-.
[16] Santhosh, C., Velmurugan, V., Jacob, G., Jeong, SK., Grace, AN., & Bhatnagar, A. Role of nanomaterials in water treatment applications: A review, Chem. Eng. J., 2016, 306, 1116–1137.
[17] Li, GH., Zhao, YM., Lv, MQ., Shi, Y., & Cao, D. Super hydrophilic poly(ethylene terephthalate) (PET)/poly(vinyl alcohol) (PVA) composite fibrous mats with improved mechanical properties

prepared via electrospinning process, Colloids Surf. a-Physicochem. Eng. Asp., 2013, 436, 417–424.

[18] Veleirinho, B., Rei, MF., & Lopes-da-Silva, JA. Solvent and concentration effects on the properties of electrospun poly(ethylene terephthalate) nanofiber mats, J. Polym. Sci. Part B-Polym. Phys., 2008, 46, 460–471.

[19] Strain, IN., Wu, Q., Pourrahimi, AM., Hedenqvist, MS., Olsson, RT., & Andersson, RL. Electrospinning of recycled PET to generate tough mesomorphic fibre membranes for smoke filtration, J. Mater. Chem. A, 2015, 3, 1632–1640.

[20] Zander, NE., Sweetser, D., Cole, DP., & Gillan, M. Formation of nanofibers from pure and mixed waste streams using electrospinning, Ind. Eng. Chem. Res., 2015, 54, 9057–9063.

[21] Khorram, M., Mousavi, A., & Mehranbod, N. Chromium removal using adsorptive membranes composed of electrospun plasma-treated functionalized polyethylene terephthalate (PET) with chitosan, J. .Environ. Chem. Eng., 2017, 5, 2366–2377.

[22] Zander, NE., Gillan, M., & Sweetser, D. Recycled PET nanofibers for water filtration applications, Materials, 2016, 9, 247.

[23] Shin, C. Filtration application from recycled expanded polystyrene, J. Colloid Interface Sci., 2006, 302, 267–271.

[24] Shin, C., Chase, GG., & Reneker, DH. Recycled expanded polystyrene nanofibers applied in filter media, Colloids Surf. a-Physicochem. Eng. Asp., 2005, 262, 211–215.

[25] Hidaka, K., Iwakawa, Y., Maoka, T., Tanimoto, F., & Oku, A. Viable chemical recycling of poly (carbonate) as a phosgene equivalent illustrated by the coproduction of bisphenol A and carbohydrate carbonates, J. Mater. Cycles Waste Manage., 2009, 11, 6–10.

[26] Isık, T., & Demir, MM. Medical waste treatment via waste electrospinning of PS, Fiber Polym., 2018, 19, 767–774.

[27] Matulevicius, J., Krugly, E., & Kliucininkas, L. Utilisation of thermoplastic polymer waste for nanofiber air filter production2015.

[28] Aluigi, A., Rombaldoni, F., Tonetti, C., & Jannoke, L. Study of Methylene Blue adsorption on keratin nanofibrous membranes, J. Hazard. Mater., 2014, 268, 156–165.

[29] Aluigi, A., Vineis, C., Tonin, C., Tonetti, C., Varesano, A., & Mazzuchetti, G. Wool keratin-based nanofibres for active filtration of air and water, J. Biobased Mater. Bio., 2009, 3, 311–319.

[30] Aluigi, A., Corbellini, A., Rombaldoni, F., & Mazzuchetti, G. Wool-derived keratin nanofiber membranes for dynamic adsorption of heavy-metal ions from aqueous solutions, Text. Res. J., 2013, 83, 1574–1586.

[31] Woranuch, S., Pangon, A., Puagsuntia, K., Subjalearndee, N., & Intasanta, V. Starch-based and multi-purpose nanofibrous membrane for high efficiency nanofiltration, RSC Adv., 2017, 7, 35368–35375.

[32] Yao, ZT., Ji, XS., Sarker, PK. et al. A comprehensive review on the applications of coal fly ash, Earth-Sci. Rev., 2015, 141, 105–121.

[33] Saud, PS., Pant, B., Park, M. et al. Preparation and photocatalytic activity of fly ash incorporated TiO2 nanofibers for effective removal of organic pollutants, Ceram. Int., 2015, 41, 1771–1777.

[34] Rajak, A., Mm, Munir., & Abdullah, M. Khairurrijal. Photocatalytic activities of electrospun TiO2/styrofoam composite nanofiber membrane in degradation of waste water, Mater. Sci. Forum, 2015, 827, 7–12.

[35] Aykut, Y. Enhanced field electron emission from electrospun co-loaded activated porous carbon nanofibers, ACS Appl. Mater. Int., 2012, 4, 3405–3415.

[36] Wang, MX., Huang, ZH., Bai, Y., Kang, FY., & Inagaki, M. Porous carbon nanofibers: Preparation and potential applications, Curr. Org. Chem., 2013, 17, 1434–1447.

[37] Song, KL., Wu, QL., Zhang, Z. et al. Fabricating electrospun nanofibers with antimicrobial capability: A facile route to recycle biomass tar, Fuel, 2015, 150, 123–130.

[38] Song, KL., Wu, QL., Zhang, Z. et al. Porous carbon nanofibers from electrospun biomass tar/polyacrylonitrile/silver hybrids as antimicrobial materials, ACS Appl. Mater. Int., 2015, 7, 15108–15116.

[39] Kampeerapappun, P. Preparation characterization and antimicrobial activity of electrospun nanofibers from cotton waste fibers, Chiang Mai J. Sci., 2012, 39, 712–722.

[40] Park, SW., Kim, JC., Dar, MA., Shim, HW., & Kim, DW. Superior lithium storage in nitrogen-doped carbon nanofibers with open-channels, Chem. Eng. J., 2017, 315, 1–9.

[41] Jiang, Y., Zhang, Y., Yan, XM., Tian, ML., Xiao, W., & Tang, HL. A sustainable route from fly ash to silicon nanorods for high performance lithium ion batteries, Chem. Eng. J., 2017, 330, 1052–1059.

[42] Huang, FL., Xu, YF., Peng, B. et al. Coaxial electrospun cellulose-core fluoropolymer-shell fibrous membrane from recycled cigarette filter as separator for high performance lithium-ion battery, ACS Sustain. Chem. Eng., 2015, 3, 932–940.

[43] Yang, W., Shi, ZR., Guo, H., Guo, J., Lei, X., & Yue, LG. Study on preparation of nanocarbon fibers from wheat-straw based on electrostatic spinning method and its application in supercapacitor, Int. J. Electrochem. Sci., 2017, 12, 5587–5597.

[44] Jin, J., Yu, SJ., Shi, ZQ., Wang, CY., & Chong, CB. Lignin-based electrospun carbon nanofibrous webs as free-standing and binder-free electrodes for sodium ion batteries, J. Power Sources, 2014, 272, 800–807.

[45] Hubbell, JA. Biomaterials in tissue engineering, Bio-Technol., 1995, 13, 565–576.

[46] Martin, P. Wound healing - Aiming for perfect skin regeneration, Science, 1997, 276, 75–81.

[47] Streuli, C. Extracellular matrix remodelling and cellular differentiation, Curr. Opin. Cell Biol., 1999, 11, 634–640.

[48] Morganti, P., Fusco, A., Paoletti, I. et al. Anti-inflammatory, immunomodulatory, and tissue repair activity on human keratinocytes by green innovative nanocomposites, Materials, 2017, 10, 843–855.

[49] Padamwar, MN., & Pawar, AP. Silk sericin and its applications: A review, J. Sci. Ind. Res., 2004, 63, 323–329.

[50] Eslah, S., Tavanai, H., & Morshed, M. Electrospinning and characterization of poly (vinyl alcohol)-sericin nanofibers as a potential for tissue engineering applications, J. Text. Inst., 2016, 107, 949–957.

[51] Jin, GR., Prabhakaran, MP., Kai, D., Annamalai, SK., Arunachalam, KD., & Ramakrishna, S. Tissue engineered plant extracts as nanofibrous wound dressing, Biomaterials, 2013, 34, 724–734.

[52] Vashisth, P., Sharma, M., Nikhil, K. et al. Antiproliferative activity of ferulic acid-encapsulated electrospun PLGA/PEO nanofibers against MCF-7 human breast carcinoma cells, 3 Biotech., 2015, 5, 303–315.

[53] Persin, Z., Ravber, M., Kleinschek, KS., Knez, Z., Skerget, M., & Kurecic, M. Bio-nanofibrous mats as potential delivering systems of natural substances, Text. Res. J., 2017, 87, 444–459.

[54] Guzman-Puyol, S., Heredia-Guerrero, JA., Ceseracciu, L. et al. Low-cost and effective fabrication of biocompatible nanofibers from silk and cellulose-rich materials, ACS Biomater. Sci. Eng., 2016, 2, 526–534.

[55] Ramphul, H., Bhaw-Luximon, A., & Jhurry, D. Sugar-cane bagasse derived cellulose enhances performance of polylactide and polydioxanone electrospun scaffold for tissue engineering, Carbohydr. Polym., 2017, 178, 238–250.

[56] Correia, DM., Ribeiro, C., Ferreira, JCC. et al. Influence of electrospinning parameters on poly (hydroxybutyrate) electrospun membranes fiber size and distribution, Polym. Eng. Sci., 2014, 54, 1608–1617.
[57] Martinez-Sanz, M., Lopez-Rubio, A., Villano, M. et al. Production of bacterial nanobiocomposites of polyhydroxyalkanoates derived from waste and bacterial nanocellulose by the electrospinning enabling melt compounding method, J. Appl. Polym. Sci., 2016, 133, 42486–42500.
[58] Canadas, RF., Cavalheiro, J., Guerreiro, JDT. et al. Polyhydroxyalkanoates: Waste glycerol upgrade into electrospun fibrous scaffolds for stem cells culture, Int. J. Biol. Macromol., 2014, 71, 131–140.
[59] Letnik, I., Avrahami, R., Rokem, JS., Greiner, A., Zussman, E., & Greenblatt, C. Living composites of electrospun yeast cells for bioremediation and ethanol production, Biomacromolecules, 2015, 16, 3322–3328.
[60] Fk, Mahar., Mehdi, M., Ua, Qureshi. et al. Dyeability of recycled electrospun polyethylene terephthalate (PET) nanofibers: Kinetics and thermodynamic study, J. Mol. Liq., 2017, 248, 911–919.
[61] Khan, WS., Asmatulu, R., Davuluri, S., & Dandin, VK. Improving the economic values of the recycled plastics using nanotechnology associated studies, J. Mater. Sci. Technol., 2014, 30, 854–859.
[62] Dong, BY., Jiang, HQ., Manolache, S., Wong, ACL., & Denes, FS. Plasma-mediated grafting of poly(ethylene glycol) on polyamide and polyester surfaces and evaluation of antifouling ability of modified substrates, Langmuir, 2007, 23, 7306–7313.
[63] Santos, RPO., Rodrigues, BVM., Ramires, EC., Ruvolo, AC., & Frollini, E. Bio-based materials from the electrospinning of lignocellulosic sisal fibers and recycled PET, Ind. Crops Prod., 2015, 72, 69–76.

Mirja Palo, Ezgi Özliseli, Didem Sen Karaman, Karin Kogermann

11 Electrospun biocomposite fibers for wound healing applications

Abstract: Wound healing is a complex and dynamic process consisting of devitalized and missing cellular structure replacements and regeneration of tissue layers. Each wound requires individual diagnosis and care in order to remove the barriers for the healing process. It is of relevance that wound dressings are not only to protect from microbial contamination, physical protection and moisture retention, but also to support the normal wound healing and provide important biological molecules and/or active substances for the wound. There is a growing interest in the development of electrospun biocomposite fibers as wound dressings due to their tunable and multifunctional properties. The main advantages of electrospun biocomposite fibers in wound healing process include their resemblance to natural extracellular matrix, high surface-area-to-volume ratio, tunable porosity, sufficient gas exchange and possibility to include different active substances and living cells into the fibers. Electrospinning technology enables to produce ultrafine fibers from different materials by using various setups. Both solution (blend, coaxial and emulsion) electrospinning and green solvent-free electrospinning approaches can be used for producing biocompatible fibers for wound care. Materials suitable for wound dressings are expected to meet the requirements of the wound and to be biocompatible. Multifunctional biocomposite fiber dressings enable to combine the physical and biochemical properties of natural and synthetic polymers together with active substances that are beneficial for wound healing. Furthermore, biocomposite fibers are able to interact with the wound, and their properties can be modified according to the need. This chapter reviews electrospun biocomposite fibers as novel materials for the preparation of wound dressings. The relevant methods and materials for the preparation, and the properties and methods for the analysis of electrospun fibers will be discussed in more detail by considering their suitability for wound care. Data sources will include mainly peer-reviewed literature, reports, patents, and guidelines where available.

Keywords: green electrospinning, biocomposite, nanofibers, microfibers, wound care, tissue regeneration

Mirja Palo, Ezgi Özliseli, Didem Sen Karaman, Pharmaceutical Sciences Laboratory, Åbo Akademi University, Turku, Finland
Karin Kogermann, Institute of Pharmacy, University of Tartu, Estonia

https://doi.org/10.1515/9783110581393-011

11.1 Introduction

Nonhealing wounds are a huge problem for the society due to increased healthcare costs and a significant burden to patients. It has been estimated that 1.5–2 million people in Europe suffer from acute or chronic wounds [1]. The number of patients suffering from chronic wounds is increasing due to population aging and insufficient treatment options. Furthermore, chronic wounds take considerable time to heal and hence are much more expensive to treat.

The dynamic process of wound healing follows a complex sequence of events, which can be explained under four different phases: hemostasis, inflammation, proliferation and tissue remodeling [2]. Especially during the final steps of wound healing, a massive tissue regeneration takes place, and the process becomes challenging. In that wound healing stage, tissue engineering approaches are needed to be implemented to provide better healing. Therefore, tremendous research efforts have been done to implement tissue engineering for wound healing and skin regeneration [3]. The delivery of stem cells within the native matrix is one of the leading approaches that can be combined with the delivery of growth factors or therapeutics to ensure effective regeneration of the appendages as well within the same wound care product.

Currently, there are several traditional and advanced wound care products on the market. In that light, the advanced wound care market is expected to reach $12,454 million by 2022 [4]. Several products and treatment options already on the market require specialists (nurses and doctors) for their appropriate administration. Available commercial engineering therapies for wound healing can be classified as follows: (i) acellular matrix (i.e., biological matrix and biologically processed matrix); (ii) cellular products (i.e., epidermal and dermal/epidermal–dermal composites); and (iii) growth factor products (i.e., recombinant platelet derived). However, none of the existing therapies are multifunctional, which can provide diagnosis and healing simultaneously to support better self-treatment by patients [5]. Therefore, smart wound care products are direly needed to overcome the challenges and give signals to patients when changes take place in the wound [5]. This enables to apply a new dressing and/or seek help from clinicians in time.

Electrospinning of nanofiber mats has been proposed as an excellent technology to produce fibrous advanced wound care products [6, 7]. An example of electrospun nanofiber mats is shown in Figure 11.1 together with possible electrospun fiber structures.

Biocomposite fibers can be prepared from synthetic or natural, degradable or nondegradable polymers depending on the intended use. Biopolymers have environmental-friendly characteristics and more suitable lifetime for wound healing and tissue regeneration applications. Biocomposite fibers are usually made of two or more materials combined together to create a new and effective material in terms of quality and production process [8]. Hence biocomposite fibers can be designed to meet the requirements of complex wound healing process. In those applications,

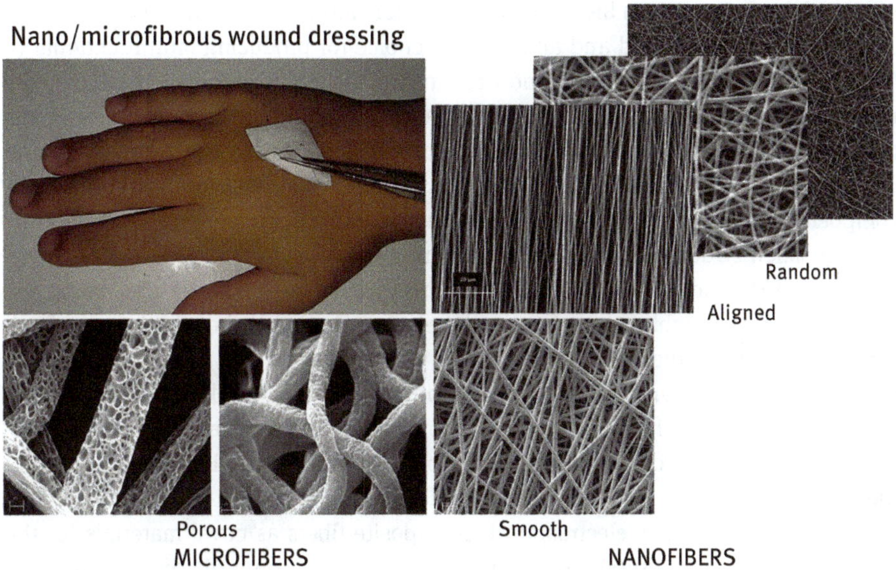

Figure 11.1: Electrospun fiber structures, and an example of electrospun nanofibrous wound dressing. Scanning electron micrograph of aligned fibers obtained from Contipro a.s., made on 4SPIN LAB, polymer: poly(vinyl alcohol) (PVA).

biocompatible and safe natural polymer-based composites are used. They have also good mechanical properties, thermal stability, permeability, optical transparency and desired degradation rate [9].

The main advantages of biocomposite fibers in wound healing process include: (i) resemblance of natural extracellular matrix (ECM); (ii) high surface-area-to-volume ratio; (iii) tunable porosity; (iv) sufficient gas exchange; and (v) possibility to include active pharmaceutical ingredients (APIs) and even living cells into fibers [10]. All these properties make electrospun biocomposite fiber mats an ideal platform for wound healing treatments [11–13].

Biocomposite fibers can be obtained by several methods out of which electrospinning is a versatile and widely used technology for the production of fibers with unique features [14]. Although electrospinning technology is easily scalable [15], not many products are on the market today. One of the main reasons is the lack of standardized production and analysis methods for the electrospun biocomposite fibers. The manufacturing of biocomposite fibers is challenging at a larger commercial scale and under Good Manufacturing Practice (GMP). Another major bottleneck is to develop specific methods for the analysis of electrospun nanofiber formulations (e.g., for testing biological activity), since not all the methods suitable for the analysis of semisolid or solid topical formulations are applicable. Moreover, appropriate choices have to be made on which electrospinning setup and technology to use for the preparation of electrospun biocomposite fibers for wound care. Blend electrospinning is the mostly used method and an excellent choice for producing multifunctional fibers. However, to be used as wound dressings, water-soluble polymer mats usually need some post-electrospinning treatment and insoluble polymers require solvents that are not environmental-friendly for the production. Post-treatment methods and incompatibilities with the solvents may reduce the activity and functionality of biocomposite fibers. Alternatively, emulsion and coaxial electrospinning allow producing core–shell structured fibers while sheltering the bioactivity of substances within the core. More and more solvent-free electrospinning methods are also being explored. The latter enables to perform green electrospinning and avoid the use of solvents. The main challenges in solvent-free electrospinning are the larger diameter due to high viscosity and low conductivity of the precursor, a relatively complicated setup (e.g., a heater is needed in melt electrospinning and nitrogen is used in ultraviolet (UV)-curing electrospinning), the lack of systematic theoretical studies [16] and the risk of losing the bioactivity during the electrospinning.

This chapter reviews electrospun biocomposite fibers as novel materials for the preparation of wound dressings. Specific emphasis is put onto the discussion about green electrospinning options for wound healing applications found in the current literature. The relevant properties and analysis methods for revealing their suitability for biotechnological applications, especially wound healing, will be discussed in more detail. Data sources will include mainly peer-reviewed literature, reports, patents, and guidelines where available.

11.2 Fabrication of electrospun biocomposite fibers for wound dressings

One of the widely used technologies for producing biocomposite fibers is electrospinning. Besides electrospinning, fibrous biocomposites can be prepared by several other methods, such as melt blowing, phase separation, self-assembly, and template synthesis. The uniqueness of electrospinning lies in the fabrication of (ultra)fine fibers with a high length-to-width (aspect) ratio with a relatively simple and cost-effective apparatus. For successful electrospinning, several materials, process and environmental parameters need to be optimized and best solutions should be selected for each case. The structure of electrospun fibers/mats depends largely on these parameters.

11.2.1 Methods

Biocomposite fiber mats for wound dressings can be prepared by several electrospinning setups or in combination with other techniques. The biocomposite fibers divide largely into single-phase (blends) or dual-phase (core–shell and core–sheath) systems [17]. These biocomposite fiber mats can be obtained by several different electrospinning setups (Figure 11.2), including:

- Blend electrospinning
- Coaxial electrospinning
- Emulsion electrospinning
- Melt electrospinning
- Post-electrospinning surface immobilization

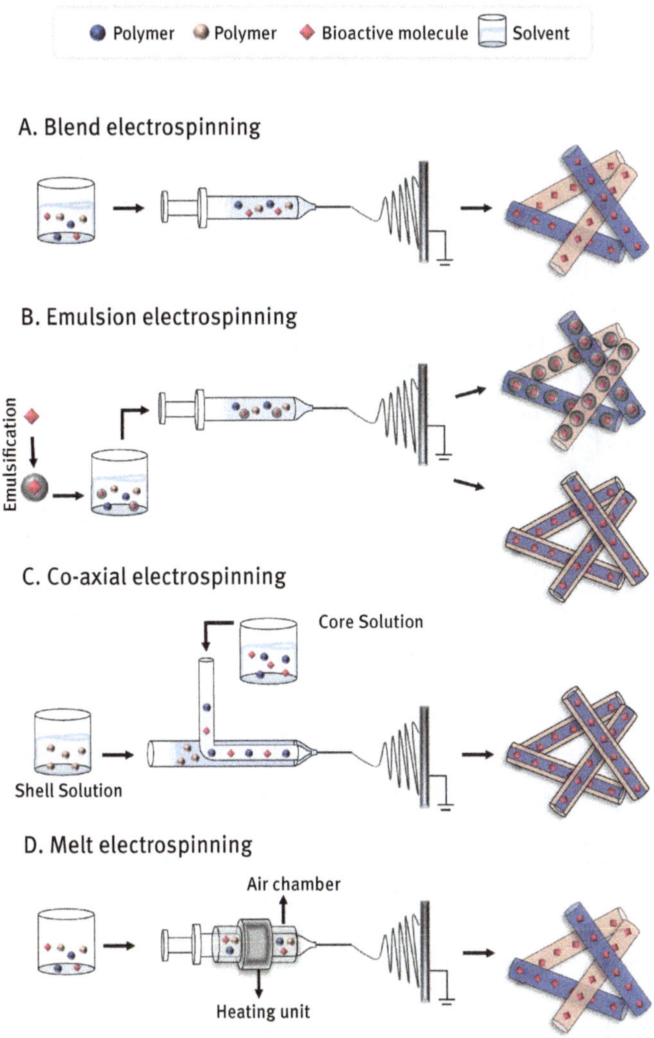

Figure 11.2: Electrospinning setups used for the preparation of biocomposite fibers.

The basic electrospinning setup consists of a flow-regulating material dispenser, the container for the polymer material, high-voltage power supply and grounded fiber collector. In terms of green electrospinning, each method has its own advantages and limitations that affect the processability of the materials and the structural integrity of the electrospun fibers [18, 19]. At the same time, post-processing and/or surface functionalization of electrospun fibers allow modifying the properties of the fiber mats and enable to conduct solvent-free green electrospinning.

Blend electrospinning

The blend electrospinning is the most common configuration for biomedical applications. The electrospun fibers are produced from polymer solutions with or without other additives. The process parameters and material properties that affect the outcome of electrospinning have been reviewed previously [20–23]. The influential process variables are as follows: applied voltage, solution feeding rate, the gauge diameter of the nozzle, the fiber collector type and the distance between the nozzle of the spinneret and the collector. Other factors that influence the fiber formation are related to the properties of the polymer solution and electrospinning environment. The combination of optimized settings allows obtaining defect-free solid fibers from a continuous jet of the polymer solution. The alignment, packing density and macroporosities of fibrous mats can be adjusted by changing the aforementioned process parameters [24–26].

The main challenge for environmental-friendly blend electrospinning of biocomposite fibers is to find common solvents for all the ingredients. Thus, emulsion electrospinning has been proposed to be an alternative method. Nevertheless, unwanted interactions and phase separation between materials may occur within unstable systems. Compared to other solvent-based methods, the advantage of blend electrospinning with nontoxic solvents and biocompatible polymers is the higher cytocompatibility for wound healing applications.

Biocomposite fibers can be prepared as single-phase systems with the conventional single spinneret setup. However, blend electrospinning has gone through a lot of advancements over the last decade. New design setups have been developed with the aim to scale up the electrospinning process for higher speed and production yield [15, 27]. By obtaining several polymer jets at the same time through a multinozzle spinneret [28] or a needleless setup [29], the amount of produced fibers is multiplied accordingly. Several other advancements for needle-free or free-surface electrospinning have been reported, involving ultrasound-assisted electrospinning [30, 31], Nanospider™ technology, ball-electrospinning and many others [32–34]. In addition, some helpful add-ons to the existing methods have been developed to increase the fiber production. An example is a dual power source electrospinning that is a combination of two single spinneret systems [35]. Recently, LeCorre-Bordes et al.

developed a shear spinneret for increasing the feeding rate of viscous polymer solutions [36]. On the other hand, the spinneret with an electroblowing add-on enhances the production of fibers from polymer solutions with slowly evaporating solvents [37].

Alternatively, portable instruments for smaller scale production are useful for more personalized wound care, for example, at the bedside or at home. These instruments are suitable for direct electrospinning of polymer solutions onto the skin [38–40]. The small-sized and the battery-powered apparatuses can provide good quality at high production flexibility.

Coaxial electrospinning

Coaxial electrospinning is a method for producing dual-phase biocomposite fibers. Functionalized electrospun core–shell fibers are fabricated through encapsulating bioactive and/or other functionalizing agents into the core or the shell part of the fibers [41]. The dual-phase systems are obtained by a two-part syringe setup, consisting of a smaller inner capillary that is mounted into a larger outer needle. Continuous fiber formation is obtained by controlling the flow rate of the electrospinning solutions in both phases separately. The flow rate of the core material(s) is typically much slower compared to the shell polymer solution [42–45]. For example, Song et al. demonstrated that the distribution of drug-loaded gelatin nanospheres in the core–shell fibers can be controlled through the feeding rate ratio between the core and shell solutions [42]. Furthermore, encapsulated vancomycin-loaded nanospheres showed good sustained antibacterial effect without any cytotoxicity.

The spinnability of the shell polymer is the driving factor for obtaining defect-free fibers, whereas the core solution needs to be viscous enough to stay intact within the shell solution. Typically, the physical phase separation in the core–shell fibers is a result of using immiscible solvents. However, clearly distinguishable core–shell structures with core-defined drug loading can also be fabricated by using the same solvent [44].

Coaxial electrospinning gives possibilities to control the release rate of encapsulated therapeutics. At the same time, the shell of the fibers can provide physical protection from environmental factors and unwanted interactions, and hinder premature release of molecules from the core structure. For example, Zhang et al. constructed complex electrospun biocomposites for vascular regeneration with growth factors [45]. Besides the incorporation of growth factors into the inner chitosan-based hydrogel structure, dual-release fibers were manufactured by loading the active ingredients into the inner and outer layers of the composite fibers. Zupančič et al. presented a novel type of core–shell nanofibers with a blended core structure [44]. A ciprofloxacin-loaded polymer mixture with poly(vinyl alcohol) (PVA) and poly(methyl methacrylate) (PMMA) was used as the blended core and PMMA as the shell of the composites. The release of the API was controlled through modifying the polymer ratio in the blended core of the biocomposite fibers at higher degree compared to a conventional monolithic core [44].

The main disadvantage of coaxial electrospinning is the dual-nozzle spinneret. The risk for nozzle clogging is high, although the flow rate of the solutions can be optimized with two separate syringe pumps. As mentioned, the feed rate is highly important for uniform structure and particle distribution within the core–shell fibers. To simplify the instrument configuration, a needleless coaxial electrospinning has been developed [46]. This uncommon setup can provide an approximately fourfold increase in the production yield.

Emulsion electrospinning

Emulsion electrospinning is another method for producing core–shell or core–sheath (encapsulated blends) biocomposite fibers. As the name indicates, compared to blend electrospinning with homogeneous solutions, emulsion electrospinning uses emulsions or colloidal solutions with immiscible fluids (e.g., water and organic phase). In principle, oil-in-water or water-in-oil fluid blends are applicable. Prior to electrospinning, the dual-phase system is emulsified through high-shear mixing. The biocomposite fibers are formed due to the phase separation of the materials after electrospinning.

Emulsion electrospinning is beneficial for the stabilization and controlled release of encapsulated hydrophilic compounds in the polymeric core of the fibers [18, 19, 47–50]. The encapsulation of hydrophobic molecules in a hydrophilic shell is less common. In the fabrication of core–shell composites with hydrophilic shell, the relative humidity and the polymer concentration influence significantly the fiber morphology and the encapsulation efficiency [51].

The emulsion electrospinning uses a conventional single-nozzle spinneret. The system is simpler compared to the coaxial electrospinning; however, the control over the fiber composition is lacking [48]. In addition, an uncontrolled denaturation of proteins may occur in contact with the organic solvents and/or during the process [50, 52].

The emulsion electrospinning is limited in terms of using organic solvents for the shell-forming polymers [19]. The electrospinning with benign solvent systems has emerged only in the recent years. Chen et al. presented the fabrication of cross-linked dexamethasone-loaded core–shell biocomposite nanofibers with water-soluble polymers: silk fibroin as the core and poly(ethylene oxide) (PEO) as the shell [53]. The authors pointed out that higher drug loading in the fibers was achieved by dissolving the hydrophobic drug molecules in plant oils. The results demonstrated that the core–shell biocomposite fibers showed superior anti-inflammatory effect and cell viability compared to the conventional blend-type nanofibers.

Melt electrospinning

Melt electrospinning is a solvent-free method for producing fiber mats from thermoplastic polymers. It is a one-step production method for polymers and their blends without any toxic solvents [54–57]. Hence, it allows to conduct green electrospinning. The incorporation of drugs and bioactive molecules is limited by their tolerance to high temperature applied during the process. However, melt electrospinning is highly advantageous mainly due to the avoidance of toxic solvents and the higher production yield. Furthermore, it gives means for producing fibers directly onto cells or tissue [58, 59]. The biocomposite fibers containing polymer blends and/or other active molecules are obtained from solid mixtures of the ingredients. In pharmaceutics, it provides an alternative for developing drug delivery systems (DDSs) with enhanced dissolution behavior for thermostable drugs with poor water solubility [56, 57].

The higher viscosity and low electrical conductivity of the polymer solution dictate the physical properties of the fibers. In wound healing, the larger fiber diameter (≥500 µm) of melt electrospun fibers can be a limiting factor for cell adhesion [54, 55]. However, the fiber alignment and the macroporosity in the melt electrospun fiber mats can be controlled by using patterned and/or rotating collectors. This has great impact on the fiber mats designed for tissue engineering during treatment. For example, Brown et al. (2014) demonstrated that poly(ε-caprolactone) (PCL) fibers mats with defined porosity and a three-dimensional (3D) structure can be obtained by using patterned electroconductive collectors or by controlling the movement of the collector plate (i.e., melt electrospinning writing) [60].

The melt electrospinning setup requires a specific heating unit for keeping a constant temperature for melting the polymer, which makes the instrumental setup more complex compared to blend/emulsion electrospinning. The fiber formation is highly affected by the temperature control in the nozzle of the spinneret. Similar to solution electrospinning, the material feeding systems and collectors vary a lot between the different melting-based instruments, but there are also several heating unit configurations (e.g., electrical, circulating fluid or oil) available.

Post-electrospinning surface immobilization

Post-electrospinning adsorption is currently the most common method for preparing functionalized nanofiber mats by loading them with drugs or other active agents. For example, dipping the nanofiber mats into peptide or growth factor solutions enables to produce bioactive biocomposite fiber mats for wound healing applications. Furthermore, the surface of the fibers can be functionalized by coatings [61, 62], chemical reactions [63, 64] and plasma treatment [65, 66]. The latter two approaches enable to modify the surface characteristics, while preserving the structure and morphology of fibers. This consequently is of importance for achieving better bioactivity of the fiber mats [65, 66]. Furthermore, the combination of different methods can help to tailor the structure and functionality of biocomposite fiber mats even more. For example, robocasting technique [67] and hydrogel formation [68] have been applied to obtain certain structural characteristics to the electrospun mats.

Common post-processing of biocomposite fiber mats typically includes also cross-linking of the fibers [62, 69–72]. The cross-linking of the fibers improves the stability and mechanical strength of the biocomposite fiber mats and enables to preserve the mats for a determined time. It is mostly used to make water-soluble matrices water insoluble, and thus increase their stability at high humidity and/or in contact with physiological fluids. Sterilization of the mats is also considered as a post-electrospinning modification that requires separate attention. For wound healing applications, the sterility of the mats needs to be verified. However, some sterilizations may change the behavior and properties of the biocomposite nanofiber mats [73].

11.2.2 Materials

Biocomposites are composite materials made from natural fibers and petroleum-derived nonbiodegradable polymer epoxies or biopolymers. Biopolymer and synthetic fibers, such as glass and carbon blended composites, are also counted as biocomposites. Biocomposites derived from plant-derived fibers (natural/biofiber) and crop/bioderived plastics (biopolymer/bioplastic) are likely more eco-friendly, and such biocomposites are sometimes termed "green composites" [74–76]. These biocomposites can be easily disposed of or composted at the end of their life without harming the environment, which is not possible with only synthetic fiber-based polymer composites [77].

The components of biocomposites and their properties vary depending on the aimed applications. For instance, in wound healing applications, which is the main focus of this chapter, the employed biocomposite fibers are expected to promote the cell–biomaterial interactions, adhesion, permit sufficient transport of nutrients and oxygen, biodegrade at a controllable rate, be nontoxic and have nonantigenic and also appropriate mechanical properties [78, 79].

Natural polymers

Natural polymers can be classified as proteins (silk, collagen, gelatin, fibrinogen, elastin, keratin, actin and myosin), polysaccharides (cellulose, amylose, dextran, chitin and glycosaminoglycans) or polynucleotides [80]. Natural polymers that are mainly used in wound healing applications together with their principal sources are summarized in Figure 11.3.

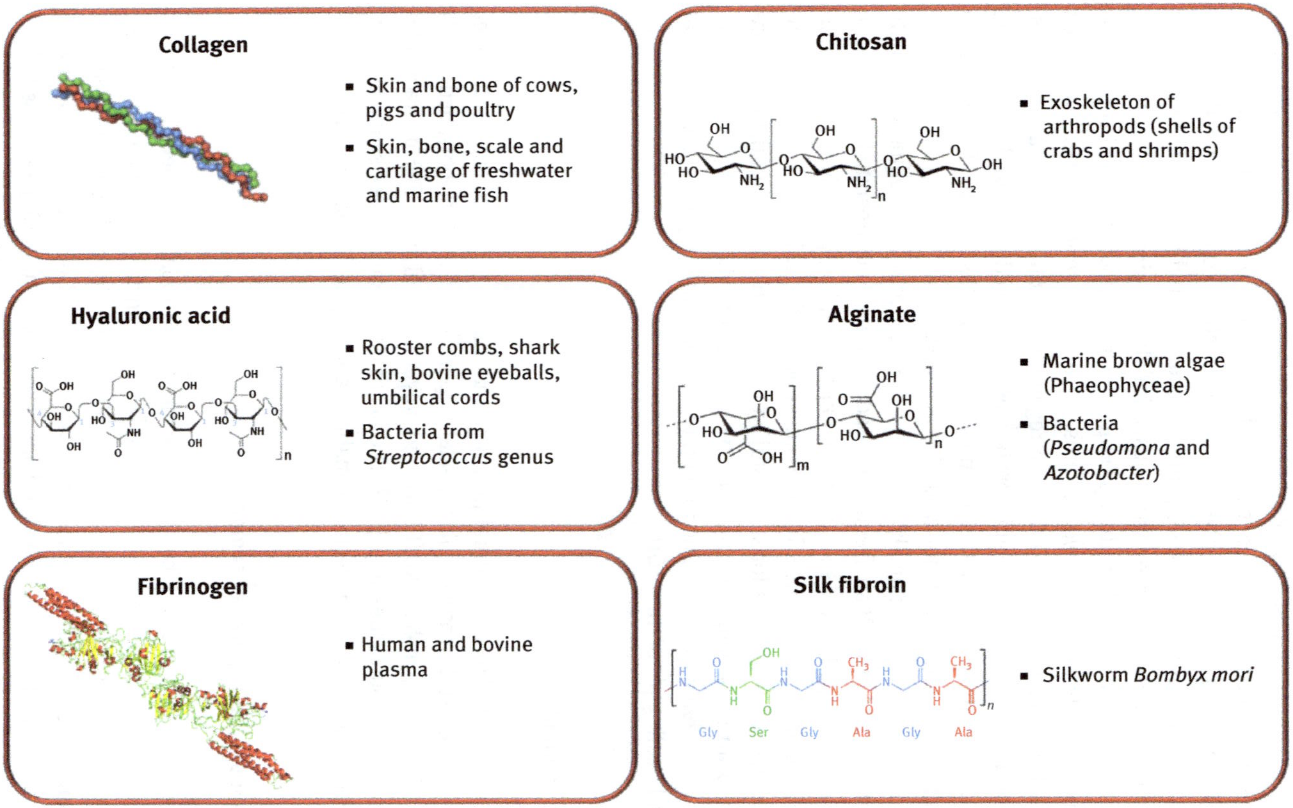

Figure 11.3: Natural polymers and polysaccharides employed in the preparation of biocomposite fibers for wound healing applications with their main sources. Adapted from Ref. [81] with permission from the Royal Society of Chemistry.

Naturally occurring proteins and polysaccharides are extensively used as the constituents of biocomposite fibers due to a high level of biocompatibility and biomimicry. The proteins that form fibers in nature, such as collagen [82–84], fibrinogen [85] and silk fibroin [42, 86–89], are especially suitable for wound dressing preparation, and these have been widely used for electrospinning of fiber mats [81]. For example, silk is a natural product produced by the silkworm during cocoon formation and consists mainly of two proteins, fibroin and sericin, which have been widely used in wound healing and tissue regeneration among other applications. Fibroin and sericin have remarkable properties to be used in wound remodeling. Chao et al. demonstrated that tigecycline loading on the sericin/PVA composite fibers could serve as ideal mats for wound healing [90].

In addition to proteins, namely chitosan [91–96], chitin [97], cellulose, alginate and hyaluronic acid [98, 99] are frequently used to prepare electrospun biocomposite fiber wound dressings [100]. As an additional approach, it is possible to change/improve the properties of polymers by different modifications. For example, chitosan has several desired properties for successful wound healing; however, the disadvantage of chitosan is its natural origin and low solubility in different solvents. Hence, Antunes et al. have used L-arginine to modify chitosan and improve its bactericidal and wound healing properties [101].

Synthetic polymers

Although high abundance of natural polymers that fulfills the requirements for fiber mat preparations, still synthetic biodegradable polymers like PCL, poly(lactic acid) (PLA), poly(lactic-*co*-glycolic acid) (PLGA) and polyethylene glycol (PEG) are frequently used materials in wound dressings [102]. The major attraction of those synthetic polymers is depicted as their better electrospinnability and physical properties [86, 103].

Natural and synthetic polymer composites

For forming electrospun fibers by the above-mentioned biomacromolecules, adding reinforcing additives or processing is required [104]. For instance, alginate or collagen fibers alone do not represent good electrospinning properties and adequate mechanical resistance. Therefore, those biopolymers are hybridized with the incorporation of synthetic polymers [105–107], nanoparticles (NPs) [105, 108, 109] or cross-linkers [110–112] with in situ or post-blending approaches [81]. More and more composite mats consisting of various synthetic, synthetic/natural or natural polymers are used [113], and less single polymer-derived approaches are used due to the lack of desired physicochemical or regenerative properties. There are several studies in the recent literature that have electrospun polymer combinations for wound care (Table 11.1).

Table 11.1: Electrospinning of biocomposite fibers for wound healing applications.

Natural polymer	Synthetic polymer	ES method	Solvent/surfactant	API	Application	Ref.
Chitosan	PVP PEO	Blend Post-treatment: CL with genipin	Acetic acid (90% w/v) (chitosan); distilled water (PVP, PEO)	Graphene oxide	Skin grafts for wound healing	[114]
Chitosan	PCL	Emulsion Post-treatment: collagen coating; CL with NHS/EDC	Chl/methanol (PCL); acetic acid (90% w/v) (chitosan)	-	Bilayered wound dressing Skin substitute for acute burn wounds	[115]
Chitosan, grafted	PPC	Blend	HFIP/TFA	Curcumin	Wound dressing	[92]
Chitosan Gelatin	SMPU	Blend Post-treatment: CL with GTA	Formic acid	-	Smart wound dressing	[116]
Collagen	PCL	Coaxial	HFIP (collagen); methylene dichloride (PCL)	Chitosan-Ag-NPs Vitamin A palmitate	Wound dressing	[117]
Collagen	PCL	Blend	HFIP	-	Wound dressing	[118]
Gelatin	PCL	Blend	TFE	QAS	Wound dressing	[119]

(continued)

Table 11.1 (continued)

Natural polymer	Synthetic polymer	ES method	Solvent/surfactant	API	Application	Ref.
Gelatin	PCL	Emulsion Posttreatment: CL with GTA	Acetic acid (25% v/v) (gelatin); Chl/methanol (PCL) Surfactant: Span80	Ketoprofen	Wound dressing	[120]
Gelatin	PCL	Blend Posttreatment: EDC in ethanol (20 mM)	Acetic acid	CeO_2 NPs	Wound dressing	[121]
Gelatin	PHB	Blend Posttreatment: collagen coating	HFIP	OSA	Wound dressing	[61]
Gelatin	PU	Blend Posttreatment: CL with GTA	Formic acid (gelatin), DMF (PU)	Amoxicillin	Bilayered wound dressing	[122]
Gelatin	PVA	Blend Posttreatment: CL with GTA	Formic acid (88% w/v) (gelatin); water (PVA)	-	Bilayered wound dressing on HeraDerm®	[123]
GT	PCL	Blend	Acetic acid (90% v/v)	Curcumin	Wound dressing	[124]
GT	PLGA	Blend Coaxial	Blend: HFIP Coaxial: water (GT); HFIP (PLGA)	Tetracycline	Wound dressing	[125]

GT	PVA; PCL	Dual nozzle	Distilled water (GT/PVA); Chl/DMF (PCL)	-	Skin substitute	[126]
Na-alginate	*PVA*	*Blend*	*Distilled water*	*Ciprofloxacin*	*DDS (patch) for wound healing*	[103]
Starch	PCL	Coaxial	DMSO (starch); DMF (PCL)	-	Wound dressing	[127]
Starch	PVA	Blend	Water/ethanol	-	Wound dressing	[128]
Silk	*PEG*	*Emulsion Posttreatment: immersion in ethanol, distilled water*	*Deionized water Surfactant: CTAB, n-octadecane, Span 80*	-	*Wound dressing*	*[129]*
SF	*PEO*	*Blend*	*Ultrapure water*	*Manuka honey*	*Wound dressing*	[86]
SF	*PEO*	*Blend Posttreatment: ethanol vapor*	*Ultrapure water*	*Grape seed extract*	*Wound dressing*	[88]
SF	*PVA*	*Blend*	*Distilled water*	*rhEGF*	*Wound dressing*	[130]
SF	*PVA*	*Blend Posttreatment: ethanol vapor*	*Distilled water*	*EGF, bFGF, AMP-LL-37*	*Wound dressing*	[131]
Sericin	*PVA*	*Blend*	*Deionized water*	*Tigecycline*	*Wound dressing*	[90]
–	PDEGMA/PLCL	Blend	HFIP	Ciprofloxacin	Wound dressing	[132]
–	PNIPAAm/PLCL	Blend	HFIP	Ciprofloxacin	Wound dressing	[133]

(continued)

Table 11.1 (continued)

Natural polymer	Synthetic polymer	ES method	Solvent/surfactant	API	Application	Ref.
–	PLGA PVA	Emulsion	HFIP (PLGA); water (PVA)	*Aloe vera* extract/rhEGF	Wound dressing	[134]
–	PCL PEO	Blend	Chl/methanol	Chloramphenicol	Wound dressing	[135]
–	*PCL-b-MPEG* *PEO*	*Emulsion*	*Water*	–	*Biomedical*	[136]

Note: Examples summarized from recent literature.
AMP-LL-37, antimicrobial peptide LL-37; API, active pharmaceutical ingredient; Chl, chloroform; CL, cross-linking; CTAB, hexadecyl trimethylammonium bromide; DDS, drug delivery system; DMF, dimethylformamide; DMSO, dimethyl sulfoxide; EDC, 1-ethyl-3-(3-dimethylaminopropyl)-carbodiimide; EGF, epidermal growth factor; ES, electrospinning; EtOAc, ethyl acetate; GT, gum tragacanth; GTA, glutaraldehyde; HFIP, 1,1,1,3,3,3-hexafluoro-2-propanol; rhEGF, recombinant human epidermal growth factor; NHS, *N*-hydroxysuccinimide; NPs, nanoparticles; PCL, poly(ε-caprolactone); PCL-b-MPEG, poly [(ε-caprolactone)-block-(methoxypolyethylene glycol)] block copolymer; PDEGMA, poly(di(ethylene glycol) methyl ether methacrylate); PEG, poly(ethylene glycol); PHB, poly(3-hydroxybutyric acid); PLCL, poly(L-lactic acid-*co*-ε-caprolactone); PLGA, poly(lactic-*co*-glycolic acid); PNIPAAm, poly(*N*-isopropylacrylamide); PPC, poly(propylene carbonate); PU, polyurethane; SF, silk fibroin; SMPU, shape memory polyurethane; TFA, trifluoroacetic acid; TFE, 2,2,2-trifluoroethanol; QAS, trimethoxysilylpropyl octadecyldimethyl ammonium chloride; OSA, ostholamide. The work with nontoxic solvents has been given in italics.

Both synthetic and natural polymers have their advantages and disadvantages. Synthetic polymers are low cost, easy and tunable in production; however, they lack bioactive surfaces and have low hydrophilicity. On the other hand, natural polymers are highly biocompatible and biodegradable, but they do not offer the mechanical strength and tunable processing ability of the synthetic polymers [137, 138]. Biocomposite nanofibers consist of a combination of natural and synthetic polymers to enhance the mechanical and biological properties of the nanofibers and their use has gained attention lately. For example, Sultana et al. have developed PCL-based composites with poly(ethylene glycol) methyl ether (PEGME) or chitosan electrospun fibrous mats. The authors showed that PCL–PEGME mats had larger fibrous structure, which consequently induced the expression of genes responsible for the differentiation of endothelial cells [139].

Multifunctional composite nanofibrous mats for smart wound healing have been electrospun and proven to have the desired mechanical behavior, water vapor transmission ratio, surface wettability, antibacterial activity, cytocompatibility and hemostatic properties [116]. These mats consisted of blended natural protein chitosan and polysaccharide gelatin, and a synthetic shape memory polyurethane; the antimicrobial activity was obtained by post-electrospinning modification with silver nitrate solution [116]. Komur et al. have shown that PCL can be incorporated into starch nanofibers, where increasing the starch content improved the cell viability [127]. In this case, both advantages of natural and synthetic polymers are preserved making these good wound dressing materials. It has been shown that the presence of hydrophilic polymer improves the hydrophilicity of nanofibrous mat. While the uniform distribution of PCL in nanofibers enhances the strength and elongation of the PVA-gum tragacanth-PCL (blend-hybrid) fiber mat [126]. Biocomposite fibers using coaxial electrospinning enable to coat hydrophobic synthetic polymers (PCL) with hydrophilic natural polymers (gelatin) and enhance the fibroblast adhesion and proliferation important for wound healing [140].

Therapeutic agents – small molecules and macromolecules

Electrospun fiber matrix provides flexible platform for the therapeutic delivery due to its tunable characteristics, such as porosity and high surface area. The nanoscale size and other specific properties of the fibers enable high therapeutic loading mass transfer through the matrix. Variety of biomolecules (e.g., proteins and genes), namely, bovine serum albumin, lysozyme, epidermal growth factor (EGF), alkaline phosphatase are needed to provide an effective wound healing (Figure 11.4) with the aid of bioactive fiber mat [141].

In the literature, researchers have incorporated bioactive molecules with different design strategies in order to provide effective therapies. For instance, it is known that incorporating growth factors within or onto electrospun fiber mats can

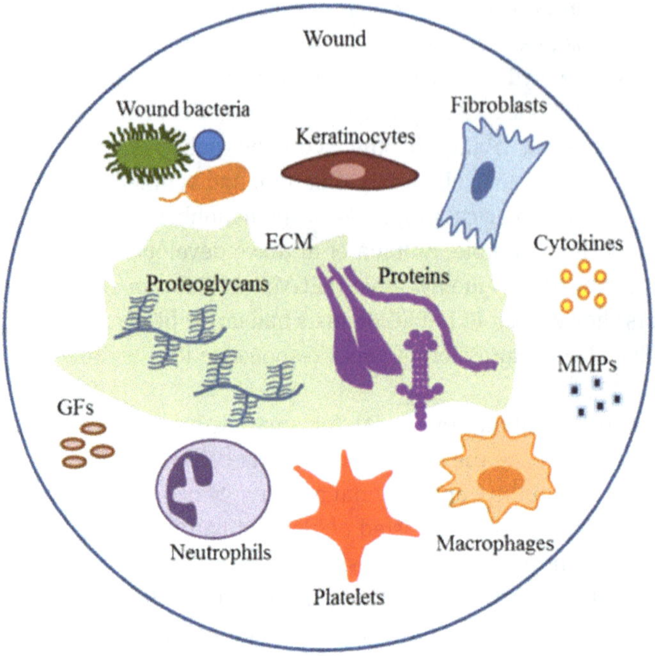

Figure 11.4: Schematics of bioactive molecules important for wound healing which can be used to design and prepare biocomposite fibers. Adapted from Ref. [6]. Copyright Wiley-VCH Verlag GmbH & Co. KgaA. Reproduced with permission. ECM- extracellular matrix; GFs-Growth factors; MMPs-metalloproteinases.

stimulate the stem cell recruitment, and thus cell proliferation and differentiation [141]. Therefore, the incorporation of the growth factors (such as vascular endothelial growth factor, basic fibroblast growth factor (bFGF) and platelet-derived growth factor-BB) has been frequently used [35, 45, 142]. These bioactive molecules can be introduced prior to the electrospinning or can be supplemented later on the electrospun fibers with surface modification [6]. Often bioactive molecules are electrospun with synthetic polymers or copolymers to improve their electrospinnability, biocompatibility as well as biodegradability. In addition to biomolecules delivery, electrospun mats are also commonly assigned for the therapeutics delivery [143, 144] (e.g., antibiotics [145], crude plant extracts [146], essential oils [147], plant-derived chemical components [144] and NPs [148]). This strategy is widely used in particular to release the active substances directly at the injury site to accelerate the wound healing [145].

Antiseptics, antibiotic and anti-inflammatory drugs are incorporated into electrospun fibers for wound healing in order to prevent or treat the wound infection or reduce the local inflammation in the wound, respectively. For instance, in ciprofloxacin-loaded sodium alginate/PLGA electrospun fibrous mats alginate is used to increase the hydrophilicity and to obtain a burst release of an antibiotic for wound healing [149]. Ciprofloxacin is used as an antibiotic in wound care, although for chronic wounds the use of local antibiotics should be less frequent due to possible resistance problems [150]. Thus, the use of antimicrobial substances, such as iodine or silver, in wound dressings is more recommended with chronic wounds and has been shown to have an effect against microbial biofilms [150].

Biodegradable nanofibers from poly(3-hydroxybutyrate-*co*-3-hydroxyvalerate) (PHBV) as a natural and thermoplastic aliphatic polyester synthesized by bacteria were investigated as wound dressing material with curcumin loading to prevent bacterial growth in chronic wound area [151]. The authors showed that these nanofibers were nontoxic and stimulated the cell growth in mouse fibroblasts [151]. Additionally, Dai et al. have also shown that the anti-inflammatory effect of curcumin can be utilized for wound healing in curcumin/gelatin-blended nanofiber mats [152].

Cell-based therapy

Cell-based therapies have encountered much interest during last years to enhance the skin wound healing [153]. These therapies enable to use suitable materials that support the normal wound healing and have the needed biocompatibility, biodegradability as well as safety. One of the first issues to solve is the selection of the cell origin; thus, encapsulation requires an appropriate source of functional cells. In this regard, the potential use of allogeneic versus xenogeneic cells has provoked important social and ethical debates [154]. The most used and studied polymer to prepare biocomposite fibers to be used as matrices for cell delivery is PCL. PCL-based nanofiber mats have proved to have the needed properties for cell delivery for wound healing applications. PCL and gelatin fibrous biocomposites exhibit significantly higher cell proliferation, greater cell spreading and integration within the nanofiber mats [153]. Several methods enable to encapsulate living cells into alginate beads, such as air flow, inkjet printing and microfluidic devices [155]. Recent developments in the electrospinning technology have led to the possibility that cells can be also successfully electrospun [156]. Biocomposite fiber mats containing living cells are a good opportunity to help the wounds to heal specifically when skin dermis is already damaged. Cell therapy is one of the methods that has been tested for the treatment of chronic wound infections and also for burn therapy [157]. There are several advantages that other therapeutic methods cannot achieve. However, the delivery of living cells requires to use electrospinning technology that enables also to maintain the viability of the cells [158]. Therefore, coaxial and emulsion electrospinning approaches are the most suitable and used ones.

In Figure 11.5, different setups have been summarized, which enable to electrospin living cells while preserving their viability, and even multiple-cell types can be electrospun into different layers. It has been proposed that the number of layers can be increased by increasing the number of needles used or by using a single needle which will spin a given cell type for the creation of that respective layer [159].

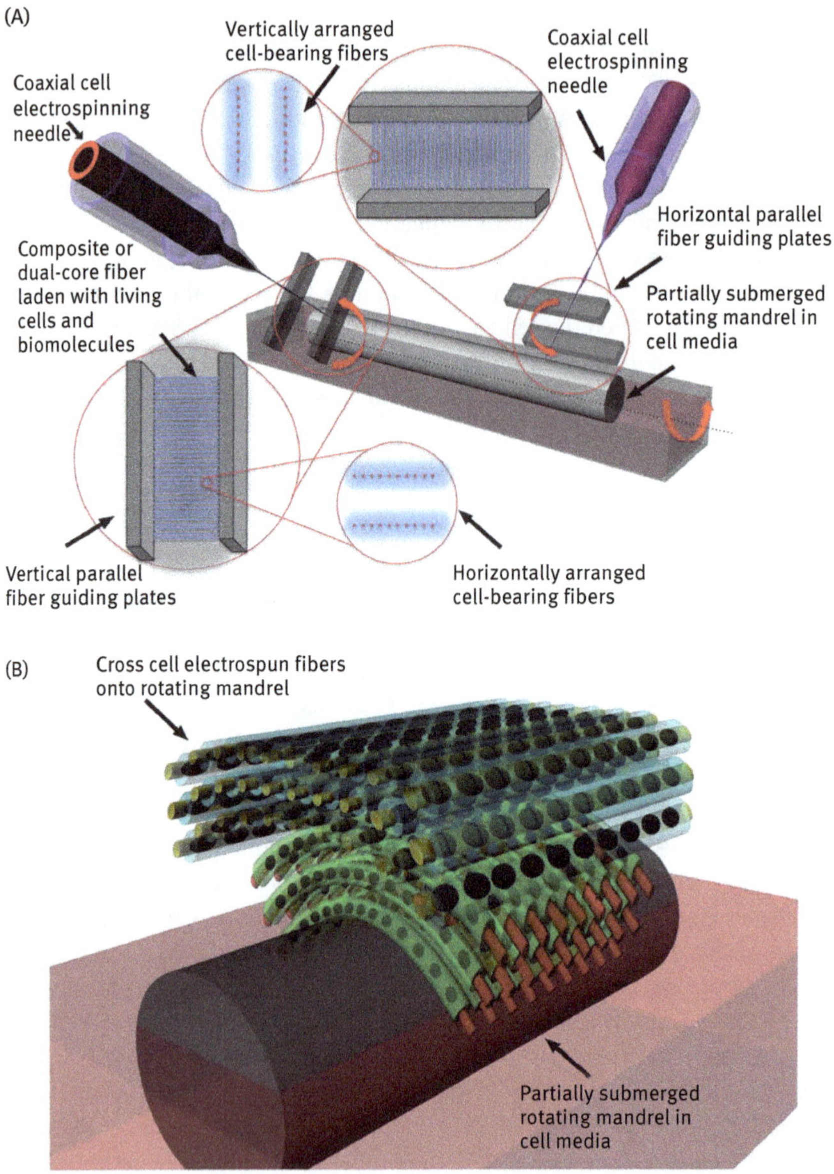

Figure 11.5: Schematic images (A) demonstrating the cell electrospinning equipment arrangement explored for generating cross-stitched living scaffolds for fabricating living vessel architectures and (B) a cross section of the finally generated cell-bearing construct. Reproduced from Ref. [159] with permission from the Royal Society of Chemistry.

Solvents for electrospinning biocomposite fibers

Solvents used for the preparation of biocomposite fibers need to be volatile and with low toxicity, in order to obtain a homogeneous solution of the components. Solvent largely determines the surface tension, electrical conductivity and viscosity of the polymer solution. For coaxial electrospinning, the selection is wider, since the core does not necessarily need to be a solution, but rather a dispersion. Electrospinning solutions need to be electrospinnable, which means that the solvent and polymer selection is one of the most important decisions [22, 160]. The use of less toxic solvents and solvents suggested as safe according to the U.S. Food and Drug Administration solvent recommendations [161] enables to conduct eco-friendly electrospinning to prepare safe and effective wound dressings.

According to Table 11.1, most often still quite toxic solvents are used for electrospinning of biocomposite fibers. After fabrication, the residual solvent inside the fibers should be removed by freeze-drying or drying under vacuum. It is one of the requirements for biomedical applications that the amount of residual solvent inside electrospun mats is less than the exposure limits regulated by international health authorities and institutes [160]. However, emulsion electrospinning does enable to use less toxic solvents in combination with distilled water. For green electrospinning applications, highly toxic organic fluorinated solvents are not used. In this case, distilled water is the main solvent to be used or melt electrospinning technique is selected for conducting solvent-free electrospinning. Green electrospinning may also involve the recycling of the solvents, if this could be integrated into the process, as it is currently done for the preparation of biocomposite films [162]. However, solvent recovery is a big challenge for the industrial-level production of ultrathin fibers due to volatile organic solvents. Novel approaches, such as UV-curing electrospinning [163], supercritical CO_2-assisted electrospinning [164], anion-curing electrospinning, thermocuring electrospinning or other solvent-free methods [165] could be used for the preparation of wound dressings without the use of solvents. While their use for biocomposite fibers consisting of bioactive substances/cells without losing the bioactivity still needs to be determined. Wound dressings prepared using solvent-free electrospinning or green electrospinning without any organic solvents (eco-friendly) have been listed in Table 11.2.

However, it must be stated that the actual borderline between green or not green electrospinning is not that strict. For example, Castilla-Casadiego et al. have discussed about green electrospinning of biomedical scaffolds, but they used acetic acid and water in order to prepare collagen and hydroxyapatite composite fibers [83]. The authors have also proposed that we could still discuss about green electrospinning with acetic acid, if the residual solvent amount is negligible and not toxic to the eukaryotic cells; thus, supporting the cell adhesion on the matrix and making it suitable for tissue regeneration. Most of the dressings are prepared by using water-soluble polymers that do not provide prolonged release of drugs or other

Table 11.2: Biocomposite fiber mats prepared using green electrospinning for wound healing applications.

Natural polymer	Synthetic polymer	ES method	Solvent	API	Application	Ref.
SF (20% w/v)	PEO (2% w/v)	Blend	Ultrapure water	Manuka honey (0–70% w/v)	Wound dressing	[86]
SF (20% w/v)	PEO (2% w/v)	Blend Post-treatment: ethanol vapor	Ultrapure water	Grape seed extract (0–5% w/v)	Wound dressing Tissue engineering Skin care	[88]
SF (30% w/v)	–	Blend	Ultrapure water	Vitamin B5 (4% w/w)	Skin care	[166]
SF (25% w/v)	–	Blend Post-treatment: ethanol vapor	Distilled water	Vitamin E (2–8% w/w)	Skin care	[167]
Sericin (2% w/w)	PVA (6% w/w)	Blend	Deionized water	Tigecycline (0.15% w/w)	Wound dressing	[90]
DAC (5–10% w/v)	PVA (0.75–7.5% w/v)	Blend	Water	–	Tissue engineering	[168]
Na-alginate (2% w/w)	PVA (8% w/w)	Blend	Distilled water	Ciprofloxacin (3.2% (w/v)	Transdermal patch	[103]

Note: Examples are summarized from the recent literature.
API, active pharmaceutical ingredient; DAC, "dialdehyde cellulose" (periodate-oxidized cellulose); ES, electrospinning; PEO, poly(ethyleneoxide); PVA, poly (vinyl alcohol); SF, silk fibroin.

bioactive substances. Therefore, even though green electrospinning can be performed during the production, additional cross-linking/curing steps are needed in order to provide prolonged release. Fan et al. have developed pantothenic acid-loaded silk fibroin composite nanofibrous mats using a green electrospinning technique together with ethanol-based post-spin treatment (Figure 11.6.).

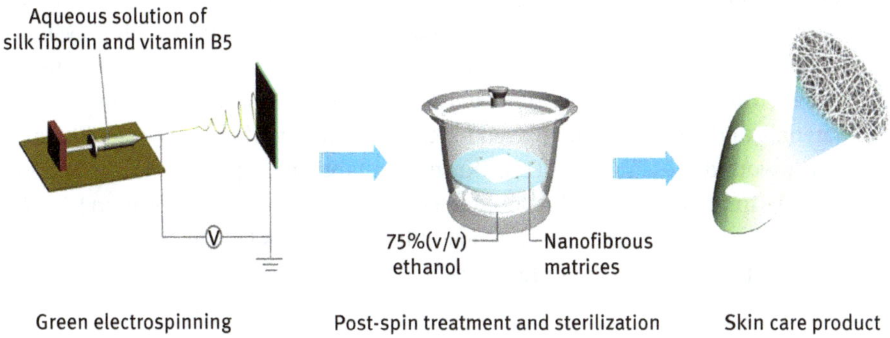

Figure 11.6: Green process of electrospinning pantothenic acid/silk fibroin composite nanofibers-based personal skin care products. Reprinted from Ref. [166]. Copyright (2014), with permission from Elsevier.

Since there are different cross-linking options available [16], this allows eco-friendly manufacturing of novel wound dressings. The created biocomposite materials should be able to combine the merits of both for attaining the desired biological and physicochemical characteristics for the aimed application.

11.3 Modular design options with electrospun biocomposite fibers for wound healing

It has been proposed that future wound dressings are multifunctional, biodegradable medical devices incorporating regenerative engineering, infection prevention, treatment and control, stimulatory aspects and sensors to provide real-time feedback [169]. This means that experts from different fields and interdisciplinary collaborations are needed in order to design functional wound dressings considering the clinical as well as material-related factors important for normal wound healing. Advanced wound dressings are not only to cover the wound but interact with the wound environment and help the wound to heal. An ideal wound dressing should mimic the structure and biological function of the native ECM proteins, which provide support and regulate cellular activities [170].

Consequently, more and more types of modern wound dressings involving drugs (i.e., antimicrobial, anti-inflammatory, anesthetics), bioactive molecules (i.e., proteins and peptides), nutrients and diagnostic reagents are developed in order to provide modular designs with electrospun fibers for wound healing [117]. Therefore, novel design strategies are often used with layered structures and combinations of different materials as well as preparation methods. In recent reports, scientists have been able to present modular design options with electrospun biocomposite fibers. Electrospinning technology enables to design modular biocomposite fiber systems and achieve the needed functionality and quality which cannot be reached by using these single unit building blocks separately. As discussed in previous sections, various materials and methods can be used for the preparation of biocomposite fibers as fiber mats, wound dressings or DDSs. During the fabrication process, the polymeric materials are functionalized and tailored to comply with the final application [17, 171, 172]. The core–shell structures are the most interesting ones for the encapsulation of functional biological molecules or living cells into the fibers [173].

11.3.1 Design strategies of functionalized electrospun biocomposite fibers

Electrospun fibers for wound healing applications can be functionalized by adding active compounds into the polymer blend and by using multiple drugs with different release kinetics simultaneously [174–176]. These electrospun fiber mats can contain small molecule drugs [82, 91, 103], extract from natural sources [86, 88], silver NPs [91, 94, 95], proteins, peptides and gene vectors. Recent studies have also reported the loading of gold NPs [177]; mesoporous silica NPs [178] and carbon nanotubes (CNTs) [179] into the biocomposite nanofibers via blend electrospinning. Hence, a large variety of materials can be used to produce electrospun fibers in the pursuit of medical fabrics in wound dressing. For wound healing, all these substances have a specific role, and therefore functionalized fibers are obtained.

New wound dressings do not consist of purely passive, biocompatible absorbent polymers. They rather contain functional materials that aim to actively encourage the healing processes and minimize the risk of infection. For instance, Wei et al. presented core–shell structures (vitamin A palmitate releasing core and surrounding shell with Ag-NPs) in order to provide heal-promoting and antibacterial activity simultaneously [117]. By this way, multifunctional nanofibrous dressings that show a promising potential for clinical applications could be prepared. For infection diagnosis, there are a large number of systems that could be applied for wound monitoring, such as pH, bacterial metabolites, wound biomarkers and temperature. Smart wound dressings with intelligent electrochemical sensors have been developed, which monitor the pH in the wound and give signal to the patient [5].

Novel nanofiber designs are used to improve the properties of nanofibrous wound dressings [92]. Different NPs-in-nanofiber systems have been developed [180, 181]. Nanofibrous electrospun grafts have been investigated for multiple applications in wound healing due to its versatile and controllable grafting parameters, ability to produce continuous fibers, low cost and easy application [182, 183]. Especially the tissue engineering stage of wound healing is challenging, since the designed wound dressings should mimic the ECM, which plays a major role in cell attachment, alignment, migration, proliferation and differentiation [137]. In this regard, ultrafine nanofiber mats with submicron and nanoscale fiber diameters that relate to the fiber mat characteristics of the natural ECM can be achieved with electrospinning.

Different strategies and methods enable to produce and mimic the real *in vivo* tissues; electrospinning of biocomposite fibers is one of such methods [141, 153]. As mentioned earlier, normal wound healing is a dynamic process that consists of four main and overlapping stages that need to be kept in mind when designing bioactive wound dressings [184, 185]. It is utmost important to diagnose the wound, understand the properties of wounds and then design the wound dressing accordingly. Hence, dressing material–wound interactions at the molecular level need to be revealed during the characterization of the wound dressings. Furthermore, the analysis in not only *in vitro* but also *in vivo*-mimicking conditions enables to make conclusions about the properties and functionality of wound dressings in biorelevant conditions. Active wound dressings using biocomposite fiber mats are not only to protect from microbial contamination, physical protection and moisture retention, but also to support normal wound healing and provide important biological molecules and/or active substances (e.g., antimicrobial agents) for the wound.

The development and properties of electrospun nanofibers for the wound healing application have been reviewed in recent literature [47, 144, 186, 187]. It is interesting that there are many papers and also several patents available about electrospinning technology for the preparation of fibrous wound dressings. However, not many clinically relevant products are yet on the market, but several medical devices and most likely drug preparations are under the development, and it is expected that these will be launched within the next few years. One of the first electrospun medical devices for wound healing application on the market is Restrata Wound Matrix [188, 189]. The product has been proposed to be successfully used for the treatments of partial and full thickness wounds, chronic wounds (e.g., ulcers) and severe wounds caused by trauma or surgery [189]. Interestingly, this electrospun matrix consists of only biocompatible synthetic polymers (i.e., polyglactin 910 and polydioxanone). Indeed, fully synthetic wound matrix materials have the advantage of better control over the material and mechanical properties. Thus, the risk for disease transmission is low; however, they commonly lack native tissue architecture and suitable biocompatibility profiles compared to biologic matrices [188].

The design of biocomposite electrospun nanofibers as wound dressings should meet the requirements of wound healing process. Even more difficult is this task due to the fact that it is not even known yet what exactly is happening during the wound healing at the molecular level, and new discoveries elucidating mechanisms of physiologic wound repair are constantly being reported [190]. Specific strategies need to be developed for chronic wounds, since all the normal wound healing stages are disturbed and need to be adjusted according to the wound case by case or the dressing should be designed to respond to the wound environment [191]. There are different strategies proposed for the development of novel wound dressings consisting of biocomposite fibers. Some of these strategies and electrospinning technologies for wound dressing development are discussed.

Three-dimensional structured electrospun biocomposite fibers

Conventional electrospinning techniques produce two-dimensional (2D) densely packed nanofiber membranes that restrict the cellular penetration and thus limit their use for *in situ* dermal tissue. On the other hand, 3D matrices composed of electrospun nanofibers promote cellular infiltration and could serve as an improved synthetic matrix for dermal repair/regeneration. Among the existing approaches in the literature, Shing et al. fabricated 3D electrospun PCL nanofiber mats with a larger pore size compared with traditional 2D nanofiber matrices, which improved dermal fibroblast infiltration and proliferation throughout the nanofiber and the secretion of ECM proteins from the cells, showing potential for dermal tissue engineering [192]. Another novel approach that enables to produce 3D structures has been reported by Chen et al. [193]. In their study, liquid collector was used to produce electrospun 3D fibrous mats. Materials selected were biodegradable synthetic polymer PLGA and the surface was chemically modified with collagen type I (natural ECM protein). Hence, biologically active matrix was obtained, which provided both structural support and biological recognition for cell growth and recruitment which are important for the wound healing [193]. This means that more prolonged and controlled release of growth factors or APIs can be obtained. For example, Xie et al. incorporated different growth factors into chitosan/PEO nanofibers [180]. These systems were tested in full-thickness rat skin wound models, and it was proved that NPs/nanofiber wound dressings significantly accelerated the wound healing process by promoting angiogenesis, increasing re-epithelialization and controlling granulation tissue formation [180]. For more complex chronic wounds, novel strategies are being searched. For example, Yang et al. incorporated polyplexes of basic fibroblast growth factor-encoding plasmid (pbFGF) with poly(ethylene imine) into electrospun fibers with a core–sheath structure, and PEG was included into the fiber sheath to allow a sustained release of pbFGF for 4 weeks. *In vitro* tests on mouse embryo fibroblasts revealed that pbFGF-loaded fibrous mats enhanced cell proliferation [194].

Nanoparticles-in-nanofiber systems

Several nano-in-nano systems have been prepared also by emulsion electrospinning [42, 195, 196] and coaxial electrospinning [117]. Figure 11.7 shows a schematic picture about NPs-in-nanofibers system. For example, Spano et al. encapsulated inorganic nanocrystals and human antibodies into nontoxic core–shell composites with polypropylene carbonate, polyethylene and sericin protein [196]. The authors concluded that the polymer ratios in the systems tune the degradation of fibers and the release kinetics of the loaded molecules/particles. Another recent study in the literature was carried out by Ali et al. [181]. In their study, they prepared differently designed NPs of Pluronic nanomicelles and PLGA NPs with phenytoin loading (Figure 11.7). Furthermore, they aimed also to control the drug release profiles by additional coating with lecithin and by altering the electrospinning parameters to obtain NPs incorporated into electrospun biocomposite nanofibers. The prepared nanocarriers attained a high drug entrapment efficiency exceeding 94% and showed promising sustained release profiles compared to free API. Results also demonstrated that nanofibers incorporating the optimized lecithin-coated phenytoin-loaded PLGA NPs could be the most promising candidate for efficient wound healing. These NPs-in-nanofiber systems conferred a well-controlled and sustained release of the drug over 9 days (Figure 11.7). Moreover, they showed the best re-epithelization and healing quality during the *in vivo* study with minimal inflammatory and necrotic cells formation. Hence, electrospun NPs-in-nanofiber systems show huge potential in preparing electrospun biocomposites with prolonged release.

In addition, NP-based dressings are used frequently. These include NPs incorporated into polymer matrix [197]. Yang et al. have proposed a novel strategy to fight against wound infections and developed 6-aminopenicillanic acid-coated gold NPs loaded into gelatin/PCL nanofibers [198]. Rath et al. have utilized the antibiotic effect of silver NPs with biocompatible and bioadhesive natural protein collagen in order to prepare electrospun nanofiber mats [148]. *In vivo* study has demonstrated that NP-incorporated nanofibers accelerate the wound healing by increasing the cell growth in fibroblast cells and inducing collagen deposition. Recently, Liao et al. have proposed a novel and interesting approach through combining PVA with CNTs and EGF by electrospinning on gauze [199]. CNTs were used as a carrier material coated with EGF, and PVA was used as a modifier and dispersant for CNTs. The authors showed that accelerated skin wound healing was obtained due to the specific structure and composition of the biocomposites; thus, no cytotoxicity was revealed.

One of the interesting approaches that enable to modify the release rates of active substances from biocomposite fibers is the use of liposomes. Liposomes alone have a short retention time and washout from the site of application, but the incorporation into nanofibers enables to develop local DDSs [200] and help in the delivery of other bioactive molecules important for wound healing. Antibacterial and antimicrobial agents incorporated into the liposomes and incorporated into the

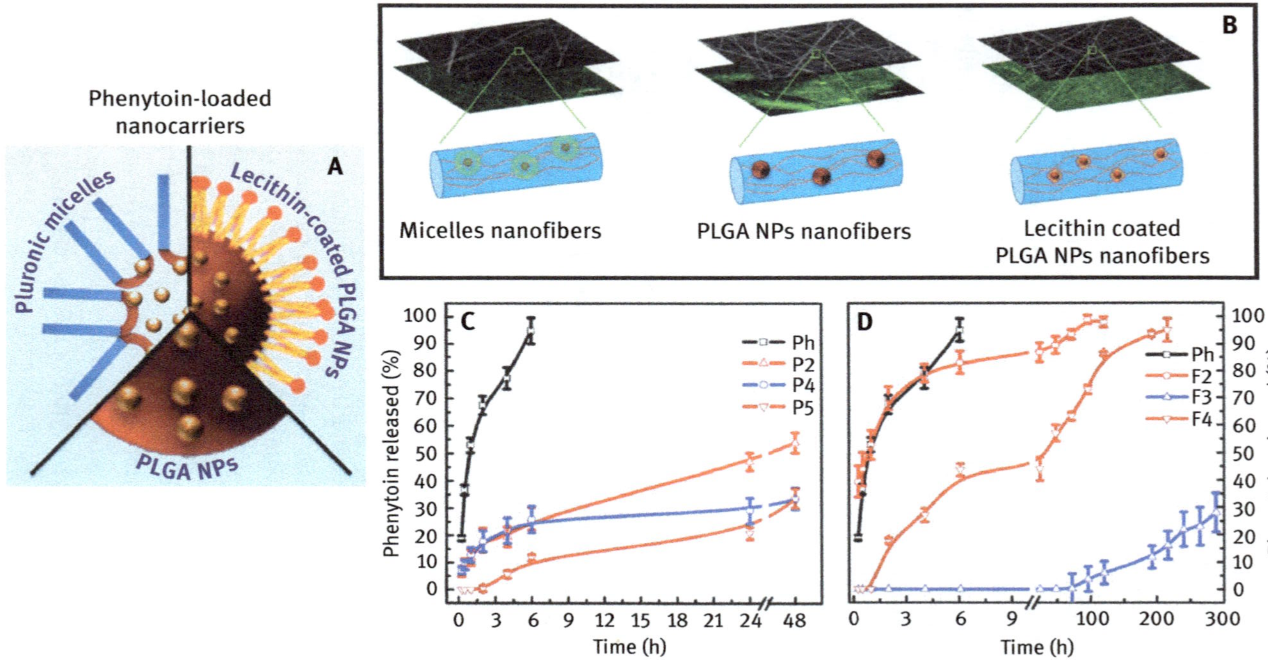

Figure 11.7: (A) Phenytoin (Ph)-loaded nanocarriers (micelles, poly(lactic-*co*-glycolic) acid nanoparticles (PLGA NPs), lecithin-coated PLGA NPs). (B) Nanocarriers loaded into nanofibers (micelles in nanofibers, PLGA NPs in nanofibers, lecithin-coated PLGA NPs in nanofibers). (C) Cumulative release profiles of Ph from micelles (P2), PLGA-NPs (P4), lecithin-coated PLGA-NPs (P5). (D) Cumulative release profiles of Ph from micelles in nanofibers (F2), PLGA NPs in nanofibers (F3), lecithin-coated PLGA NPs in nanofibers (F4). Adapted with permission from Ref. [181]. Copyright (2018) American Chemical Society.

nanofibers have been shown by several authors [201–204]. The method has been shown to be applicable for wound infections and wound healing treatment.

Staffa et al. (2017) have developed a liposomes-in-nanofiber systems using needleless electrospinning technology [200]. The authors incorporated several growth factors into nanofiber scaffold by using PCL as a polymer and chloroform/ethanol as a solvent mixture for electrospinning. Bioactive liposome-loaded nanofiber mats were obtained with controlled and more sustained release of growth factors compared to growth factor-loaded nanofibers.

Layered systems

An interesting bilayer fiber mat with PCL–chitosan nanofibrous membrane on 3D cotton-wool-like fluffy PCL–chitosan scaffold was developed by Pal et al. [115]. The authors showed that different solvents used for the preparation of polymer mixture resulted in an emulsion system, and consequent electrospinning allowed producing biocomposite fibers for wound healing applications. Furthermore, the nanofiber surface was impregnated with collagen I to improve the biofunctionality of the matrices. The developed bilayered mats consisted of core–shell structured fibers and the scaffold retained its 3D structure after collagen percolation within it. Such bilayered biofunctional 3D scaffold promoted efficient fibroblast cell attachment, infiltration and proliferation from initial days of cell seeding, eventually covering the entire fiber structure, while keratinocytes grew only on the surface of the nanofibrous membrane. The mats showed faster healing, re-epithelialization and mature dermal tissue development, and suitability as skin substitute for healing acute burn wounds (Figure 11.8(A)) [115].

Mirzaei et al. (2013) have patented a multilayered electrospun matrix, which was developed as a wound dressing [205]. The latter comprises three electrospun layers and consists of natural origin extract *Melilotus officinalis* developed for local wound healing. The outer hydrophilic polymeric layer acts as a support layer. The middle layer comprises a genipin cross-linked reservoir layer loaded with a herbal extract of *M. officinalis* and is cross-linked to control a release of the extract in an aqueous environment. The third layer is a wound contacting inner layer comprising of a noncross-linked chitosan-based nanofibrous layer that is hydrophilic in nature [205]. No results about its efficacy were provided in the initial patent application.

Besides the fabrication of homogeneous fiber mats with uniform distribution of the materials, double-layered [91] or sandwich-type [82] structures can be prepared as functionalized wound dressings. Croisier et al. have prepared polysaccharide-based nanofibers (chitosan and hyaluronic acid) with a multilayered structure by combining electrospinning and layer-by-layer deposition techniques [206]. In their study, authors have shown that a multilayered structure is achieved by alternating the deposition of the positively charged chitosan with the deposition of a negatively

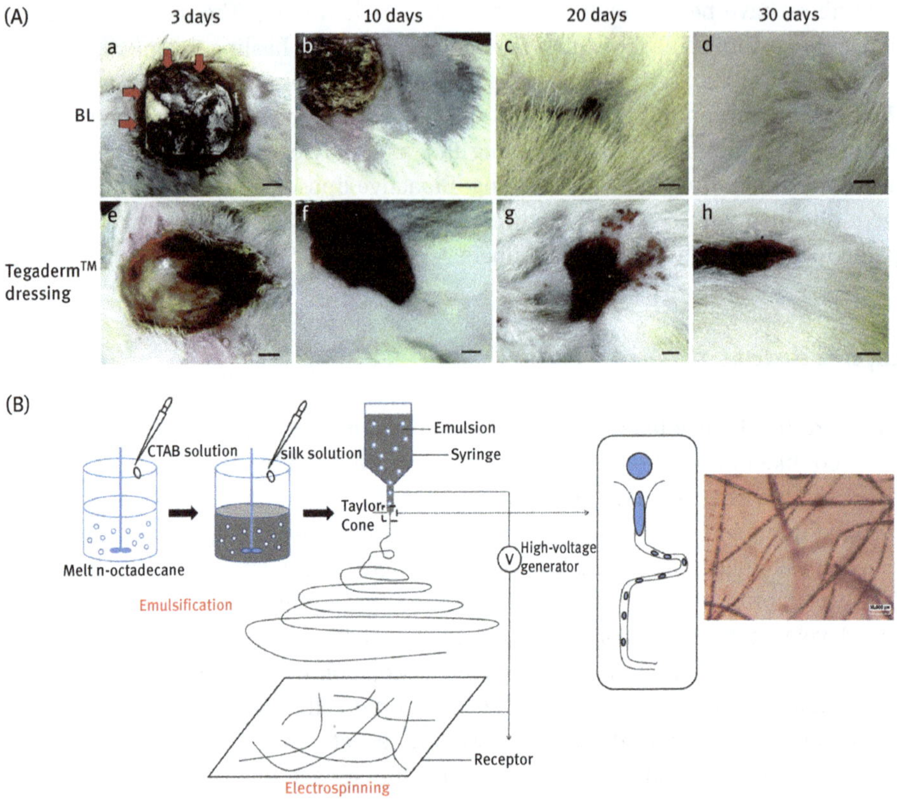

Figure 11.8: (A) Wound healing effects of the bilayered scaffold (BL) compared to Tegaderm™ standard dressing as control in rat burn model. (a-–h) Optical photographs showing progression of healing of the burn wound treated with the scaffold and control after 3, 10, 20 and 30 days. Adapted from Ref. [115] with permission from the Royal Society of Chemistry. (B) Emulsion electrospinning of composite fibers [129]. Reprinted from Ref. [129]. Copyright © 2017, John Wiley and Sons. Reproduced with permission.

charged polyelectrolyte. The mats had both the mechanical resistances typical to the core material PCL and also desired surface properties, biocompatibility and hydrophilicity of chitosan.

Complex and combined systems

Emulsion electrospinning has been successfully used in order to produce novel biocomposite fiber materials to be used in wound healing applications using environment-friendly electrospinning (Figure 11.8(B)). Zhao et al. have electrospun *n*-octadecane and silk and revealed that the developed material has the capacity to store or release the thermal energy through the reversible phase transitions of *n*-octadecane within the fiber [129]. These composites have shown to exhibit the capability to regulate the interior temperature as the ambient temperature alters; therefore, this material has a huge potential to be used as a wound dressing material.

A more complex system consisting of various components important for wound healing, for example, silk fibroin blended with PVA and functionalized with growth factors and LL-37 antimicrobial peptide have been electrospun recently by Chouhan et al. [131]. This bioactive wound dressing enabled to regulate ECM deposition, which resulted in faster and complete repair of chronic diabetic cutaneous wounds. Ravichandran et al. have prepared a combination of nanofibrous mats comprising PLA/poly(aspartic acid)/collagen with stem cell therapy along with the inclusion of bFGF to promote skin regeneration [170].

Among other approaches, nanofiber constructs using dendrimer conjugates have been reported. Dongargaonkar et al. have electrospun blends of gelatin and gelatin–dendrimer conjugates (branched star-shaped polyamidoamine dendrimer) [207]. Blends of gelatin and gelatin–dendrimer conjugates were electrospun together with silver acetate into nanofiber constructs and further converted into semi-interpenetrating networks with photoreactive PEG diacrylate. According to the authors, these constructs showed good wound dressing properties and antimicrobial activity.

Aligned fibers

Usually nanofibers are collected as nonwoven or randomly arranged structures, due to the "whipping instability" of the electrospinning jet and the use of planar stationary plate. However, when the electrospinning conditions are modified, either with the collector or by the manipulation of electric field, then aligned nanofibers are formed [14]. Alignment of fibers has been proposed as one of the key properties when tissue scaffolds and wound dressings are developed. Uniaxial and radial alignments improve the cell attachment compared to random orientation [208]. Uniaxially aligned nanofiber mats promote cell migration along one specific direction, and radially aligned fibers promote cell migration from the surrounding tissue to the center having improved properties for wound healing (e.g., shorter healing time) [209]. Therefore, novel strategies allow obtaining structures that have superior properties compared to the traditional structures used in dressings. Xie et al. have shown that electrospun PCL nanofibers can be aligned radially to generate fiber mats potentially useful as dural substitutes (Figure 11.9) [209]. For wound dressing applications, on the other hand, the resemblance of randomly oriented fibers with the ECM creates a highly porous scaffold for cellular infiltration and vascularization which is important for wound repair. Dressings consisting of complex engineered biocomposite fiber layers of different geometries most likely enable to take into account several mentioned features and improve the properties even more.

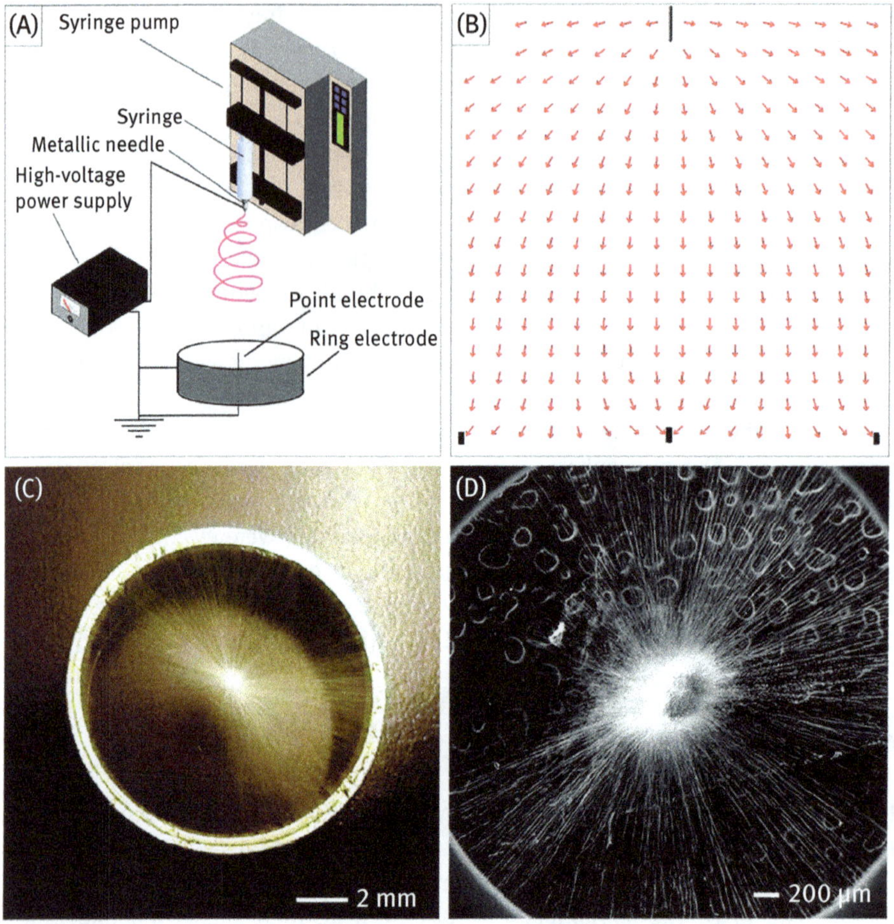

Figure 11.9: (A) Electrospinning setup for generating fiber mats consisting of radially aligned nanofibers. (B) Electric field strength vectors calculated for the region between the spinneret and the collector. (C) Photograph of a scaffold of radially aligned nanofibers directly deposited on the ring collector. (D) Scanning electron microscopy image showing the radial alignment for the nanofibers in the scaffold. The circular features behind the nanofibers were dents on the conductive carbon tape. Reprinted with permission from Ref. [209]. Copyright (2010) American Chemical Society.

11.4 Characterization of biocomposite fibers

Several characterization methods are employed to investigate the physicochemical properties of electrospun biocomposite fibers for wound healing applications. The commonly used techniques to characterize electrospun fiber mats and their biological activity will be discussed in this part by also considering their pitfalls in practice.

Some of the most used physical parameters that are provided include: structure characterization, morphology analysis and mechanical properties of biocomposite fiber mats. When biocomposites are prepared for wound healing application, then tests that really illustrate the performance and efficacy of the electrospun fiber mats need to be conducted. These tests include cell culture testing as well as *ex vivo/in vivo* tests. Later phases also *in vivo* human trials are conducted.

11.4.1 Biocomposite fiber structure characterization

For the evaluation of the morphology of the electrospun fibers, imaging methods are commonly used. The typical imaging methods are optical microscopy in the visible range, scanning electron microscopy, transmission electron microscopy and atomic force microscopy. The main advantage of the imaging methods is that the structure can be directly visualized at various places of the sample. However, in order to conduct quantitative comparison between the different samples, the image analysis with the additional statistical analysis is needed [210]. Additionally, the quantitative topographic properties of the total electrospun fiber mat, such as surface roughness, can be obtained directly from the scanning white light interferometry profiles [211]. Recent developments in the X-ray computed tomography have increased its feasibility in the 3D structure profiling of composites together with image-based modeling [212]. Furthermore, fluorescent and confocal microscopy techniques are applied to visualize the cell attachment and growth on the fibrous mats. For imaging the inner morphology of the fibers, additional preparation processes are used. For example, cross-sectional visualization can be obtained by cutting of fibers in liquid N_2 with a surgical blade [213].

The electrospun fibers with a diameter in nano- to micrometer scale leads to the formation of fiber mats with a high surface area to mass ratio. Hence the surface area can be increased by creating pores and/or dimples on the surface of the fibers. For the determination of the surface area of electrospun fibers, the pore size and porosity of the fiber mats can be identified by the following methods: mercury porosimetry [135, 214, 215], capillary flow porometry [216, 217], gas sorption [218], pycnometry [214] or gravimetric techniques [26, 219]. During wound healing, the favorable environment for the adhesion and growth of different cell types is dependent on the total porosity and the pore size distribution [214]. In DDSs, the porosity and packing density of the fibers in the mats affect the drug release kinetics, since it is influenced by the liquid penetration rate and the available surface area [26, 215].

The mechanical properties of the fiber mats are determined typically by tensile test, where the tensile strength and elastic modulus can be obtained from the stress–strain curve. Additionally, in a compression test, the compression or burst strength and elongation of the fiber mats can be determined [214, 220]. Furthermore, image-based modeling can be applied to estimate the elastic modulus and tensile properties of interconnected fiber meshes [221]. The test conditions and the pretreatment of the samples vary depending on the used instruments. The samples are usually cut out into rectangular-shaped strips of various sizes. Alternatively, the dumbbell-shaped samples have been measured [35]. As expected, the measurement environment needs to be adjusted to the storage and handling conditions. The mechanical testing at controlled temperature and humidity allows for evaluating the properties in biorelevant conditions [179]. Furthermore, the physical strength of the biocomposite fiber mats could be determined after wetting or immersion in water or physiological fluids [35, 222].

11.4.2 Solid-state properties of electrospun biocomposite fibers

The solid-state characterization of DDSs is highly important for the evaluation of their therapeutic performance. The crystallinity of the polymers in the fibers affects the mechanical strength, whereas the solid state of the active ingredients dictates their thermodynamic stability, release rate, appearance and bioavailability. The most used approaches for studying the solid-state properties of materials are thermal methods, X-ray diffraction (XRD) and spectroscopic methods [e.g., infrared (IR), Raman spectroscopy and solid-state nuclear magnetic resonance].

The analytical methods for evaluating the thermal behavior of electrospun fibers include differential scanning calorimetry (DSC) and thermogravimetric analysis (TGA). The DSC is suitable for determining the solid-state properties of the composition, encapsulation efficiency of crystalline compounds and the interactions between the materials (e.g., plasticizing effect and hydrogen bonding). The analyses of the composites enable to appoint suitable ratios of polymers to avoid phase separation [153] and the amount of cross-linking agents for the improved physical stability [72, 223]. DSC is widely used for learning about the solid state of the APIs in the fibers [56, 224]. However, one must keep in mind that the crystalline fractions under the minimum detection limit may still exist. Raman mapping has been suggested to be an alternative method with higher sensitivity for detecting small quantities of crystalline materials [56]. On the other hand, TGA is mainly used to evaluate the stability and solvent evaporation of formulations. If the thermal behavior of the components is distinguishable from each other, the relative content (loading) of the materials can be identified. For example, TGA was used to quantify the content of hydrophobic compounds that were encapsulated into the core structure of fibers by emulsion electrospinning [195]. TGA has also been exploited to evaluate the weight loss of starch-based nanofibers in a wound-mimicking environment over the period of 28 days [128]. The recorded weight loss due to increased temperature enables to evaluate the effect of polymer mixtures, additives (e.g., drugs and biomolecules), posttreatment cross-linking and occurring interactions on the stability of the electrospun fiber mats [69, 103].

XRD is a well-established method for determining the crystalline structure of materials and their mixtures. Crystalline solids have a well-defined molecular order and specific reflections in diffractograms, which are lacking for the materials in the amorphous state (an amorphous halo is recorded on the diffraction pattern). XRD has been typically used to detect the changes in the crystallinity of initial materials after electrospinning [225–227]. The encapsulation efficiency of active agents in core–shell fibers can be obtained through the ascertainment of crystallization limits [228]. As expected, the XRD is widely applied for confirming the formation of amorphous solid dispersions after blend electrospinning [135, 224, 225], melt electrospinning [56, 57] and even high-speed electrospinning [229].

The IR spectroscopy enables to study the composition of the electrospun fibers. Typically, the attenuated total reflectance accessory is used for the surface characterization of the fiber mats (penetration depth of few micrometers) without any sample pretreatment. This method has been used successfully to detect polymer–polymer or polymer–active compound interactions, such as hydrogen bonding, after electrospinning [44, 135, 177, 227]. Furthermore, the effect of surface functionalization [230] and cross-linking can be detected, which provides valuable information about the mechanical strength and toxicity (residual cross-linkers) of the matrices [71, 72, 128, 231]. IR spectra provide information about the drug distribution and the degradation, for example, after melt electrospinning [57]. IR spectroscopy can also be applied to confirm the removal of pore-forming agents [214] or changes to the fiber mats after the immersion in cell medium [231].

Raman spectroscopy is considered as a complementary method to IR spectroscopy. It can be applied for the identification and quantification of molecular structures of chemical entities and the occurring changes after electrospinning, post-treatment and subsequent storage. The advanced method of Raman scattering microspectroscopy mapping can be used to evaluate the distribution of drugs or other Raman-active molecules within the electrospun fibers at high method sensitivity [56, 135]. Furthermore, near-IR [232, 233] and Raman spectroscopy [234–236] handheld devices and probes for in-line process monitoring are available to control the quality of the manufacturing and the final products.

11.4.3 Other relevant properties

Recent literature reveals that electrospun fiber mats provide an excellent platform for sustaining the release of drugs and other active substances (e.g., growth factors). Differently designed electrospun fiber mats are specifically suitable for hydrophilic small molecule drugs that are highly loaded in fiber formulations [237]. For wound healing application, specifically for wound infection prevention and treatment, this approach is of relevance and provides better treatment outcome [135, 238]. Antimicrobial agents and antimicrobial peptides release can be modified by modifying the electrospun layer thickness [215] or by different dressing design [239, 240]. It has been proposed that in addition to the control of drug release from electrospun nanofibers by drug diffusion, dissolution and/or physical desorption rates, diameters of fibers (lengths of diffusion barrier), and/or polymer degradation/erosion rates, also smart electrospun fiber mats can be designed [241]. These mats are able to respond to changes in environmental parameters including pH value, temperature, light, electrical field, and magnetic field, which can tailor the drug release rate [241].

Release studies are usually conducted using modified dissolution studies in small vials by immersing the fiber mats in buffered solutions. The API concentrations are analyzed usually by high-performance liquid chromatography or UV-Visible spectrophotometry. Novel approaches that enable to avoid the drawbacks of conventional dissolution testing have also been tested, such as fully automated fiberoptics-based dissolution testing system for *in situ* monitoring of drug release from electrospun fiber mats [242] and UV-imaging technique [243].

In addition to storage stability studies that are performed with every dosage form/DDS and medical device to guarantee the quality of the product, wettability and surface charge are also measured in order to understand the properties of biocomposite fibers as wound dressings and predict their *in vivo* behavior. Moreover, not only the electrospun fiber and *in vivo* tissue interactions, but also bacteria and electrospun fiber mat interactions can be predicted from these parameters. It has been shown that fiber wettability, surface charge and chemistry largely affect the ability of bacterial cells to attach and proliferate throughout the nanofiber mats [244]. This is of real importance while developing antimicrobial wound dressings.

Usually wettability is measured using conventional contact angle measurement. Surface charge measurements have been conducted both with Zetasizer (Zeta Potential analysis is a technique for determining the surface charge) as well as newly developed approaches, such as streaming current measurement in a microfluidic device [245]. Since electrospun biocomposite fiber mats are developed as wound dressings, also all other tests relevant for wound healing in general need to be investigated (i.e., fluid handling capability, dehydration rate, vertical wicking, absorption behavior and pH values of the dressings, bioadhesiveness, etc.) [246]. Among other properties, oxygen and water vapor permeability are the most important ones, since these parameters are known to have superior values for electrospun fibers compared to other dressings [247]. Even oxygen sensors have been developed using nanofiber mats applicable for sensing oxygen levels [248] important for wound healing [249]. Several recent papers (Table 11.1) have characterized electrospun nanofiber mats using the above-mentioned techniques and proposed their suitability for the management of wound repair [250, 251].

11.4.4 Electrospun fibers–biological environment interactions

The interaction of biocomposite fibers and biological systems is correlated with their physicochemical properties. Biocomposite fibers intended to be used in wound healing should possess well-organized structure that allows cellular adhesion, sufficient penetration for growth and new tissue formation [252]. However, inappropriate immune responses in the host can limit its proper functioning which might be caused by the incompatibility of composite material with the host. Hence, the physicochemical and structural properties of those electrospun biocomposite fibers have a great impact on the behavior of the materials in a biological environment.

In general, the interactions of any biocomposite materials can be examined by evaluating their biocompatibility, biodegradation and biological activity. Biocompatibility is defined by the International Union of Pure and Applied Chemistry as the "ability to be in contact with a living system without producing an adverse effect" [253]. A plethora of tests are needed to deem the biocomposite material as biocompatible which usually starts by considering the intended patient contact and duration of the contact. In the case of using biocomposite fibers for wound healing even though the endpoint and readout can differ, similar investigations are needed to verify the biocompatibility of fibers. The biocompatibility of materials that in the final product are intended for the clinical use should be assessed according to the provided guidance by the authorities [i.e., International Organization for Standardization (ISO) series of guidance documents for biocompatibility, ISO 10,993]. Biocompatibility investigations should cover the appropriateness of the materials assured with the cytotoxicity, sensitization, irritation/intracutaneous creativity, systemic toxicity, genotoxicity, implantation and hemocompatibility investigations [254].

Regarding the biodegradability and bioresorbability, biocomposite fibers are expected to decompose naturally but their degraded products remain inside the human body. For bioresorbable materials, they will degrade after a certain period of time of implantation, and nontoxic products will be produced during their metabolism and/or elimination. With reference to the definitions of these two phenomena different degradation mechanisms are possible. Especially in the biological environment, water plays an important role in the degradation of those materials and water penetration through the biocomposite matrix accompanied by swelling which is followed with the scission process due to hydrolysis [255]. In addition to hydrolytic degradation, biodegradation takes place as the continuation of the process. In biodegradation usually, enzymes play an important role, which is called enzymolysis or enzymatic hydrolysis.

Degradation factors and control methods are mainly focusing on the following aspects: (i) the constituents of the composite; (ii) the crystallinity of the constituents (polymers or reinforcing ingredient); (iii) the influence of surrounding environment; and (iv) the morphology of the fibers. The degradation process can happen at the outermost layer of fibers that is interacting with the surrounding environment or throughout the whole volume of the composite fibers in a simultaneous manner. A straightforward measure of composite material is a mass loss that can alter the morphology, and physical and chemical properties of the materials. In the literature, the degradation investigations start usually by measuring the wettability and swelling of the materials further continued by following the morphology and mechanical properties of the materials [256]. Degradation investigations need to be carried out with biological relevant media; however, simultaneous degradation follow-up methods are not that common.

The biocompatibility and biodegradation investigations of the biocomposite fibers are needed within the biomimetic 3D systems, which can be easily provided already with the advantageous 3D features of biocomposite fiber matrix. By this way *in vivo* complexity can be adequately re-created, which enables to predict the effects when the materials are taken for preclinical investigations. Currently numerous studies are ongoing to provide biomimetic 3D models (i.e., 3D wound models) for the advanced screening for the wound dressings, alternative to the traditional *in vitro* and *in vivo* studies [257, 258].

Biological activity of the biocomposite fibers is explored via cell adhesion, migration, proliferation, differentiation and vascularization to attain a proper integration between cells and mat for synthesizing a new tissue [259] or exploring the safety of wound dressings. Biocompatibility tests can be discussed under two distinct focuses regarding the testing conditions, *in vitro* or *in vivo*. *In vitro* cytotoxicity and genotoxicity examinations are preferable prior to *in vivo* tests since the newly developed mats can be examined outside the body and the results are more reproducible, facilitating clinical evaluation of a medicinal product that can be replicated and estimated. Bioactive molecule functionality, cytotoxic and genotoxic effect evaluations are required to assess the safety of the fiber mats intended for direct body contact or permanent implantation and validate cytocompatibility and risk of effect on genetic material of cells. For antimicrobial wound dressings, antimicrobial activity studies are performed with biocomposite fibers using bacterial liquid cultures or disk diffusion assays and relevant bacteria. Furthermore, antibiofilm activity of biocomposite fiber mats is being studied [135, 260]. Advancements are explored to test the activity also in more biorelevant conditions mimicking wound environment [261] or in *in vivo* animal wound models [262, 263]. These enable to understand whether developed antimicrobial biocomposite fiber dressings have superior antimicrobial efficiency compared to commercial wound dressings. Several studies have shown that differently designed electrospun fiber mats are able to effectively kill bacteria [203, 264–266] and reduce bacterial biofilm formation [135, 260, 267].

Colorimetric assay-based cell viability tests are continually being carried out for cell viability, proliferation and cytotoxicity due to their easy application. However, genotoxic effects also raise concern even though the biocomposite fibers may be cytocompatible in *in vitro* assessment. To investigate genotoxic effect, several different DNA damage detection methods have been performed [268]. *In vitro* biocompatible studies enable to reveal the details of the cellular and molecular interactions and should be carried out prior to *in vivo* tests to eliminate the possible harm to humans and animals.

The most straightforward and commonly used *in vivo* biocompatibility test is applied by constructing a wound in a controlled manner on an animal skin and inserting wound dressings on the wounded area. Afterward, wound area is monitored for multiple weeks and wound size reduction is calculated [90]. In order to assess the functional relationships of these biological processes in tissue regeneration, biological marker analysis of the proliferation and reorganization of the cells within the fiber mats have been used traditionally. For instance, integrin markers for cell adherence, fibronectin marker for dermal fibroblast presence, keratin marker for keratinocytes and collagen markers for new tissue formation are among the most common biological indicators widely used [269].

In vivo biocompatibility tests are generally employed on histological samples, which require tissue fixation; therefore, these do not provide noninvasive monitoring of tissue regeneration, which is essential for complex and dynamic process of wound healing. More and more alternative biocompatibility evaluation techniques have been introduced to the tissue engineering with the state-of-art imaging systems. Hyperspectral imaging and second harmonic imaging techniques have been recognized in the tissue engineering area to allow super-resolution imaging of 3D features in living cell environment [153, 257].

11.5 Conclusions

Electrospinning is a promising technology to produce fiber mats from nanometer- to micrometer-sized fibers. Hence, it has been shown to be the only fiber forming method which is easily scalable from the laboratory level to the industrial level. Electrospun biocomposite fiber mats have huge potential to be used as advanced wound dressings due to several structural properties, such as high surface area, microporosity, ability to load drugs and other biomolecules into the fibers, and their fibrous structure for promoting proliferation and migration of cells due to the resemblance to the natural ECM. Moreover, when using suitable materials/biopolymers together with biologically active molecules important for wound healing (including living cells) in eco-friendly or solvent-free electrospinning, the wound healing is even more supported. It is of importance to design advanced wound dressings and to select the most suitable electrospinning technique case by case. Several available design options enable to develop wound dressings with desired properties and functionality; thus, these electrospun fibers enable to adjust to the wound environment and function according to the need. Furthermore, incorporated signaling systems allow monitoring the wound condition and detect changes in the wound. Proper characterization of the electrospun biocomposite fibers reveals the relevant physicochemical and biological properties of the wound dressings and enables to predict their *in vivo* behavior (safety/efficacy).

11.6 Future aspects with biocomposite fibers as wound dressings

The amount of research publications and patents published within last years has shown that there is enough evidence that supports the efficacy of using electrospun biocomposite fiber mats for wound healing applications. Together with the development of novel electrospun wound dressings and their production methods there is a huge need to develop also suitable testing methods that enable to understand the safety, efficacy and quality of the dressings. Not all method currently used for other dosage forms are suitable as discussed during the chapter, since electrospun nanofiber mats have specific material-related and structure-related properties that determine their functionality. For instance, the design of electrospun fibers functioning as responsive with electrical stimulation, mechanical stress and pulsed magnetic field can help to obtain more effective healing. In addition to those novel designs, the produced electrospun fibers especially the short fibers can be used as ink for printing 3D skin tissue constructs composed of dermal fibroblasts, and keratinocytes that can promote the wound healing more efficiently with the appropriate 3D construct. The specific challenges remain unsolved, which requires multidisciplinary approaches and collaboration between experts from different fields. Only this enables to develop also GMP guidelines and regulatory legislation and move the technology closer to industry and the products closer to the market. Unmet clinical need for new wound treatments requires innovative wound care products that help the patients with nonhealing wounds.

Acknowledgments: This work was financially supported by the personal research project no. PUT1088 (Dr K. Kogermann). The Estonian Science Foundation and Estonian Ministry of Education and Research are also thanked for funding.

Abbreviations

2D	Two-dimensional
3D	Three-dimensional
API	Active pharmaceutical ingredient
CNTs	Carbon nanotubes
DDS	Drug delivery system
ECM	Extracellular matrix
GMP	Good Manufacturing Practice

HPLC	High-performance liquid chromatography
IR	Infrared
ISO	International Organization for Standardization
NPs	Nanoparticles
PAMAM	Polyamidoamine
pbFGF	Basic fibroblast growth factor-encoding plasmid
PCL	Poly(ε-caprolactone)
PEG	Polyethylene glycol
PEGME	Poly(ethylene glycol) methyl ether
PEO	Poly(ethylene oxide)
pFGF	Basic fibroblast growth factor
PHBV	Poly(3-hydroxybutyrate-co-3-hydroxyvalerate)
PLA	Poly(lactic acid)
PLGA	Poly(lactic-*co*-glycolic acid)
PMMA	Poly(methyl methacrylate)
PVA	Poly(vinyl alcohol)
TGA	Thermogravimetric analysis
UV	Ultraviolet
XRD	X-ray diffraction

References

[1] Lindholm, C., & Searle, R. Wound management for the twenty-first century: combining effectiveness and efficiency, Int. Wound. J., 2016, 13(S2), 5–15.

[2] Orgill, DP., & Blanco, C. Biomaterials for Treating Skin Loss, Cambridge, Woodhead Publishing Limited, United Kingdom, 2009.

[3] Ho, J., Walsh, C., Yue, D., Dardik, A., & Cheema, U. Current advancements and strategies in tissue engineering for wound healing: a comprehensive review, Adv. Wound. Care., 2017, 6(6), 191–209.

[4] Malik, MA. Advanced wound care market. allied market research, 2016. (Accessed March 29, 2018 https://www.alliedmarketresearch.com/advanced-wound-care-market

[5] McLister, A., McHugh, J., Cundell, J., & Davis, J. New developments in smart bandage technologies for wound diagnostics, Adv. Mater., 2016, 28(27), 5732–5737.

[6] Abrigo, M., McArthur, SL., & Kingshott, P. Electrospun nanofibers as dressings for chronic wound care: advances, challenges, and future prospects, Macromol. Biosci., 2014, 14(6), 772–792.

[7] Huang, ZM., Zhang, YZ., Kotaki, M., & Ramakrishna, S. A review on polymer nanofibers by electrospinning and their applications in nanocomposites, Compos. Sci. Technol., 2003, 63(15), 2223–2253.

[8] Suhaily, SS., Kahlil, AH., Nadirah, WW., & Jawaid, M. Bamboo Based Biocomposites Material, Design and Applications. Y Mastai, ed. Materials Science – Advanced Topics, InTechOpen, 2013, 489–517. DOI: 10.5772/56057

[9] Bramhill, J., Ross, S., & Ross, G. Bioactive nanocomposites for tissue repair and regeneration: a review, Int. J. Environ. Res. Public Health, 2017, 14(1), 66.

[10] Hu, X., Liu, S., Zhou, G., Huang, Y., Xie, Z., & Jing, X. Electrospinning of polymeric nanofibers for drug delivery applications, J. Control Release, 2014, 185, 12–21.

[11] Dhivya, S., Padma, VV., & Santhini, E. Wound dressings – a review, BioMedicine, 2015, 5(4), 24–28.

[12] Jones, V., Grey, JE., & Harding, KG. Wound dressings, BMJ, 2006, 332, 777–780.
[13] Zahedi, P., Rezaeian, I., Ranaei-Siadat, S-O., Jafari, S-H., & Supaphol, P. A review on wound dressings with an emphasis on electrospun nanofibrous polymeric bandages, Polym. Adv. Technol., 2010, 21, 77–95.
[14] Li, D., & Xia, Y. Electrospinning of nanofibers: reinventing the wheel?, Adv. Mater., 2004, 16(14), 1151–1170.
[15] Persano, L., Camposeo, A., Tekmen, C., & Pisignano, D. Industrial upscaling of electrospinning and applications of polymer nanofibers: a review, Macromol. Mater. Eng., 2013, 298(5), 504–520.
[16] Zhang, B., Yan, X., He, H-W., Yu, M., Ning, X., & Long, Y-Z. Solvent-free electrospinning: opportunities and challenges, Polym. Chem., 2017, 8(2), 333–352.
[17] Sundaramurthi, D., Krishnan, UM., & Sethuraman, S. Electrospun nanofibers as scaffolds for skin tissue engineering, Polym. Rev., 2014, 54(2), 348–376.
[18] Krishnan, R., Sundarrajan, S., & Ramakrishna, S. Green processing of nanofibers for regenerative medicine, Macromol. Mater. Eng., 2013, 298(10), 1034–1058.
[19] Agarwal, S., & Greiner, A. On the way to clean and safe electrospinning – green electrospinning: emulsion and suspension electrospinning, Polym .Adv. Technol., 2011, 22(3), 372–378.
[20] Patil, J V., Mali, SS., Kamble, AS., Hong, CK., Kim, JH., & Patil, PS. Electrospinning: a versatile technique for making of 1D growth of nanostructured nanofibers and its applications: an experimental approach, Appl. Surf. Sci., 2017, 423, 641–674.
[21] Thenmozhi, S., Dharmaraj, N., Kadirvelu, K., & Kim, HY. Electrospun nanofibers: new generation materials for advanced applications, Mater. Sci. Eng. B Adv. Funct. Solid-State Mater., 2017, 217, 36–48.
[22] Haider, A., Haider, S., & Kang, IK. A comprehensive review summarizing the effect of electrospinning parameters and potential applications of nanofibers in biomedical and biotechnology, Arab. J. Chem.,2015. 11(8), December 2018,1165–1188. https://doi.org/10.1016/j.arabjc.2015.11.015
[23] Bhardwaj, N., & Kundu, SC. Electrospinning: a fascinating fiber fabrication technique, Biotechnol. Adv., 2010, 28(3), 325–347.
[24] Castilla-Casadiego, DA., Ramos-Avilez, H V., Herrera-Posada, S. et al. Engineering of a stable collagen nanofibrous scaffold with tunable fiber diameter, alignment, and mechanical properties, Macromol. Mater. Eng., 2016, 301(9), 1064–1075.
[25] Liverani, L., & Boccaccini, A. Versatile production of poly(epsilon-caprolactone) fibers by electrospinning using benign solvents, Nanomaterials, 2016, 6, 75.
[26] Soliman, S., Sant, S., Nichol, JW., Khabiry, M., Traversa, E., & Khademhosseini, A. Controlling the porosity of fibrous scaffolds by modulating the fiber diameter and packing density, J. Biomed. Mater. Res. Part A, 2011, 96A(3), 566–574.
[27] Tan, R., Yang, X., & Shen, Y. Robot-aided electrospinning toward intelligent biomedical engineering, Robot Biomimetics, 2017, 4, 17.
[28] Zhou, FL., Gong, RH., & Porat, I. Mass production of nanofibre assemblies by electrostatic spinning, Polym. Int., 2009, 58(4), 331–342.
[29] Wang, X., Niu, H., Wang, X., & Lin, T. Needleless electrospinning of uniform nanofibers using spiral coil spinnerets, J. Nanomater., 2012, 2012, 785920, 9. http://dx.doi.org/10.1155/2012/785920
[30] Nieminen, HJ., Laidmäe, I., Salmi, A. et al. Ultrasound-enhanced electrospinning, Sci. Rep., 2018, 8, 4437.
[31] Laidmäe, I., Nieminen, H., Salimi, A., et al. Device and method to produce nanofibers and constructs thereof. WO patent WO/2016/151191. 2016.

[32] Molnar, K., & Nagy, ZK. Corona-electrospinning: needleless method for high-throughput continuous nanofiber production, Eur. Polym. J, 2016, 74, 279–286.
[33] Niu, H., & Lin, T. Fiber generators in needleless electrospinning, J. Nanomater., 2012, 2012, 725950.
[34] Liu, Y., & He, J-H. Bubble electrospinning for mass production of nanofibers, Int. J. Nonlinear Sci. Numer. Simul., 2007, 8(3), 393–396.
[35] Lai, HJ., Kuan, CH., Wu, HC. et al. Tailored design of electrospun composite nanofibers with staged release of multiple angiogenic growth factors for chronic wound healing, Acta Biomater., 2014, 10(10), 4156–4166.
[36] LeCorre-Bordes, D., Tucker, N., Huber, T., Buunk, N., & Staiger, MP. Shear-electrospinning: extending the electrospinnability range of polymer solutions, J. Mater. Sci., 2016, 51(14), 6686–6696.
[37] Pokorny, M., Rassushin, V., Wolfova, L., & Velebny, V. Increased production of nanofibrous materials by electroblowing form blend of hyaluronic acid and polyethylene oxide, Polym. Eng. Sci., 2016, 56(8), 932–938.
[38] Yan, X., Yu, M., Zhang, L-H. et al. A portable electrospinning apparatus based on a small solar cell and a hand generator: design, performance and application, Nanoscale, 2016, 8(1), 209–213.
[39] Xu, S-C., Qin, C-C., Yu, M. et al. A battery-operated portable handheld electrospinning apparatus, Nanoscale, 2015, 7(29), 12351–12355.
[40] Mouthuy, PA., Groszkowski, L., & Ye, H. Performances of a portable electrospinning apparatus, Biotechnol. Lett, 2015, 37(5), 1107–1116.
[41] Zamani, M., Prabhakaran, MP., & Ramakrishna, S. Advances in drug delivery via electrospun and electrosprayed nanomaterials, Int. J. Nanomed., 2013, 8, 2997–3017.
[42] Song, J., Klymov, A., Shao, J. et al. Electrospun nanofibrous silk fibroin membranes containing gelatin nanospheres for controlled delivery of biomolecules, Adv. Healthc. Mater., 2017, 6(14), 1700014.
[43] Rahimi Tanha, N., & Nouri, M. An experimental study on the coaxial electrospinning of silk fibroin/poly (vinyl alcohol)–salicylic acid core-shell nanofibers and process optimization using response surface methodology, J. Ind. Text., 2017, 0(0), 1–20.
[44] Zupančič, Š., Sinha-Ray, S., Sinha-Ray, S., Kristl, J., & Yarin, AL. Controlled release of ciprofloxacin from core-shell nanofibers with monolithic or blended core, Mol. Pharm., 2016, 13(4), 1393–1404.
[45] Zhang, H., Jia, X., Han, F. et al. Dual-delivery of VEGF and PDGF by double-layered electrospun membranes for blood vessel regeneration, Biomaterials, 2013, 34(9), 2202–2212.
[46] Vysloužilová, L., Buzgo, M., Pokorný, P. et al. Needleless coaxial electrospinning: a novel approach to mass production of coaxial nanofibers, Int. J. Pharm., 2017, 516(1–2), 293–300.
[47] Wang, J., & Windbergs, M. Functional electrospun fibers for the treatment of human skin wounds, Eur. J. Pharm. Biopharm., 2017, 119, 283–299.
[48] McClellan, P., & Landis, WJ. Recent applications of coaxial and emulsion electrospinning methods in the field of tissue engineering, Biores. Open Access, 2016, 5(1), 212–227.
[49] Ingavle, GC., & Leach, JK. Advancements in electrospinning of polymeric nanofibrous scaffolds for tissue engineering, Tissue Eng. Part B Rev., 2014, 20(4), 277–293.
[50] Jiang, H., Wang, L., & Zhu, K. Coaxial electrospinning for encapsulation and controlled release of fragile water-soluble bioactive agents, J. Control Release, 2014, 193, 296–303.
[51] Camerlo, A., Bühlmann-Popa, AM., Vebert-Nardin, C., Rossi, RM., & Fortunato, G. Environmentally controlled emulsion electrospinning for the encapsulation of temperature-sensitive compounds, J. Mater. Sci., 2014, 49(23), 8154–8162.

[52] Rho, KS., Jeong, L., Lee, G. et al. Electrospinning of collagen nanofibers: Effects on the behavior of normal human keratinocytes and early-stage wound healing, Biomaterials, 2006, 27(8), 1452–1461.

[53] Chen, W., Li, D., El-Shanshory, A. et al. Dexamethasone loaded core-shell SF/PEO nanofibers via green electrospinning reduced endothelial cells inflammatory damage, Colloids Surf. B Biointerfaces, 2015, 126, 561–568.

[54] Brown, TD., Dalton, PD., & Hutmacher, DW. Melt electrospinning today: An opportune time for an emerging polymer process, Prog. Polym. Sci., 2016, 56, 116–166.

[55] Zhang, L-H., Duan, X-P., Yan, X. et al. Recent advances in melt electrospinning, RSC Adv., 2016, 6(58), 53400–53414.

[56] Balogh, A., Drávavölgyi, G., Faragó, K. et al. Plasticized drug-loaded melt electrospun polymer mats: characterization, thermal degradation, and release kinetics, J. Pharm. Sci., 2014, 103(4), 1278–1287.

[57] Nagy, ZK., Balogh, A., Drávavölgyi, G. et al. Solvent-free melt electrospinning for preparation of fast dissolving drug delivery system and comparison with solvent-based electrospun and melt extruded systems, J. Pharm. Sci., 2013, 102(2), 508–517.

[58] Qin, C-C., Duan, X-P., Wang, L. et al. Melt electrospinning of poly(lactic acid) and polycaprolactone microfibers by using a hand-operated wimshurst generator, Nanoscale, 2015, 7(40), 16611–16615.

[59] Dalton, PD., Klinkhammer, K., Salber, J., Klee, D., & Möller, M. Direct in vitro electrospinning with polymer melts, Biomacromolecules, 2006, 7(3), 686–690.

[60] Brown, TD., Edin, F., Detta, N., Skelton, AD., Hutmacher, DW., & Dalton, PD. Melt electrospinning of poly(ε-caprolactone) scaffolds: phenomenological observations associated with collection and direct writing, Mater. Sci. Eng. C, 2014, 45, 698–708.

[61] Kandhasamy, S., Perumal, S., Madhan, B. et al. Synthesis and fabrication of collagen-coated ostholamide electrospun nanofiber scaffold for wound healing, ACS Appl. Mater. Interfaces, 2017, 9(10), 8556–8568.

[62] Sridhar, R., Sundarrajan, S., Vanangamudi, A., Singh, G., Matsuura, T., & Ramakrishna, S. Green processing mediated novel polyelectrolyte nanofibers and their antimicrobial evaluation, Macromol. Mater. Eng., 2014, 299, 283–289.

[63] Ulrica, Edlund., Martina Källrot, A., & Albertsson, A-C. Single-step covalent functionalization of polylactide surfaces, J. Am. Chem. Soc., 2005, 127(24), 8865–8871.

[64] Zhu, A., Zhang, M., Wu, J., & Shen, J. Covalent immobilization of chitosan/heparin complex with a photosensitive hetero-bifunctional crosslinking reagent on PLA surface, Biomaterials, 2002, 23, 4657–4665.

[65] Yan, D., Jones, J., Yuan, XY. et al. Plasma treatment of electrospun PCL random nanofiber meshes (NFMs) for biological property improvement, J. Biomed. Mater. Res. Part A, 2013, 101A(4), 963–972.

[66] Liu, W., Zhan, J., Su, Y. et al. Effects of plasma treatment to nanofibers on initial cell adhesion and cell morphology, Colloids Surf. B Biointerfaces, 2014, 113, 101–106.

[67] Ortega, I., Dew, L., Kelly, AG., Chong, CK., MacNeil, S., & Claeyssens, F. Fabrication of biodegradable synthetic perfusable vascular networks via a combination of electrospinning and robocasting, Biomater. Sci., 2015, 3(4), 592–596.

[68] Lee, HJ., & Koh, WG. Hydrogel micropattern-incorporated fibrous scaffolds capable of sequential growth factor delivery for enhanced osteogenesis of hMSCs, ACS Appl. Mater. Interfaces, 2014, 6(12), 9338–9348.

[69] Dhand, C., Venkatesh, M., Barathi, VA. et al. Bio-inspired crosslinking and matrix-drug interactions for advanced wound dressings with long-term antimicrobial activity, Biomaterials, 2017, 138, 153–168.

[70] Wang, X., Pellerin, C., & Bazuin, CG. Enhancing the electrospinnability of low molecular weight polymers using small effective cross-linkers, Macromolecules, 2016, 49(3), 891–899.
[71] Mirzaei, E., Faridi-Majidi, R., Shokrgozar, MA., & Paskiabi, FA. Genipin cross-linked electrospun chitosan-based nanofibrous mat as tissue engineering scaffold, Nanomed. J., 2014, 1(3), 137–146.
[72] Giebel, E., & Greiner, A. Water-stable nonwovens composed of electrospun fibers from aqueous dispersions by photo-cross-linking, Macromol. Mater. Eng., 2012, 297(6), 532–539.
[73] Valente, TAM., Silva, DM., Gomes, PS., Fernandes, MH., Santos, JD., & Sencadas, V. Effect of sterilization methods on electrospun poly(lactic acid) (PLA) fiber alignment for biomedical applications, ACS Appl. Mater. Interfaces, 2016, 8(5), 3241–3249.
[74] Bharath, KN., & Basavarajappa, S. Applications of biocomposite materials based on natural fibers from renewable resources: a review, Sci. Eng. Compos. Mater., 2016, 23(2), 123–133.
[75] Ramnath, V., Sekar, S., Sankar, S., Sankaranarayanan, C., & Sastry, TP. Preparation and evaluation of biocomposites as wound dressing material, J. Mater. Sci. Mater. Med., 2012, 23(12), 3083–3095.
[76] John, MJ., & Thomas, S. Biofibres and biocomposites, Carbohydr. Polym., 2008, 71(3), 343–364.
[77] Gurunathan, T., Mohanty, S., & Nayak, SK. A review of the recent developments in biocomposites based on natural fibres and their application perspectives, Compos. Part A Appl. Sci. Manuf., 2015, 77, 1–25.
[78] Dhandayuthapani, B., Yoshida, Y., Maekawa, T., & Kumar, DS. Polymeric scaffolds in tissue engineering application: a review, Int. J. Polym. Sci., 2011, 2011, 290602.
[79] Sezer, AD., & Cevher, E. Biopolymers as Wound Healing Materials: Challenges and New Strategies. R Pignatello, ed. *Biomaterials Applications for Nanomedicine*, InTechOpen, 2011, 383–414. DOI:10.5772/25177
[80] Alexander, H., Brunski, JB., Cooper, SL. et al. Classes of Materials Used in Medicine. BD Ratner, AS Hoffman, FJ Schoen, & JE Lemons, eds. *Biomaterials Science: An Introduction to Materials in Medicine*, San Diego, Elsevier Inc, 1996, 37–130. https://doi.org/10.1016/B978-0-08-050014-0.50007-9
[81] Mele, E. Electrospinning of natural polymers for advanced wound care: towards responsive and adaptive dressings, J. Mater. Chem. B, 2016, 4(28), 4801–4812.
[82] Reshmi, CR., Menon, T., Binoy, A., Mishra, N., Elyas, KK., & Sujith, A. Poly(L-lactide-co-caprolactone)/collagen electrospun mat: potential for wound dressing and controlled drug delivery, Int. J. Polym. Mater. Polym. Biomater., 2017, 66(13), 645–657.
[83] Castilla-Casadiego, DA., Maldonado, M., Sundaram, P., & Almodovar, J. “Green” electrospinning of a collagen/hydroxyapatite composite nanofibrous scaffold, MRS Commun., 2016, 6(4), 402–407.
[84] Zhong, S., Teo, W., Zhu, X., Beuerman, R., Ramakrishna, S., & Yung, L. An aligned nanofibrous collagen scaffold by electrospinning and its effects on in vitro fibroblast culture, J. Biomed. Mater. Res. Part A, 2006, 79A(3), 456–463.
[85] McManus MC, Boland ED, Simpson DG, Barnes CP, Bowling GL Electrospun fibrinogen: Feasibility as a tissue engineering scaffold in a rat cell culture model, J. Biomed. Mater. Res. Part A, 2007, 81A(2), 299–309.
[86] Yang, X., Fan, L., Ma, L. et al. Green electrospun Manuka honey/silk fibroin fibrous matrices as potential wound dressing, Mater. Des., 2017, 119, 76–84.

[87] Zhu, J., Luo, J., Zhao, X., Gao, J., & Xiong, J. Electrospun homogeneous silk fibroin/poly (ε-caprolactone) nanofibrous scaffolds by addition of acetic acid for tissue engineering, J. Biomater. Appl., 2016, 31(3), 421–437.
[88] Lin, S., Chen, M., Jiang, H. et al. Green electrospun grape seed extract-loaded silk fibroin nanofibrous mats with excellent cytocompatibility and antioxidant effect, Colloids Surf. B Biointerfaces, 2016, 139, 156–163.
[89] Ju, HW., Lee, OJ., Lee, JM. et al. Wound healing effect of electrospun silk fibroin nanomatrix in burn-model, Int. J. Biol. Macromol., 2016, 85, 29–39.
[90] Chao, S., Li, Y., Zhao, R. et al. Synthesis and characterization of tigecycline-loaded sericin/ poly(vinyl alcohol) composite fibers via electrospinning as antibacterial wound dressings, J. Drug. Deliv. Sci. Technol., 2018, 44, 440–447.
[91] Hassiba, AJ., Zowalaty ME, El., Webster, TJ. et al. Synthesis, characterization, and antimicrobial properties of novel double layer nanocomposite electrospun fibers for wound dressing applications, Int. J. Nanomed., 2017, 12, 2205–2213.
[92] Mei, L., Fan, R., Li, X. et al. Nanofibers for improving the wound repair process: The combination of a grafted chitosan and an antioxidant agent, Polym. Chem., 2017, 8(10), 1664–1671.
[93] Ahmed, S., & Ikram, S. Chitosan based scaffolds and their applications in wound healing, Achiev. Life Sci., 2016, 10, 27–37.
[94] Liu, YY., Liu, YY., Liao, N., Cui, F., Park, M., & Kim, HY. Fabrication and durable antibacterial properties of electrospun chitosan nanofibers with silver nanoparticles, Int. J. Biol. Macromol., 2015, 79, 638–643.
[95] Abdelgawad, AM., Hudson, SM., & Rojas, OJ. Antimicrobial wound dressing nanofiber mats from multicomponent (chitosan/silver-NPs/polyvinyl alcohol) systems, Carbohydr. Polym., 2014, 100, 166–178.
[96] Matsuda, A., Kagata, G., Kino, R., & Tanaka, J. Preparation of chitosan nanofiber tube by electrospinning, J. Nanosci. Nanotechnol., 2007, 7(3), 852–855.
[97] Noh, HK., Lee, SW., Kim, JM. et al. Electrospinning of chitin nanofibers: Degradation behavior and cellular response to normal human keratinocytes and fibroblasts, Biomaterials, 2006, 27(21), 3934–3944.
[98] Liu, X., Lin, T., Gao, Y. et al. Antimicrobial electrospun nanofibers of cellulose acetate and polyester urethane composite for wound dressing, J. Biomed. Mater. Res. – Part B Appl. Biomater., 2012, 100B, 1556–1565.
[99] Ji, Y., Ghosh, K., Shu, XZ. et al. Electrospun three-dimensional hyaluronic acid nanofibrous scaffolds, Biomaterials, 2006, 27(20), 3782–3792.
[100] Dong, Y., Liao, S., Ngiam, M., Chan, CK., & Ramakrishna, S. Degradation behaviors of electrospun resorbable polyester nanofibers, Tissue Eng. Part B, 2009, 15(3), 333–351.
[101] Antunes, BP., Moreira, AF., Gaspar, VM., & Correia, IJ. Chitosan/arginine–chitosan polymer blends for assembly of nanofibrous membranes for wound regeneration, Carbohydr. Polym., 2015, 130, 104–112.
[102] Garcia-Orue, I., Pedraz, JL., Hernandez, RM., & Igartua, M. Nanotechnology-based delivery systems to release growth factors and other endogenous molecules for chronic wound healing, J. Drug. Deliv. Sci. Technol., 2017, 42, 2–17.
[103] Kataria, K., Gupta, A., Rath, G., Mathur, RB., & Dhakate, SR. In vivo wound healing performance of drug loaded electrospun composite nanofibers transdermal patch, Int. J. Pharm., 2014, 469(1), 102–110.
[104] Dias, J., Granja, P., & Bártolo, P. Internal crosslinking evaluation of gelatin electrospinning fibers with 1,4-butanediol diglycidyl ether (BDDGE) for skin regeneration, Front Bioeng.

Biotechnol. Conf. Abstr. 10th World Biomater. Congr., 2016, 896. doi:10.3389/conf.FBIOE.2016.01.00896

[105] Sambudi, NS., Sathyamurthy, M., Lee, GM., & Park S, Bin. Electrospun chitosan/poly (vinyl alcohol) reinforced with CaCO3 nanoparticles with enhanced mechanical properties and biocompatibility for cartilage tissue engineering, Compos. Sci. Technol., 2015, 106, 76–84.

[106] Sui, G., Yang, X., Mei, F. et al. Poly-L-lactic acid/hydroxyapatite hybrid membrane for bone tissue regeneration, J. Biomed. Mater. Res. – Part A, 2007, 82A(2), 445–454.

[107] Huang, L., Nagapudi, K., Apkarian, PR., & Chaikof, EL. Engineered collagen – PEO nanofibers and fabrics, J. Biomater. Sci. Polym. Ed., 2001, 12(9), 979–993.

[108] De Silva, RT., Mantilaka, MMMGPG., Goh, KL., Ratnayake, SP., Amaratunga, GAJ., & De Silva, KMN. Magnesium oxide nanoparticles reinforced electrospun alginate-based nanofibrous scaffolds with improved physical properties, Int. J. Biomater., 2017, 2017, 1391298.

[109] Kim, H., Che, L., Ha, Y., & Ryu, W. Mechanically-reinforced electrospun composite silk fibroin nanofibers containing hydroxyapatite nanoparticles, Mater. Sci. Eng. C, 2014, 40, 324–335.

[110] Dias, JR., Baptista-Silva, S., Oliveira, CMT d. et al. In situ crosslinked electrospun gelatin nanofibers for skin regeneration, Eur. Polym. J., 2017, 95, 161–173.

[111] Tamimi, E., Ardila, DC., Haskett, DG. et al. Biomechanical comparison of glutaraldehyde-crosslinked gelatin fibrinogen electrospun scaffolds to porcine coronary arteries, J. Biomech. Eng., 2015, 138(1), 11001.

[112] Bonino, CA., Krebs, MD., Saquing, CD. et al. Electrospinning alginate-based nanofibers: From blends to crosslinked low molecular weight alginate-only systems, Carbohydr. Polym., 2011, 85, 111–119.

[113] Butcher, AL., Offeddu, GS., & Oyen, ML. Nanofibrous hydrogel composites as mechanically robust tissue engineering scaffolds, Trends Biotechnol., 2014, 32(11), 564–570.

[114] Mahmoudi, N., Eslahi, N., Mehdipour, A. et al. Temporary skin grafts based on hybrid graphene oxide-natural biopolymer nanofibers as effective wound healing substitutes: pre-clinical and pathological studies in animal models, J. Mater. Sci. Mater. Med., 2017, 28(5), 73.

[115] Pal, P., Dadhich, P., Srivas, PK., Das, B., Maulik, D., & Dhara, S. Bilayered nanofibrous 3D hierarchy as skin rudiment by emulsion electrospinning for burn wound management, Biomater. Sci., 2017, 5(9), 1786–1799.

[116] Tan, L., Hu, J., Huang, H., Han, J., & Hu, H. Study of multi-functional electrospun composite nanofibrous mats for smart wound healing, Int. J. Biol. Macromol., 2015, 79, 469–476.

[117] Wei, Q., Xu, F., Xu, X. et al. The multifunctional wound dressing with core-shell structured fibers prepared by coaxial electrospinning, Front Mater. Sci., 2016, 10(2), 113–121.

[118] Sun, L., Gao, W., Fu, X. et al. Enhanced wound healing in diabetic rats by nanofibrous scaffolds mimicking the basketweave pattern of collagen fibrils in native skin, Biomater. Sci., 2018. Jan 30, 6(2), 340–349. doi: 10.1039/c7bm00545h

[119] Shi, R., Geng, H., Gong, M. et al. Long-acting and broad-spectrum antimicrobial electrospun poly (ε-caprolactone)/gelatin micro/nanofibers for wound dressing, J. Colloid Interface Sci., 2018, 509, 275–284.

[120] Basar, AO., Castro, S., Torres-Giner, S., Lagaron, JM., & Turkoglu Sasmazel, H. Novel poly(ε-caprolactone)/gelatin wound dressings prepared by emulsion electrospinning with controlled release capacity of Ketoprofen anti-inflammatory drug, Mater. Sci. Eng. C, 2017, 81, 459–468.

[121] Naseri-Nosar, M., Farzamfar, S., Sahrapeyma, H. et al. Cerium oxide nanoparticle-containing poly (ε-caprolactone)/gelatin electrospun film as a potential wound dressing material: In vitro and in vivo evaluation, Mater. Sci. Eng. C, 2017, 81, 366–372.

[122] Tan, L., Hu, J., & Zhao, H. Design of bilayered nanofibrous mats for wound dressing using an electrospinning technique, Mater Lett, 2015, 156, 46–49.

[123] Yao, C-H., Lee, C-Y., Huang, C-H., Chen, Y-S., & Chen, K-Y. Novel bilayer wound dressing based on electrospun gelatin/keratin nanofibrous mats for skin wound repair, Mater. Sci. Eng. C, 2017, 79, 533–540.
[124] Ranjbar-Mohammadi, M., & Bahrami, SH. Electrospun curcumin loaded poly(ε-caprolactone)/gum tragacanth nanofibers for biomedical application, Int. J. Biol. Macromol., 2016, 84, 448–456.
[125] Ranjbar-Mohammadi, M., Zamani, M., Prabhakaran, MP., Bahrami, SH., & Ramakrishna, S. Electrospinning of PLGA/gum tragacanth nanofibers containing tetracycline hydrochloride for periodontal regeneration, Mater. Sci. Eng. C, 2016, 58, 521–531.
[126] Zarekhalili, Z., Bahrami, SH., Ranjbar-Mohammadi, M., & Milan, PB. Fabrication and characterization of PVA/Gum tragacanth/PCL hybrid nanofibrous scaffolds for skin substitutes, Int. J. Biol. Macromol., 2017, 94, 679–690.
[127] Komur, B., Bayrak, F., Ekren, N. et al. Starch/PCL composite nanofibers by co-axial electrospinning technique for biomedical applications, Biomed. Eng. Online, 2017, 16(1), 40.
[128] Waghmare, VS., Wadke, PR., Dyawanapelly, S., Deshpande, A., Jain, R., & Dandekar, P. Starch based nanofibrous scaffolds for wound healing applications, Bioact. Mater., 2017, 3(3), September 2018, 255-266. https://doi.org/10.1016/j.bioactmat.2017.11.006
[129] Zhao, L., Luo, J., Li, Y., Wang, H., Song, G., & Tang, G. Emulsion-electrospinning n-octadecane /silk composite fiber as environmental-friendly form-stable phase change materials, J. Appl. Polym. Sci., 2017, 134(47), 45538.
[130] Chouhan, D., Chakraborty, B., Nandi, SK., & Mandal, BB. Role of non-mulberry silk fibroin in deposition and regulation of extracellular matrix towards accelerated wound healing, Acta Biomater., 2017, 48, 157–174.
[131] Chouhan, D., Janani, G., Chakraborty, B., Nandi, SK., & Mandal, BB. Functionalized PVA-silk blended nanofibrous mats promote diabetic wound healing via regulation of extracellular matrix and tissue remodelling, J. Tissue Eng. Regen. Med., 2018 Mar, 12(3), e1559-e1570. doi: 10.1002/term.2581.
[132] Li, H., Williams, GR., Wu, J. et al. Thermosensitive nanofibers loaded with ciprofloxacin as antibacterial wound dressing materials, Int. J. Pharm., 2017, 517, 135–147.
[133] Li, H., Williams, GR., Wu, J., Wang, H., Sun, X., & Zhu, L-M. Poly(N-isopropylacrylamide)/poly (l-lactic acid-co-ε-caprolactone) fibers loaded with ciprofloxacin as wound dressing materials, Mater. Sci. Eng. C., 2017, 79, 245–254.
[134] Garcia-Orue, I., Gainza, G., Gutierrez, FB. et al. Novel nanofibrous dressings containing rhEGF and Aloe vera for wound healing applications, Int. J. Pharm., 2017, 523(2), 556–566.
[135] Preem, L., Mahmoudzadeh, M., Putrinš, M. et al. Interactions between chloramphenicol, carrier polymers, and bacteria–implications for designing electrospun drug delivery systems countering wound infection, Mol. Pharm., 2017, 14(12), 4417–4430.
[136] Bubel, K., Grunenberg, D., Vasilyev, G., Zussman, E., Agarwal, S., & Greiner, A. Solvent-free aqueous dispersions of block copolyesters for electrospinning of biodegradable nonwoven mats for biomedical applications, Macromol. Mater. Eng., 2014, 299, 1445–1454.
[137] Wang, X., Ding, B., & Li, B. Biomimetic electrospun nanofibrous structures for tissue engineering, Mater. Today., 2013, 16(6), 229–241.
[138] Jayarama Reddy, V., Radhakrishnan, S., Ravichandran, R. et al. Nanofibrous structured biomimetic strategies for skin tissue regeneration, Wound Repair Regen., 2013, 21, 1–16.
[139] Sultana, T., Amirian, J., Park, C., Lee, SJ., & Lee, B-T. Preparation and characterization of polycaprolactone–polyethylene glycol methyl ether and polycaprolactone–chitosan electrospun mats potential for vascular tissue engineering, J. Biomater. Appl., 2017, 32(5), 648–662.

[140] Zhao, P., Jiang, H., Pan, H., Zhu, K., & Chen, W. Biodegradable fibrous scaffolds composed of gelatin coated poly(ε-caprolactone) prepared by coaxial electrospinning, J. Biomed. Mater. Res. Part A, 2007, 83A(2), 372–382.
[141] Ji, W., Sun, Y., Yang, F. et al. Bioactive electrospun scaffolds delivering growth factors and genes for tissue engineering applications, Pharm. Res., 2011, 28(6), 1259–1272.
[142] Montero, RB., Vial, X., Nguyen, DT. et al. BFGF-containing electrospun gelatin scaffolds with controlled nano-architectural features for directed angiogenesis, Acta Biomater., 2012, 8(5), 1778–1791.
[143] Zhang, W., Ronca, S., & Mele, E. Electrospun nanofibres containing antimicrobial plant extracts, Nanomaterials, 2017, 7(2), 42.
[144] Chen, S., Liu, B., Carlson, MA., Gombart, AF., Reilly, DA., & Xie, J. Recent advances in electrospun nanofibers for wound healing, Nanomedicine, 2017, 12(11), 1335–1352.
[145] Cheng, F., Gao, J., Wang, L., & Hu, X. Composite chitosan/poly(ethylene oxide) electrospun nanofibrous mats as novel wound dressing matrixes for the controlled release of drugs, J. Appl. Polym. Sci., 2015, 132(24), 42060.
[146] Sadri, M., Arab-Sorkhi, S., Vatani, H., & Bagheri-Pebdeni, A. New wound dressing polymeric nanofiber containing green tea extract prepared by electrospinning method, Fibers Polym., 2015, 16(8), 1742–1750.
[147] Rieger, KA., & Schiffman, JD. Electrospinning an essential oil: cinnamaldehyde enhances the antimicrobial efficacy of chitosan/poly(ethylene oxide) nanofibers, Carbohydr. Polym., 2014, 113, 561–568.
[148] Rath, G., Hussain, T., Chauhan, G., Garg, T., & Goyal, AK. Collagen nanofiber containing silver nanoparticles for improved wound-healing applications, J. Drug Target, 2016, 24(6), 520–529.
[149] Liu, X., Nielsen, LH., Kłodzińska, SN. et al. Ciprofloxacin-loaded sodium alginate/poly (lactic-co-glycolic acid) electrospun fibrous mats for wound healing, Eur. J. Pharm. Biopharm., 2018, 123, 42–49.
[150] Lipsky, BA., & Hoey, C. Topical antimicrobial therapy for treating chronic wounds, Clin. Infect. Dis., 2009, 49(10), 1541–1549.
[151] Mutlu, G., Calamak, S., Ulubayram, K., & Guven, E. Curcumin-loaded electrospun PHBV nanofibers as potential wound-dressing material, J. Drug. Deliv. Sci. Technol., 2018, 43, 185–193.
[152] Dai, X., Liu, J., Zheng, H. et al. Nano-formulated curcumin accelerates acute wound healing through Dkk-1-mediated fibroblast mobilization and MCP-1-mediated anti-inflammation, NPG Asia Mater., 2017, 9, e368.
[153] Anjum, F., Agabalyan, NA., Sparks, HD., Rosin, NL., Kallos, MS., & Biernaskie, J. Biocomposite nanofiber matrices to support ECM remodeling by human dermal progenitors and enhanced wound closure, Sci. Rep., 2017, 7(1), 1–17.
[154] Orive, G., Hernández, RM., Gascón, AR. et al. Cell encapsulation: promise and progress, Nat. Med., 2003, 9(1), 104–107.
[155] Gasperini, L., Mano, JF., & Reis, RL. Natural polymers for the microencapsulation of cells, J R Soc. Interface., 2014, 11, 20140817.
[156] Sampson, SL., Saraiva, L., Gustafsson, K., Jayasinghe, SN., & Robertson, BD. Cell electrospinning: an in vitro and in vivo study, Small, 2014, 10(1), 78–82.
[157] You, HJ., & Han, SKJ. Cell therapy for wound healing, J. Korean Med. Sci., 2014, 29(3), 311–319.
[158] Townsend-Nicholson, A., & Jayasinghe, SN. Cell electrospinning: a unique biotechnique for encapsulating living organisms for generating active biological microthreads/scaffolds, Biomacromolecules, 2006, 7, 3364–3369.
[159] Jayasinghe, SN. Cell electrospinning: a novel tool for functionalising fibres, scaffolds and membranes with living cells and other advanced materials for regenerative biology and medicine, Analyst, 2013, 138(8), 2215–2223.

[160] Xie, J., Li, X., & Xia, Y. Putting electrospun nanofibers to work for biomedical research, Macromol. Rapid Commun., 2008, 29(22), 1775–1792.

[161] U. S. Food and Drug Administration. Q3C – Tables and List Guidance for Industry. U. S. Food and Drug Administration, 2017. (Accessed March 29, 2018 https://www.fda.gov/downloads/drugs/guidances/ucm073395.pdf

[162] Tran, CD., Duri, S., & Harkins, AL. Recyclable synthesis, characterization, and antimicrobial activity of chitosan-based polysaccharide composite materials, J. Biomed. Mater. Res. Part A, 2013, 101A(8), 2248–2257.

[163] He, H-W., Wang, L., Yan, X. et al. Solvent-free electrospinning of UV curable polymer microfibers, RSC Adv., 2016, 6(35), 29423–29427.

[164] Levit, N., & Tepper, G. Supercritical CO2-assisted electrospinning, J. Supercrit. Fluids, 2004, 31(3), 329–333.

[165] Liu, S-L., Long, Y-Z., Huang, Y-Y. et al. Solventless electrospinning of ultrathin polycyanoacrylate fibers, Polym. Chem., 2013, 4(23), 5696–5700.

[166] Fan, L., Cai, Z., Zhang, K. et al. Green electrospun pantothenic acid/silk fibroin composite nanofibers: Fabrication, characterization and biological activity, Colloids Surf. B Biointerfaces, 2014, 117, 14–20.

[167] Sheng, X., Fan, L., He, C., Zhang, K., Mo, X., & Wang, H. Vitamin E-Loaded silk fibroin nanofibrous mats fabricated by green process for skin care application, Int. J. Biol. Macromol., 2013, 56, 49–56.

[168] Hell, S., Ohkawa, K., Amer, H., Potthast, A., & Rosenau, T. “Dialdehyde cellulose” nanofibers by electrospinning as polyvinyl alcohol blends: manufacture and product characterization, J. Wood Chem. Technol., 2018, 38(2), 96–110.

[169] Manoukian, OS., Ahmad, A., Marin, C., James, R., Mazzocca, AD., & Kumbar, SG. Bioactive nanofiber dressings for wound healing. M Ågren, ed. . *Wound Healing Biomaterials – Volume 2: Functional Biomaterials*, Woodhead Publishing, 2016, 451–481. https://doi.org/10.1016/B978-1-78242-456-7.00022-2

[170] Ravichandran, R., Venugopal, JR., Sundarrajan, S., Mukherjee, S., Sridhar, R., & Ramakrishna, S. Composite poly-L-lactic acid/poly-(α,β)-DL-aspartic acid/collagen nanofibrous scaffolds for dermal tissue regeneration, Mater. Sci. Eng. C Mater. Biol. Appl., 2012, 32(6), 1443–1451.

[171] Jiang, T., Carbone, EJ., Lo, KWH., & Laurencin, CT. Electrospinning of polymer nanofibers for tissue regeneration, Prog. Polym. Sci., 2015, 46, 1–24.

[172] Braghirolli, DI., Steffens, D., & Pranke, P. Electrospinning for regenerative medicine: a review of the main topics, Drug Discov. Today, 2014, 19(6), 743–753.

[173] Kalisky, J., Raso, J., Rigothier, C. et al. An easy-to-use and versatile method for building cell-laden microfibres, Sci. Rep., 2016, 6(1), 33328.

[174] Fu, Y., & Kao, WJ. Drug release kinetics and transport mechanisms of non-degradable and degradable polymeric delivery systems, Expert Opin. Drug Deliv., 2010, 7(4), 429–444.

[175] Chew, SY., Wen, J., Yim, EKF., & Leong, KW. Sustained release of proteins from electrospun biodegradable fibers, Biomacromolecules, 2005, 6(4), 2017–2024.

[176] Peh, P., Lim, NSJ., Blocki, A. et al. Simultaneous delivery of highly diverse bioactive compounds from blend electrospun fibers for skin wound healing, Bioconjug. Chem., 2015, 26(7), 1348–1358.

[177] Manjumeena, R., Elakkiya, T., Duraibabu, D., Feroze Ahamed, A., Kalaichelvan, PT., & Venkatesan, R. “Green” biocompatible organic-inorganic hybrid electrospun nanofibers for potential biomedical applications, J. Biomater. Appl., 2015, 29(7), 1039–1055.

[178] Liverani, L., Boccardi, E., Beltran, AM., & Boccaccini, AR. Incorporation of calcium containing mesoporous (MCM-41-Type) particles in electrospun PCL fibers by using benign solvents, Polymers (Basel), 2017, 9, 487.
[179] Fernández, J., Auzmendi, O., Amestoy, H., Diez-Torre, A., & Sarasua, JR. Mechanical properties and fatigue analysis on poly(ε-caprolactone)-polydopamine-coated nanofibers and poly(ε-caprolactone)-carbon nanotube composite scaffolds, Eur. Polym. J., 2017, 94, 208–221.
[180] Xie, Z., Paras, CB., Weng, H. et al. Dual growth factor releasing multi-functional nanofibers for wound healing, Acta Biomater., 2013, 9(12), 9351–9359.
[181] Ali, IH., Khalil, IA., & El-Sherbiny, IM. Single-dose electrospun nanoparticles-in-nanofibers wound dressings with enhanced epithelialization, collagen deposition, and granulation properties, ACS Appl. Mater. Interfaces, 2016, 8(23), 14453–14469.
[182] Ngadiman, NHA., Noordin, MY., Idris, A., & Kurniawan, D. A review of evolution of electrospun tissue engineering scaffold: From two dimensions to three dimensions, Proc. Inst. Mech. Eng. Part H J. Eng. Med., 2017, 231(7), 597–616.
[183] Ramakrishna, S., Lim, T-C., & Fujihara, K. *An Introduction To Electrospinning And Nanofibers (eBook)*. World Scientific Publishing Company, 2005. https://doi.org/10.1142/5894
[184] Guo, S., & Dipietro, LA. Factors affecting wound healing, J. Dent. Res., 2010, 89(3), 219–229.
[185] Sonnemann, KJ., & Bement, WM. Wound repair: toward understanding and integration of single-cell and multicellular wound responses, Annu. Rev. Cell. Dev. Biol., 2011, 27(1), 237–263. https://doi.org/10.1142/5894
[186] Liu, M., Duan, X-P., Li, Y-M., Yang, D-P., & Long, Y-Z. Electrospun nanofibers for wound healing, Mater. Sci. Eng. C, 2017, 76, 1413–1423.
[187] Kamble, P., Sadarani, B., Majumdar, A., & Bhullar, S. Nanofiber based drug delivery systems for skin: a promising therapeutic approach, J. Drug Deliv. Sci. Technol., 2017, 41, 124–133.
[188] MacEwan, MR., MacEwan, S., Kovacs, TR., & Batts, J. What makes the optimal wound healing material? a review of current science and introduction of a synthetic nanofabricated wound care scaffold, Cureus, 2017, 9(10), e1736.
[189] MacEwan, MR., MacEwan, S., Wright, AP., Kovacs, TR., Batts, J., & Zhang, L. Comparison of a fully synthetic electrospun matrix to a bi-layered xenograft in healing full thickness cutaneous wounds in a porcine model, Cureus, 2017, 9(8), e1614.
[190] Davenport, NR., & Bement, WM. Cell repair: revisiting the patch hypothesis, Commun. Integr. Biol., 2016, 9(6), e1253643.
[191] Han, G., & Ceilley, R. Chronic wound healing: a review of current management and treatments, Adv. Ther., 2017, 34(3), 599–610.
[192] Leong, WS., Wu, SC., Ng, K., & Tan, LP. Electrospun 3D multi-scale fibrous scaffold for enhanced human dermal fibroblast infiltration, Int. J. Bioprinting., 2016, 2(1), 81–92.
[193] Chen, H., Peng, Y., Wu, S., & Tan, L. Electrospun 3D fibrous scaffolds for chronic wound repair, Materials (Basel), 2016, 9(12), 272.
[194] Yang, Y., Xia, T., Chen, F. et al. Electrospun fibers with plasmid bFGF polyplex loadings promote skin wound healing in diabetic rats, Mol. Pharm., 2012, 9(1), 48–58.
[195] Jiang, S., Ma, BC., Reinholz, J. et al. Efficient nanofibrous membranes for antibacterial wound dressing and UV protection, ACS Appl. Mater. Interfaces, 2016, 8(44), 29915–29922.
[196] Spano, F., Quarta, A., Martelli, C. et al. Fibrous scaffolds fabricated by emulsion electrospinning: from hosting capacity to in vivo biocompatibility, Nanoscale, 2016, 8(17), 9293–9303.
[197] Berthet, M., Gauthier, Y., Lacroix, C., Verrier, B., & Monge, C. Nanoparticle-based dressing: the future of wound treatment?, Trends Biotechnol., 2017, 35(8), 770–784.

[198] Yang, X., Yang, J., Wang, L. et al. Pharmaceutical intermediate-modified gold nanoparticles: against multidrug-resistant bacteria and wound-healing application via an electrospun scaffold, ACS Nano., 2017, 11(6), 5737–5745.
[199] Liao, J., Zhong, S., Wang, S. et al. Preparation and properties of a novel carbon nanotubes/poly(vinyl alcohol)/epidermal growth factor composite biological dressing, Exp. Ther. Med., 2017, 14(3), 2341–2348.
[200] Staffa, A., Vocetkova, K., Sovkova, V., Rampichova, M., Filova, E., & Amler, E. Polycaprolactone nanofiber mesh with adhered liposomes as a simple delivery system for bioactive growth factors, Transl. Med. Rep., 2017, 1(6716), 58–63.
[201] Mickova, A., Buzgo, M., Benada, O. et al. Core/shell nanofibers with embedded liposomes as a drug delivery system, Biomacromolecules, 2012, 13(4), 952–962.
[202] Zylberberg, C., & Matosevic, S. Bioengineered liposome–scaffold composites as therapeutic delivery systems, Ther. Deliv., 2017, 8(6), 425–445.
[203] Monteiro, N., Martins, M., Martins, A. et al. Antibacterial activity of chitosan nanofiber meshes with liposomes immobilized releasing gentamicin, Acta Biomater., 2015, 18, 196–205.
[204] Chandrawati, R., Olesen, MTJ., Marini, TCC. et al. Enzyme prodrug therapy engineered into electrospun fibers with embedded liposomes for controlled, localized synthesis of therapeutics, Adv. Healthc. Mater., 2017, 6, 1700385.
[205] Mirzaei, E., Majidi, RF., Sarkar, S., & Rezayat, MS. Electrospun nanofibrous wound dressing and a method of synthesizing the same, United States Patent Application, 2013/0150763 A1, 2013.
[206] Croisier, F., Atanasova, G., Poumay, Y., & Jérôme, C. Polysaccharide-coated PCL nanofibers for wound dressing applications, Adv. Healthc Mater., 2014, 3(12), 2032–2039.
[207] Dongargaonkar, AA., Bowlin, GL., & Yang, H. Electrospun blends of gelatin and gelatin–dendrimer conjugates as a wound-dressing and drug-delivery platform, Biomacromolecules, 2013, 14(11), 4038–4045.
[208] Wang, S., Zhong, S., Lim, CT., & Nie, H. Effects of fiber alignment on stem cells–fibrous scaffold interactions, J. Mater. Chem. B, 2015, 3(16), 3358–3366.
[209] Xie, J., Macewan, MR., Ray, WZ., Liu, W., Siewe, DY., & Xia, Y. Radially aligned, electrospun nanofibers as dural substitutes for wound closure and tissue regeneration applications, ACS Nano., 2010, 4(9), 5027–5036.
[210] Baheti, S., & Tunak, M. Characterization of fiber diameter using image analysis, IOP Conf. Ser. Mater. Sci. Eng., 2017, 254, 142002.
[211] Paaver, U., Heinämäki, J., Kassamakov, I. et al. Nanometer depth resolution in 3D topographic analysis of drug-loaded nanofibrous mats without sample preparation, Int. J. Pharm., 2014, 462(1–2), 29–37.
[212] Garcea, SC., Wang, Y., & Withers, PJ. X-ray computed tomography of polymer composites, Compos. Sci. Technol., 2018, 156, 305–319.
[213] Yazgan, G., Dmitriev, RI., Tyagi, V. et al. Steering surface topographies of electrospun fibers: understanding the mechanisms, Sci. Rep., 2017, 7, 158.
[214] Atila, D., Keskin, D., & Tezcaner, A. Cellulose acetate based 3-dimensional electrospun scaffolds for skin tissue engineering applications, Carbohydr. Polym., 2015, 133, 251–261.
[215] Kim, G., Yoon, H., & Park, Y. Drug release from various thicknesses of layered mats consisting of electrospun polycaprolactone and polyethylene oxide micro/nanofibers, Appl. Phys. A Mater. Sci. Process., 2010, 100(4), 1197–1204.
[216] Jin, G., Prabhakaran, MP., Kai, D., Annamalai, SK., Arunachalam, KD., & Ramakrishna, S. Tissue engineered plant extracts as nanofibrous wound dressing, Biomaterials, 2013, 34, 724–734.

[217] Li, D., Frey, MW., & Joo, YL. Characterization of nanofibrous membranes with capillary flow porometry, J. Memb. Sci., 2006, 286, 104–114.

[218] Hou, Z., Li, X., Li, C. et al. Electrospun upconversion composite fibers as dual drugs delivery system with individual release properties, Langmuir, 2013, 29, 9473–9482.

[219] Ren, L., Wang, J., Yang, FY. et al. Fabrication of gelatin-siloxane fibrous mats via sol-gel and electrospinning procedure and its application for bone tissue engineering, Mater. Sci. Eng. C, 2010, 30, 437–444.

[220] Palo, M., Kogermann, K., Laidmäe, I. et al. Development of oromucosal dosage forms by combining electrospinning and inkjet printing, Mol. Pharm., 2017, 14(3), 808–820.

[221] Siimon, K., Siimon, H., & Järvekülg, M. Mechanical characterization of electrospun gelatin scaffolds cross-linked by glucose, J. Mater. Sci. Mater. Med., 2015, 26, 37.

[222] Suchý, T., Šupová, M., Klapková, E. et al. The release kinetics, antimicrobial activity and cytocompatibility of differently prepared collagen/hydroxyapatite/vancomycin layers: microstructure vs nanostructure, Eur. J. Pharm. Sci., 2017, 100, 219–229.

[223] Giebel, E., Getze, J., Röcker, T., & Greiner, A. The importance of crosslinking and glass transition temperature for the mechanical strength of nanofibers obtained by green electrospinning, Macromol. Mater. Eng., 2013, 298(4), 439–446.

[224] Lopez, FL., Shearman, GC., Gaisford, S., & Williams, GR. Amorphous formulations of indomethacin and griseofulvin prepared by electrospinning, Mol. Pharm., 2014, 11(12), 4327–4338.

[225] Tamm, I., Heinämäki, J., Laidmäe, I. et al. Development of suberin fatty acids and chloramphenicol-loaded antimicrobial electrospun nanofibrous mats intended for wound therapy, J. Pharm. Sci., 2016, 105(3), 1239–1247.

[226] He, T., Wang, J., Huang, P. et al. Electrospinning polyvinylidene fluoride fibrous membranes containing anti-bacterial drugs used as wound dressing, Colloids Surf. B Biointerfaces, 2015, 130, 278–286.

[227] Gautam, S., Dinda, AK., & Mishra, NC. Fabrication and characterization of PCL/gelatin composite nanofibrous scaffold for tissue engineering applications by electrospinning, Mater. Sci. Eng. C, 2013, 33, 1228–1235.

[228] He, M., Xue, J., Geng, H. et al. Fibrous guided tissue regeneration membrane loaded with anti-inflammatory agent prepared by coaxial electrospinning for the purpose of controlled release, Appl. Surf. Sci., 2015, 335, 121–129.

[229] Nagy, ZK., Balogh, A., Démuth, B. et al. High speed electrospinning for scaled-up production of amorphous solid dispersion of itraconazole, Int. J. Pharm., 2015, 480(1–2), 137–142.

[230] Truong, YB., Glattauer, V., Briggs, KL., Zappe, S., & Ramshaw, JAM. Collagen-based layer-by-layer coating on electrospun polymer scaffolds, Biomaterials, 2012, 33(36), 9198–9204.

[231] Gomes, SR., Rodrigues, G., Martins, GG. et al. In vitro and in vivo evaluation of electrospun nanofibers of PCL, chitosan and gelatin: A comparative study, Mater. Sci. Eng. C, 2015, 46, 348–358.

[232] Sorak, D., Herberholz, L., Iwascek, S., Altinpinar, S., Pfeifer, F., & Siesler, HW. New developments and applications of handheld Raman, mid-infrared, and near-infrared spectrometers, Appl. Spectrosc. Rev., 2012, 47(2), 83–115.

[233] Chablani, L., Taylor, MK., Mehrotra, A., Rameas, P., & Stagner, WC. Inline real-time near-infrared granule moisture measurements of a continuous granulation–drying–milling process, AAPS Pharm. Sci. Tech., 2011, 12(4), 1050–1055.

[234] Zhang, J., Ying, Y., Pielecha-Safira, B. et al. Raman spectroscopy for in-line and off-line quantification of poorly soluble drugs in strip films, Int. J. Pharm., 2014, 475(1), 428–437.

[235] Hajjou, M., Qin, Y., Bradby, S., Bempong, D., & Lukulay, P. Assessment of the performance of a handheld Raman device for potential use as a screening tool in evaluating medicines quality, J. Pharm. Biomed. Anal., 2013, 74, 47–55.
[236] Mungikar, A., & Kamat, M. Use of in-line raman spectroscopy as a non- destructive and rapid analytical technique to monitor aggregation of a therapeutic protein, Am. Pharm. Rev., 2010, 13(7), 78.
[237] Chou, SF., Carson, D., & Woodrow, KA. Current strategies for sustaining drug release from electrospun nanofibers, J. Control Release, 2015, 220, 584–591.
[238] Elsner, JJ., Egozi, D., Ullmann, Y., Berdicevsky, I., Shefy-Peleg, A., & Zilberman, M. Novel biodegradable composite wound dressing with controlled release of antibiotics: Results in a guinea pig burn model, Burns, 2011, 37, 896–904.
[239] Kiaee, B., Etaat, M., Kiaee, B., Kiaei, S., & Javar, HA. Multilayered controlled released topical patch containing tetracycline for wound dressing, J. Silico Vitr. Pharmacol., 2016, 2, 2.
[240] Zupančič, Š., Preem, L., Kristl, J. et al. Impact of PCL nanofiber mat structural properties on hydrophilic drug release and antibacterial activity on periodontal pathogens, Eur J. Pharm. Sci., (submitted manuscript) Eur J Pharm Sci. 2018 Sep 15;122:347–358. doi: 10.1016/j.ejps.2018.07.024.
[241] Weng, L., & Xie, J. Smart electrospun nanofibers for controlled drug release: recent advances and new perspectives, Curr. Pharm. Des., 2015, 21(15), 1944–1959.
[242] Seif, S., Graef, F., Gordon, S., & Windbergs, M. Monitoring drug release from electrospun fibers using an in situ fiber-optic system, Diss. Technol., 2016, 23(2), 6–11.
[243] Tamm, I., Østergaard, J., Laidmäe, I., et al. Real-time monitoring and visualization of drug release from antimicrobial nanofibrous wound dressings by UV-imaging technique. In: 6th BBBB Conference on Pharmaceutical Sciences: Strategies to improve the quality and performance of modern drug delivery systems 2015, 88.
[244] Abrigo, M., Kingshott, P., & McArthur, SL. Bacterial response to different surface chemistries fabricated by plasma polymerization on electrospun nanofibers, Biointerphases, 2015, 10(4), 04A301.
[245] Cho, D., Lee, S., & Frey, MW. Characterizing zeta potential of functional nanofibers in a microfluidic device, J. Colloid Interface Sci., 2012, 372(1), 252–260.
[246] Mennini, N., Greco, A., Bellingeri, A., De Vita, F., & Petrella, F. Quality of wound dressings: a first step in establishing shared criteria and objective procedures to evaluate their performance, J. Wound Care, 2016, 25(8), 428–437.
[247] Jayakumar, R., Prabaharan, M., Sudheesh Kumar, PT., Nair, S V., & Tamura, H. Biomaterials based on chitin and chitosan in wound dressing applications, Biotechnol. Adv., 2011, 29(3), 322–337.
[248] Xue, R., Behera, P., Xu, J., Viapiano, M., & Lannutti, J. Polydimethylsiloxane core-polycaprolactone shell nanofibers as biocompatible, real-time oxygen sensors, Sens. Actuators B Chem., 2014, 192, 697–707.
[249] Pandit, A., & Feldman, D. Effect of oxygen treatment and dressing oxygen permeability on wound healing, Wound Repair Regen., 1994, 2(2), 130–137.
[250] Rath, G., Hussain, T., Chauhan, G., Garg, T., & Goyal, AK. Fabrication and characterization of cefazolin-loaded nanofibrous mats for the recovery of post-surgical wound Artif Cells, Nanomed. Biotechnol., 2016, 44(8), 1783–1792.
[251] Bui, HT., Chung, OH., Dela Cruz, J., & Park, JS. Fabrication and characterization of electrospun curcumin-loaded polycaprolactone-polyethylene glycol nanofibers for enhanced wound healing, Macromol. Res., 2014, 22(12), 1288–1296.
[252] Jin, L., Wang, T., Zhu, M-L. et al. Electrospun fibers and tissue engineering, J. Biomed. Nanotechnol., 2012, 8(1), 1–9.

[253] Vert, M., Hellwich, K. . et al. Terminology for biorelated polymers and applications (IUPAC Recommendations 2012), Pure Appl. Chem., 2012, 84(2), 377–410.
[254] Toxicology / Biocompatibility. Geneva Laboratories, 2018. (Accessed March 27, 2018 http://www.genevalabs.com/testing-services_toxicology-biocompatibility/
[255] Zhang, H., Zhou, L., & Zhang, W. Control of scaffold degradation in tissue engineering: a review, Tissue Eng. Part B, 2014, 20, 5.
[256] Liu, H., Wang, S., & Qi, N. Controllable structure, properties, and degradation of the electrospun PLGA/PLA-blended nanofibrous scaffolds, J. Appl. Polym. Sci., 2012, 125, E468–76.
[257] Wahabzada, M., Besser, M., Khosravani, M. et al. Monitoring wound healing in a 3D wound model by hyperspectral imaging and efficient clustering, PLoS One, 2017, 12(12), e0186425.
[258] Gholami, P., Ahmadi-Pajouh, MA., Abolftahi, N., Hamarneh, G., & Kayvanrad, M. Segmentation and measurement of chronic wounds for bioprinting, IEEE J. Biomed. Heal. Inform., 2017, 99, 1–8.
[259] Zhang, TH., Xu, SG., Zhang, DZ., & Wulamu, A. Degradation modeling research of biodegradable tissue engineering scaffold, Adv. Mater. Res., 2014, 873, 642–651.
[260] Ahire, JJ., Hattingh, M., Neveling, DP., & Dicks, LMT. Copper-containing anti-biofilm nanofiber scaffolds as a wound dressing material, PLoS One, 2016, 11(3), e0152755.
[261] Mohiti-Asli, M., Risselada, M., Jacob, M., Pourdeyhimi, B., & Loboa, EG. Creation and evaluation of new porcine model for investigation of treatments of surgical site infection, Tissue Eng. Part C Methods, 2017, 23, 11.
[262] Ranjbar-Mohammadi, M., Rabbani, S., Bahrami, SH., Joghataei, MT., & Moayer, F. Antibacterial performance and in vivo diabetic wound healing of curcumin loaded gum tragacanth/poly(ε-caprolactone) electrospun nanofibers, Mater. Sci. Eng. C, 2016, 69, 1183–1191.
[263] Liu, X., Lin, T., Fang, J. et al. In vivo wound healing and antibacterial performances of electrospun nanofibre membranes, J. Biomed. Mater. Res. Part A, 2010, 94A(2), 499–508.
[264] De Faria, AF., Perreault, F., Shaulsky, E., Arias Chavez, LH., & Elimelech, M. Antimicrobial electrospun biopolymer nanofiber mats functionalized with graphene oxide-silver nanocomposites, ACS Appl. Mater. Interfaces, 2015, 7(23), 12751–12759.
[265] Rodríguez-Tobías, H., Morales, G., & Grande, D. Improvement of mechanical properties and antibacterial activity of electrospun poly(D,L-lactide)-based mats by incorporation of ZnO-graft-poly(D,L-lactide) nanoparticles, Mater. Chem. Phys., 2016, 182, 324–331.
[266] Amina, M., Hassan, MS., Al Musayeib, NM., Amna, T., & Khil, MS. Improved antibacterial activity of HAP garlanded PLGA ultrafine fibers incorporated with CuO: Synthesis and characterization, J. Sol-Gel Sci. Technol., 2014, 71(1), 43–49.
[267] Vashisth, P., Nikhil, K., Pemmaraju, SC. et al. Antibiofilm activity of quercetin-encapsulated cytocompatible nanofibers against Candida albicans, J. Bioact. Compat. Polym., 2013, 28(6), 652–665.
[268] Keong, LC., & Halim, AS. In vitro models in biocompatibility assessment for biomedical-grade chitosan derivatives in wound management, Int. J. Mol. Sci., 2009, 10(3), 1300–1313.
[269] Shan, YH., Peng, LH., Liu, X., Chen, X., Xiong, J., & Gao, JQ. Silk fibroin/gelatin electrospun nanofibrous dressing functionalized with astragaloside IV induces healing and anti-scar effects on burn wound, Int. J. Pharm., 2015, 479(2), 291–301.

Baturalp Yalcinkaya, Fatma Yalcinkaya

12 Nanofibers in liquid filtration

Abstract: Fabrication of nanofibrous material has become the center of attraction by researchers due to their unique physical properties, for instance, high porosity and super-thin fiber diameter. Electrospinning of various polymers has gained advanced chemical properties to the nanofibrous materials such as permeability besides physical ones. During the last decade, the interest in the use of the nanofibrous material, in particular, has emerged in water treatment. However, even now, the main limitation of using the nanofibrous material, especially in liquid filtration, is its weak properties and low adhesion to the other surface. Hence, this chapter aims at introducing nanofibers for liquid separation, overcoming the mechanical properties issue of nanofiber layer and preparing them as a membrane for liquid separation.

Keywords: nanofiber, electrospinning, nanofiltration, membrane, reverse osmosis, desalination, lamination, water treatment

12.1 Introduction

Water is a primary nutrient for all living beings. Living organisms constitute about 50–95% of water. About 70% of Earth's surface is covered by water; of the 70%, 96.5% is present in seas and oceans. The remaining 1.7% is present in groundwater, 1.7% in ice caps and glaciers, and others in the air as vapor or clouds. Only 3% of the Earth's water is freshwater, and 1% of freshwater is safe for drinking [1]. Regrettably, due to intense industrialization, increase in population and overurbanization, the ratio of potable water is decreased day by day.

Many continents, countries or regions are under threat of water scarcity. Many governments are searching for new potable water alternatives due to its scarcity. However, alternatives are not always available for some countries, and therefore, desalination has been the center of interest.

Desalination is an efficient water remediation process that separates highly concentrated salty water into a freshwater stream of low concentration and is an alternative source of drinkable water available today.

Baturalp Yalcinkaya, Institute for Nanomaterials, Advanced Technologies and Innovation, Technical University of Liberec, Studentska Liberec, Czech Republic
Fatma Yalcinkaya, Institute for New Technologies and Applied Informatics, Faculty of Mechatronics, Studentska Liberec, Czech Republic

https://doi.org/10.1515/9783110581393-012

In the early 1970s, semipermeable membrane is located at the heart of the desalination process. These membranes are permeable to water but reject salts and suspended solids [2]. In the last four decades, polymer-based filter media have been used for liquid filtration, especially high-performance (with ion rejection of 99.5–99.8%) reverse osmosis (RO) membranes. In the recent years of the twentieth century, fibrous materials have been considered as filter media. Especially, nonwoven webs are used for air and liquid filtration such as gas filters, car filters, medical filtration processes and swimming pool filters.

Nonwovens include a wide variety of technology, and the classification of nonwovens related to the production process is based on the selection of specific operations in each of the three steps: dry laid, wet laid and spun laid. Drylaid is distinguished from the four different forming methods such as carded, air laid, a combination of carded and air, and electrostatic laid [3]. Nonwoven filter media are ideal for filtration applications due to its high internal surface area and porosity that provides high liquid flux and dirt load capacity. Usage of nonwovens in water treatment is one of the fastest growing segments in the separation and filtration industry. Due to its low cost and increased efficiency, nonwoven materials have found main applications such as home furnishing, geotextile fabrics, roofing, surgical gowns, supporting materials or filters. Nonwoven materials can replace other forms of media such as glass, paper and carbon. Even though the nonwoven materials have enormous properties, conventional nonwovens are still incapable of separating the submicron particle due to its large pore size.

With the new era in material science, the so-called nanotechnology, the finer fibers that have lower diameter than any conventional fiber or higher porosity of fabrics are possible to produce from polymeric materials. Nanofibers became a popular term in the past decades and are rapidly growing in the application of filtration technology. Donaldson Company, USA, has already commercialized nanofiber filter media, which consisted of 10 µm sized cellulose fibers and 250 nm sized nanofibers, the so-called Ultra-Web [4]. Usage of nanofibrous materials is still a big challenge in liquid filtration. They exhibit good filtration performance in microdimensional particles. However, when it comes to nano- or subnanodimension, unaided nanofibrous layers are failed to satisfy. It will take to next decade or two to realize if nanofibrous filter materials are the serious rescuer of drought in the world.

12.2 Nanofibers

Fiber is a raw material, which is the beginning point of all production of textile surfaces. Fibers are bunched together to create yarn, woven or nonwoven fabrics. The textile industry had one's share from nanotechnology, and the thinnest form of the fiber, the so-called nanofiber, has been produced by electrospinning method. Fibers with a diameter range d = 10–900 nm can be readily electrospun into mats [5]. There are several methods to produce nanofibers. Some of these methods are drawing [6], melt blowing [7], template synthesis [8–9], self-assembly [10–11], phase separation [12], island in the sea [13] and force spinning [14, 15]. Among all, electrospinning is the most common and mostly used method for production of nanofibers. Electrospinning is the mostly used method for industrial application.

12.3 Electrospinning

Electrospinning is the most popular and influential nanotextile fabrication technology because of its advantages such as simple setup and the spinning of a wide range of polymers. Needle electrospinning was patented by Formhals first in 1930 [16], and at present, needle electrospinning is still one of the most important methods to produce nanofibers for lab application. The biggest step in the production methods of nanofiber was the invention of roller electrospinning methods [17]. This method is unique to produce nanofibers at industrial scale and was commercialized by Elmarco under the trade name "Nanospider." Nanospider equipment can produce membranes collected from fibers ranging from 50 to 600 nm in diameter [18–20]. The quality or productivity of nanofiber fabrics is affected by parameters that are involved: condition process, solution and environmental parameters. There are numerous researches for investigating of impacts of parameters on the nanofibers quality during electrospinning. For instance, the concentration of the solution, applied voltage, the distance between the electrode, humidity of chamber and additive in solution have a direct effect on the nanofiber morphology and productivity [21–23]. So far, by using Nanospider electrospinning system, various polymeric solutions were used to produce nanofibers in various applications as shown in Table 12.1.

One of the successful application areas of nanofiber is water purification. Nanofiber membranes in membrane desalination application are discussed in the next section.

12.4 Membranes

An exact definition of membranes always becomes difficult; a general definition may be a selective barrier between two phases, the term selective being inherent to a membrane or a membrane process [48]. The one side of phase is called as feed, whereas the other is called as permeate. Species float on membranes from one phase to another under the effect of a driving force (Figure 12.1). Driving force may be pressure, temperature or differences in concentration that act on the components in the feed. Type and the structure of the membranes determine their performance and the

Table 12.1: Electrospun polymers and application using needle-free electrospinning system.

Nanofiber	Application
CS, PVA, cellulose acetate, PUR/copper oxide	Air filtration [24–27]
CS/PCL, collagen, gelatin, PCL and PLGA	Tissue engineering [28–29]
Cellulose, PA6, PVDF, PAN, PVDF/PAN, PVDF, PUR, PP	Membrane for water purification [30–35]
Gelatin, silk, PCL, CS-PEO, PA6, dextran	Biomedical applications [34, 36–40]
PVA, PA6	Acoustics [41, 42]
PVA, PVB, PUR, PAN	High-performance apparels [43–46]
PVDF	Piezo applications [47]

CS, chitosan; PVA, poly(vinyl alcohol); PUR, polyurethane; PCL, poly(caprolactone); PLGA, poly (L-lactide-*co*-glycolide); PA6, polyamide 6; PAN, polyacrylonitrile; PP, polypropylene; PVDF, polyvinylidene fluoride, PVB, polyvinyl butyral.

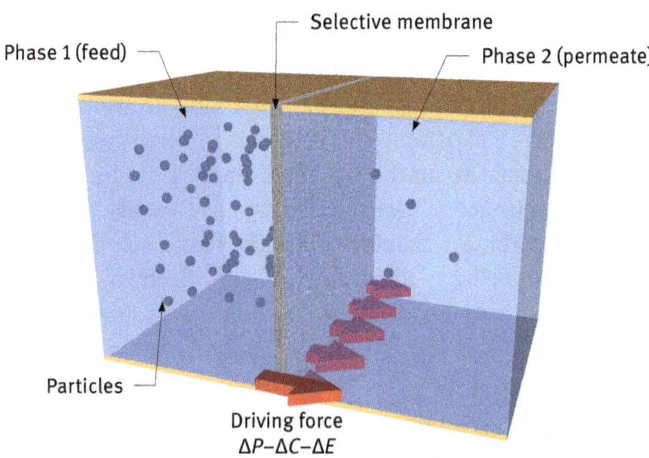

Figure 12.1: Schematic representation of a two-phase system separated by a membrane.

selectivity. In this chapter, the membranes will be investigated according to their structure, filtration method and separation properties.

12.4.1 Membrane structure

In general, a membrane can have a symmetric or asymmetric, homogeneous or heterogeneous, thick or thin structure, and transport can be active or passive. Also, membranes can be natural or synthetic, neutral or charged [48] as shown in Figure 12.2.

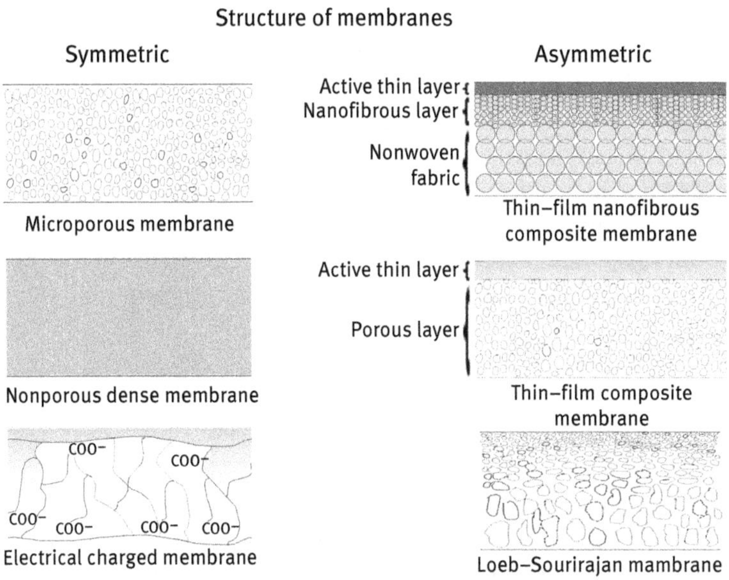

Figure 12.2: Schematic diagrams of the types of membrane.

Membranes can be classified according to their cross-sectional views to clarify the structure of membranes. If we confine ourselves to solid synthetic membranes, two kinds of the membrane may be distinguished, that is, symmetric and asymmetric membranes. The types of membrane are illustrated schematically in Figure 12.2 and described briefly later.

12.4.1.1 Symmetric membranes

The symmetric membranes are molecular, chemically or physically, homogeneous, and uniform in composition and structure. Microporous membrane has a rigid, highly voided structure with randomly distributed, interconnected pores that are in the order of 0.01–10 µm. All particles larger than the largest pores reject entirely by the microporous membranes. Nonporous, dense membranes consist of a dense film through which permeants are transported by diffusion under the driving force of pressure, concentration or electrical potential gradient. Electrically charged membranes have pore walls carrying positively, or negatively charged ions and can be dense or mostly finely microporous [49].

12.4.1.2 Asymmetric membranes

A breakthrough in industrial applications was the development of the asymmetric membranes. Asymmetric membranes have a thickness of 0.1–0.5 µm supported by a porous layer with a thickness of about 50–150 µm which has a very dense top layer or skin. Asymmetric membranes are a combination of the high selectivity of a dense membrane with the high permeation rate of a gaunt membrane. The resistance to mass transport is determined largely or entirely by the thin top layer [48]. In the composition of membranes, the top layer and sublayer originate from different polymeric materials, and each layer can be optimized independently such as in thin-film nanofibrous composite (TFNC) membranes. The layer of TFNC membranes can be a nonwoven fabric, electrospun nanofibrous material, and the thin active layer is produced individually and gathered together to create the filter media.

12.4.2 Filtration methods

The membrane filtration process is classified into two different methods: dead-end and cross-flow filtration. Feed solutions are applied perpendicular to the membrane surface in the dead-end filtration, whereas feed solutions are applied horizontally to the membrane surface in cross-flow filtration (Figure 12.3). In the dead-end filtration method, membranes particularly tend to show the fouling effect because of accumulation on the membrane surface. Eventually, the fouling effect ends up with a stop in filtration. Despite the dead end, the cross-flow filtration method has reduced fouling as far as possible and increased productivity.

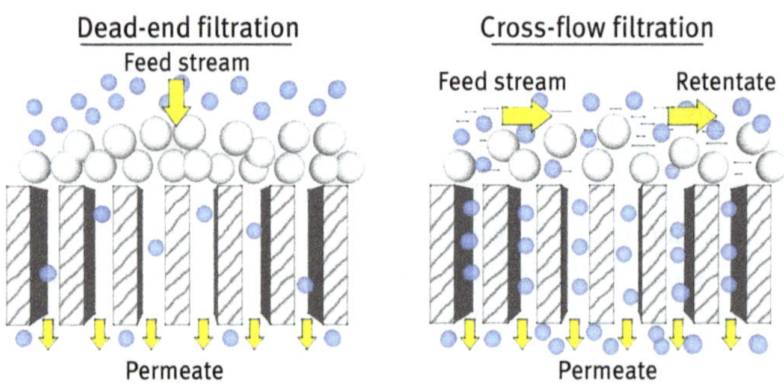

Figure 12.3: Different filtration methods.

12.4.3 Membrane separation properties

The most important properties in all fabrication systems are productivity and quality. In all kinds of filtration process, the quality is always represented by selectivity.

In liquid filtration, productivity is named by flux. The performance of a membrane is crucially related to selectivity and flux. These parameters are influenced directly by the structure of the membrane such as porosity, hydrophilicity or chemical structure. The content of feed solution and driving force also affects the performance of membranes.

12.5 Membrane transport theory

The membranes can control the rate of permeation of various species, which is one of the most important properties of them. The two models are used to describe the mechanism of permeation such as pore flow and solution diffusion models (Figure 12.4). In the pore flow model, permeants are transported by pressure-driven convective flow through pores. Separation occurs by one of the permeants (particles or impurities), which are filtered through the pores of membranes while other permeants flow through the membranes freely. The second model is the solution diffusion model. The membranes that have no pore structure, in which permeants dissolve in the membrane material, penetrate through the membrane to a concentration gradient [49].

The relative size and permeance of pores are main factors to determine differences between pore flow and solution diffusion mechanism. The solution diffusion model may be described by the Fick's law:

$$Ji = -Di\frac{dci}{dx} \tag{12.1}$$

where J_i is the rate of transfer of component i or flux and dc_i/dx is the concentration gradient of component i. D_i is called the diffusion coefficient and is a measure of the mobility of the individual molecules [49].

The pores in the membrane are tiny spaces between polymer chains by the thermal motion of the polymer chain. These pores appear and disappear on about the same timescale as the motions of the permeants traversing the membrane. On the other hand, the pore flow model may be described by Darcy's law:

$$Ji = K'ci\frac{dp}{dx} \tag{12.2}$$

where dp/dx implies the pressure gradient existing in the porous medium, c_i is the concentration of component i in the medium and K' is the coefficient reflecting the nature of the medium [49].

The pores are slightly large and fixed. The pores do not change in position or volume during permeant motion. These pores are usually connected to one another.

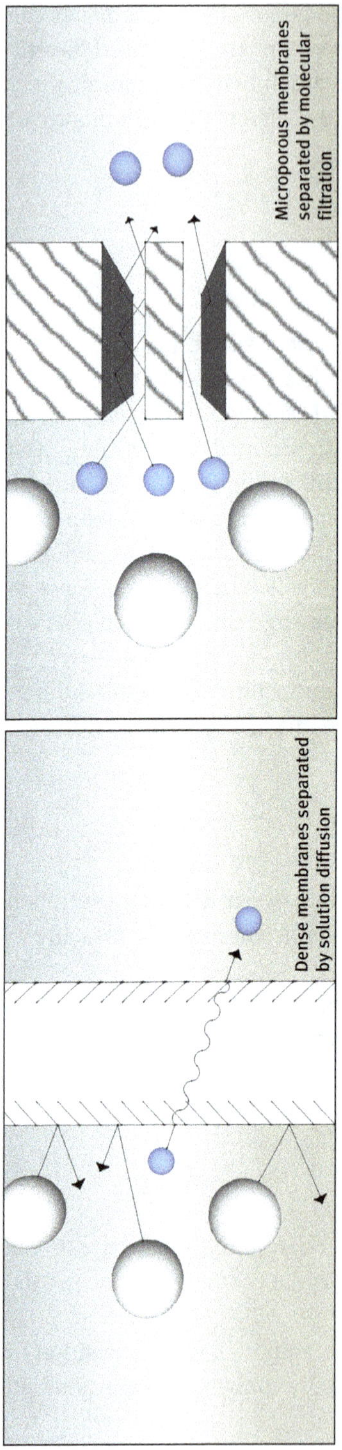

Figure 12.4: Transport of permeants by pore flow and solution diffusion.

The larger the individual pores, the more likely they are to be present long enough to produce pore flow characteristics in the membrane.

It is hard to distinguish the membranes based on their average pore size due to insufficient measurement technique of membrane pores. Therefore, the measurement of the size of molecules that permeate the membranes has been adapted and determined as the general groups of membranes (Figure 12.5).

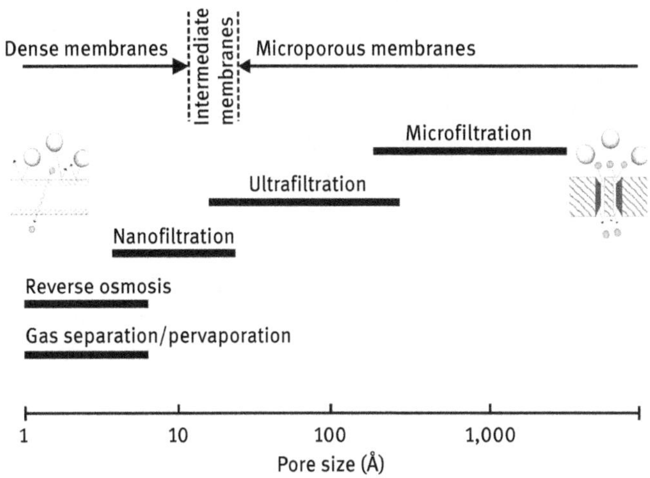

Figure 12.5: Schematic representation of the nominal pore size of membranes.

Ultrafiltration (UF) and microfiltration (MF) membranes are microporous because their pore diameter is larger than 10–15 Å and transport occurs by pore flow.

The membranes that contain a dense selective active layer with no visible pores are RO, pervaporation and polymeric gas separation membranes. The diameter of pores in these membranes is between 2 and 5 Å and the solution diffusion model describes the transport of permeants.

Nanofiltration (NF) membranes contain pores as support and dense active layer at the same time. NF membranes are placed between UF membranes due to the porous supporting structure and RO membranes clearly dense active layer. In these membrane types, pores have diameters between 5 and 15 Å [49].

12.6 Nonwoven and nanofibrous scaffolds in liquid filtration

There are several unique features of nonwoven fabrics and nanofibrous scaffolds; the surface porosity of nanofibrous scaffolds is similar to the bulk porosity, which is closer to 80% or higher. The surface pore structures are all interconnected throughout the layer, so there are no dead-end pores in nonwoven fabrics and nanofibrous scaffolds [50]. According to the pore size of the materials and driven pressure, electrospun nanofibrous scaffolds have been used in MF, UF, NF and forward osmosis (FO).

12.6.1 Nanofibers in microfiltration

In the last decade, electrospun nanofibrous scaffolds in liquid filtration have taken place at the MF level. In one of the first practices, cellulose nanofiber membranes were prepared by electrospinning as affinity membranes, whose fiber diameters are between 200 nm and 1 µm [51]. The nanofibrous scaffold was treated in NaOH solution in H_2O/ethanol to obtain regenerated cellulose (RC) nanofiber mesh. The Cibacron blue RC nanofiber membrane has a Cibacron blue content of 130 µmol g^{-1}, and capture capacity of 4 mg g^{-1} for bilirubin and 13 mg g^{-1} for bovine serum albumin (BSA). After regeneration with elution buffer, the membrane showed reusability [51]. The filtration result of the survey showed that the electrospun membranes (EMs) have lower pressure drop and higher flux compared with commercial MF membranes.

One of earlier studies was carried out by electrospinning of polyvinylidene fluoride into the membrane with a fiber diameter of approximately 380 nm [52]. The polymeric nanofibrous membranes were used to separate 1, 5 and 10 µm polystyrene particles. The EMs were successful in rejecting more than 90% of the microparticles from solution. The results led to the use of electrospun nanofibrous materials that best replace the commercial membranes as prefilters prior to UF on NF to minimize the possibility of fouling. Another study of the author has used polysulfone (PSf) electrospun nanofibrous scaffolds and observed the ability to remove microparticles from solution. The membrane was able to remove 99% of 10, 8 and 7 µm particles without any permanent fouling. However, the membrane was observed to foul irreversibly by 2 and 1 µm particles with a "cake layer" formed on the membrane surface [53].

In another study, the electrospun nylon-6 scaffold was prepared using electrospinning method with the fiber diameter ranging from 30 to 110 nm and employed as a membrane material for water filtration. EM separated all particles with sizes from 10 to 1 µm particles. The usage of submicron particles tended to the fouling effect on the membrane surface. They indicated the importance of the influence of flow patterns such as cross-flow or through the flow to obtain better filtration performance [54].

Electrospun scaffold for the MF method is conducted on PSf and polyvinylidene fluoride with high contact angle and more hydrophobic nanofibers. High hydrophilic EMs were generated by blending with several different types of surface-modifying macromolecules. The contact angle of blended EM reduced from

140 to 54 . The pure water flux of blended EM was 20% higher than that of non-blended EM [55].

The poly(trimethylene terephthalate) (PTT) nanofibers with an average diameter of 145 nm for MF also have been produced by wet-laid process instead of electrospinning. Heat application and blending with polypropylene (PP) nanofibers were exposed to enhance the mechanical properties of PTT nanofibers. The water flux and filtration efficiency have been investigated using TiO_2 suspensions with the average diameter of 100 nm. It was found that the rejection performance of membranes was above 99.6% with high flux MF media [56].

One of the latest studies about solid–liquid clarification of a lignocellulosic hydrolyzate and fermentation broth containing yeast cells has been investigated. The performance and fouling characteristics of electrospun polyimide nanofiber membrane were compared with commercial microfilters. All three microfilters [polyethersulfone (PES) and polyvinylidene fluoride] were showed lower flux, and irreversible fouling compared the PI membrane. Moreover, PI nanofibers membranes did not show any permanent decrease in flux during cell separation or solid-liquid clarification. The authors suggested the PI nanofibers membrane would have a high potential for solid–liquid clarification and cell recycle/removal operations within a biorefinery process [57].

12.6.2 Nanofibers in ultrafiltration

Another method of membrane system so-called UF that researchers have focused on high flux and low fouling effect using the electrospun nanofibrous material. In one of the study, high flux and long-term performance of UF membrane have been formed based on poly(vinyl alcohol) (PVA) electrospun nanofibrous scaffold support and PVA hydrogel coating. It was found that the 96% hydrolyzed PVA with relatively high molecular weight electrospun scaffold exhibited excellent overall mechanical performance. Results indicated that such unique hydrophilic nanofibrous composite membranes exhibited a flux rate (>130 L m^{-2} h^{-1}) significantly higher than commercial UF membranes but with similar filtration efficiency (rejection rate >99.5%) [58].

In another study, an UF membrane was produced by polyacrylonitrile electrospun nanofibrous scaffold coupled with thin top layer coating of chitosan. The electrospun polyacrylonitrile (PAN) scaffold with an average diameter of 124–720 nm. Various concentrations of PAN (4–12%) was electrospun to form a three-tier composite membrane on the nonwoven supporting material. Three-tier composite membranes exhibited flux rates that could be an order of magnitude higher than the commercial NF filter (e.g., NF 270 from Dow) after 24 h operation while they maintained excellent filtration efficiency with rejection ratios better than 99.9% [59].

Polyaniline/PSf membranes have been prepared the filtration of PANI nanofiber aqueous dispersion with PS substrate membrane. PANI nanofibers were prepared by using chemical oxidative polymerization method. The filtration performance that permeates and rejection properties of membranes were tested using the cross-flow equipment. Pure water, PEG aqueous solution, and BSA aqueous solution were used as feed solution. The results indicate that PANI/PS membranes were performed higher hydrophilicity and enhancement in permeability but no changes in rejection performance. However, the antifouling properties of PANI/PS membranes showed much better performance than PS substrate during filtration of BSA solution [60].

One of the first attempts the usage of electrospun nanofibrous in NF has been made real by Ritcharoen et al. [61] They electrospun cellulose acetate (CA) fiber mat and coated fiber surface with chitosan/sodium alginate and poly(styrene sulfonate) to obtain electrostatic multilayer. They indicated that the water flux was decreased with an increase in the number of the bilayers. For instance, they obtained that the water flux was in the range of 60 and 40 L m^{-2} h^{-1} for 15 and 25 bi-layered membranes, respectively. They have found the sodium chloride solution in flux was lower than the pure water, naturally, because of osmotic pressure. As a result of the filtration process, the rejection level of NaCl for 15 and 25 bi-layered membranes was between the range of 6 and 15%, respectively.

12.6.3 Nanofibers in nanofiltration

As mentioned above studies, neat nanofiber web itself cannot function and withstand to use in NF processes due to their low mechanical strength to withstand applied pressure. As a reason, many researchers have functionalized the surface of nanofiber webs using interfacial polymerization (IP). One of the first attempt to use porous nanofiber layer on NF process is done by electrospun PAN nanofibers. The PAN nanofibrous mid-layer was used as a support web in high flux and permeability TFNC membranes. The active barrier layer prepared by IP of piperazine, trimesoylchloride and some additives was used on the nanofibrous layer. Using magnesium sulfate ($MgSO_4$) divalent salt, the membrane performance was evaluated. It was found that the amount of piperazine was played a critical role in IP to optimize the flux and rejection performance. The TFNC membrane was showed higher permeate fluxes (21–42%) as well as higher rejection rates (2–22%) than those of TFC membranes [50]. Also, TFNC membrane was exhibited 38% permeate flux increase while maintaining rejection to that NF270 at 70 psi.

Another study has aimed an enhancement of the electrospun nanofibrous membranes performance of NF process. PES was dissolved in dimethyl formamide (DMF) and DMF/NMP (*N*-methyl-pyrrolidinone) mixture and spun onto polyethylene terephthalate (PET) nonwoven backing layer. The reason for solvent mixture selection increases the adhesion between the formed PES nanofibrous layer and PET nonwoven scaffold. The active layer was formed onto nanofibrous scaffold surface by using IP method with some ionic liquids additive to monomer solutions. Compared to commercial membrane NF-90, the TFNC membranes which were prepared with ionic liquids additive was performed two times higher flux with comparable salt rejection ratio, and comparable permeation flux and salt rejection performance as those of NF-270 [62].

One of the most important studies about polymeric nanofibrous composite membrane was reported by Kaur et al. [63]. The effect of polymeric nanofibrous structure on the thin-film composite membrane and their performance in the NF process have been investigated. In the first report, the influence of fiber diameter on the filtration performance has been investigated. Thus, various concentrations of PAN solution were electrospun onto the nonwoven scaffold. The separation performance was done using dead-end filtration cell with 2,000 ppm $MgSO_4$ feed solution. The result showed that as the fiber diameter decreased, the pore size also decreased, and the separation of salts increased. When the thickness of the electrospun layer was decreased together with lower pore size, the flux with a high salt rejection. The same author was observed the hot pressing effect on the electrospun PAN nanofibrous membrane properties and the separation of salt after IP. The reason for applying hot press is to increase the mechanical properties and adhesion between nanofibrous and nonwoven scaffolds. The NF result showed that the higher applied hot pressure membrane exhibited higher rejection and lower flux than those of TFNC membranes [64].

12.6.4 Nanofibers in forward osmosis

The technology of FO is a membrane-based separation technique, and their application areas are sustainable energy, resource recovery and water production. PES and PSf nanofibers were collected onto the PET backing nonwoven layer to prepare FO membrane using electrospinning method. The MPD (1,3-phenylene diamine) and TMC monomers were used to form an active barrier layer onto the fibrous supporting layer. This type of membrane had high water flux and low salt rejection [65].

In another study, authors used that electrospun nylon-6,6 nanofibers due to its high hydrophilic and mechanical properties compared to other nanofiber materials. To obtain high permselectivity on the FO membrane, active barrier layer was formed onto nanofibers supporting midlayer. TFC membranes were fabricated using cosolvent such as acetone to get different water permeance and salt selectivity. These membranes have low salt flux compared to the HTI cellulose acetate membrane [66].

12.7 Improved mechanical strength of nanofibers in liquid filtration

The poor mechanical strength of nanofibers and their low adhesion to a supporting layer have always been a stumbling block to prepare nanofibrous membranes in liquid filtration. So far, many researchers have been trying to solve and improve the mechanical issues of polymeric nanofibers. There are many studies about the improvement of mechanical properties of nanofibrous layer. Some studies focused on the change in chemical structure of the nanofibers while some of them developed a new mechanism for production method [67–71].

Some of the researchers were focused on increasing adhesion of interfibers to improve the mechanical strength of membranes. The mixed solvent system (DMF and NMP) were used to prepare a solution of PES, and then the PES polymer solution was electrospun using needle electrospinning. The different solvent mixed system contains various vapor pressure of solvents. Hence, the nanofiber web supported by a stronger material was still partly wetted because of the high vapor pressure of NMP. As a result, adhesion between fibers was improved. The results indicated that the mechanical strength of PES membranes was improved significantly. However, the average fiber diameter was increased directly proportional to the mechanical strength [72]. The same mixed solvent systems for electrospun PES membranes were mentioned earlier to prepare NF filter [62].

As an alternative to the mixed solvent system, EMs were exposed to the vapor of solvents to improve their mechanical strength. Using a needle electrospinning system, the PAN and PSf in DMF solutions were prepared. After spinning, the polymeric PAN and PSf nanofibers were exposed to solvent vapor for a different period. As a result, smaller pore size and fiber fusion at junction points were observed due to fiber swelling. Permeability results showed that treated membranes had lower permeability than the untreated ones, thereby enhancing the mechanical strength [73]. Some of the authors investigated the effect of chemical modification on the improvement of mechanical properties of PAN and PSf electrospun web. The chemical modification involves the coating of polydopamine (PDA), which is a hydrophilic polymer. The PDA-modified electrospun web was performed with higher strength and improved hydrophilicity [74].

Razavi et al. [75] used electrospun PAN nanofibers incorporated in an epoxy-based adhesive layer to improve the adhesive joint's mechanical performance. Compared to neat PAN nanofiber web, the epoxy-reinforced PAN nanofibers indicated that the fracture energy increased from an initial value of 0.11–0.25 N mm^{-1} for the reinforced adhesive with the maximum obtained enhancement as 127%. Even though the epoxy-based adhesives have been considered as promising and the most commonly used adhesives, these types of adhesives mainly suffer from low fracture toughness.

Kanafchian et al. [76] used melt adhesives to adhere nanofiber layer on a substrate. The strong nanofibrous composite fabrics were prepared by electrospinning PAN nanofibers onto a nonwoven substrate and incorporating into fabric system via hot press method using various temperatures. It was found that by increasing the laminating temperature to the melting point, the adhesive layer gradually melts, spreads and covers on the nanofiber web surface. When the laminating temperature arranged to melting point of the adhesive web, the nanofiber web began to be damaged. As a result, the adhesive layer completely melted and penetrated into nanofiber web and blocks its pores. In another study, it was found that above the melting point of the adhesive layer, the PAN nanofiber web began to damage at lamination temperatures and the adhesive force between layers increased by increasing the lamination temperature [77].

Yao et al. [78] investigated the effect of heat-press temperature, pressure and duration on the morphological and mechanical characteristics of the EM for desalination by membrane distillation. A PP filter layer was used as a supporting layer for the electrospun polyvinylidene fluoride–cohexafluropropylene (PH). Using heat-press treatment, an improvement of mechanical strength, permeation, salt rejection performance and liquid entry pressure was achieved on EMs.

Yoon et a. [79] prepared electrospun polyurethane (PUR) nanofiber webs and laminated on different supporting substrate fabrics using a mesh roller and PUR adhesive. First, the PUR nanofiber web was laminated to a regular polyester fabric, and then a nylon tricot-knitted fabric was laminated to the other side of the web to create a nanofiber web sandwiched between the two substrate fabrics. Results showed that composite structure design had a considerable influence on the degrees of breathability and waterproofness. The findings suggest that nanofiber webs should be incorporated into layered composite structures in such a way as to achieve a combination of high barrier and comfort performance.

In previous works [30, 80, 81], a special lamination technology with a fusing machine was used to adhere nanofibers onto a spunbond nonwoven surface. In this technology, a low melting point adhesive web is used between nanofiber and supporting layer. Under the constant heat and pressure, nanofibers adhesive web-supporting material is moving on a conveyer belt. During the heat press in a fusing equipment, adhesives melted and adhered two surfaces as shown in Figure 12.6.

In this technology, it is crucial to arrange pressure, heat and speed of membranes in the fusing machine. Overheat, pressure or slow speed of the material in a fusing machine can cause the damage of nanofibers. Moreover, the adhesive can cover all the surface of nanofibers and block the pores, which will result in a decrease in air and water permeability. Figure 12.7 shows the laminated nanofiber surface with an adhesive web. The adhesive web covers the surface partially and blocks the pores under its surface.

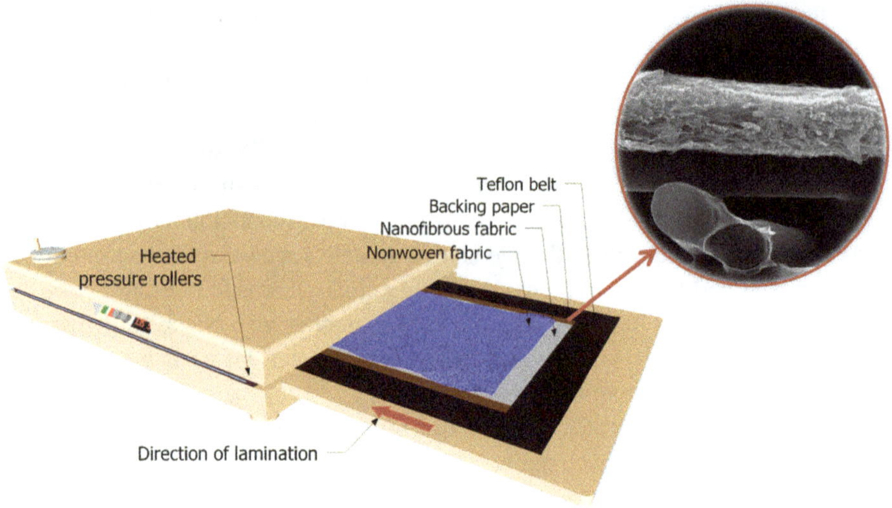

Figure 12.6: Lamination method and equipment.

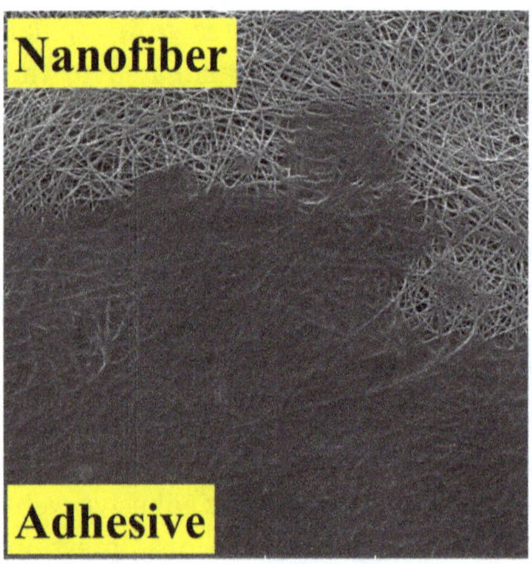

Figure 12.7: Laminated nanofiber web from the top view.

Depending on the supporting material under nanofiber layer, the mechanical strength of laminated and wetted nanofiber composite membranes could be improved 3 to 24 times higher than the single nanofiber layer [82].

12.8 Conclusion

Water desalination process is a viable method to obtain potable water supply, especially for the countries that suffer from lack of freshwater. Nowadays, RO and FO are widely applied in desalination due to their relatively high efficiency. Membranes play a main role in liquid filtration process. Hence, the application area of nanofibers is getting more attention; EMs possess some advantages involving adjustable pore size and membrane thickness, high porosity, high hydrophilicity and the large surface area that make them attractive candidates for liquid filtration.

Electrospinning technology has already been used in industrial production, and production of electrospun nanofibrous layers has become easier than ever before. Unfortunately, poor mechanical properties of electrospun layers have been delaying to find a place in the final product, especially in water treatment. The most convenient and easiest way to use electrospun nanofibrous materials is to combine nanofibrous materials with another strong material or surface. However, its easily deformable structure has limited the manipulation of nanofibrous material.

This chapter explained that the mechanical poorness of nanofiber webs was overcome by a lamination method. It is important to avoid any damage to the nanofibrous structure during the lamination process. Therefore, choosing a lamination technique is quite crucial and has a significant impact on the final structure of the supporting material.

Conflict of interest

The authors declare that they have no competing interests. This chapter is originated from author's own Ph.D. thesis.

Abbreviations

PAN	Polyacrylonitrile
PVDF	Polyvinylidene fluoride
PVA	Polyvinyl alcohol
PUR	Polyurethane
PS	Polystyrene
CA	Cellulose acetate
RC	Regenerated cellulose
BSA	Bovine serum albumin
PTT	Poly-trimethylene terephthalate
PP	Polypropylene

PANI	Polyaniline
PEG	Polyethylene glycol
PES	Polyether sulfone
NMP	*N*-Methyl-pyrrolidinone
PET	Polyethylene terephthalate
PSf	Polysulfone
MPD	1,3-Phenylene diamine
TMC	Trimesoyl chloride
PDA	Polydopamine
PH	Polyvinyline fluoride-cohexafluoropropylene
TiO_2	Titanium dioxide
TFC	Thin-film composite
TFNC	Thin-film nanofibrous composite
J_i	Flux
dc_i/dx	Concentration gradient
D_i	Diffusion coefficient
K	Coefficient of medium nature
c_i	Concentration
Å	Angstrom
MF	Microfiltration
NF	Nanofiltration
UF	Ultrafiltration
FO	Forward osmosis
RO	Reverse osmosis
NaOH	Sodium hydroxide
NaCl	Sodium hydroxide
$MgSO_4$	Magnesium sulfate
EM	Electrospun membranes
IP	Interfacial polymerization
NF-270	Commercial nanofilter
DMF	Dimethyl formamide
ΔP	Pressure differences
ΔC	Concentration gradient
ΔE	Electrical potential differences

References

[1] “Drinking water,” 2017. [Online]. Available: https://en.wikipedia.org/wiki/Drinking_water. [Accessed: 01-Jan-2017].

[2] Ribeiro, J. “Desalination technology – Survey and prospects”, European Joint Research Center, IPTS, Seville, Spain, August 1996.

[3] Jirsák, O., &Wadsworth, L. C. Nonwoven textiles, Carolina Academic Press, Durham, N.C. 1997.

[4] Grafe, T. H., & Graham, K. M., “Nanofiber Webs from Electrospinning,” 2003.

[5] Reneker, D. H., & Chun, I. Nanometre diameter fibres of polymer, produced by electrospinning, Nanotechnology, 1996, 7(3), 216–223. Sep.

[6] Ondarçuhu, T., & Joachim, C. Drawing a single nanofibre over hundreds of microns, Europhys. Lett., 1998, 42(2), 215–220. Apr.

[7] Hassan, M. A., Yeom, B. Y., Wilkie, A., Pourdeyhimi, B., & Khan, S. A. Fabrication of nanofiber meltblown membranes and their filtration properties, J. Membr. Sci., 2013, 427, 336–344. Jan.

[8] Wang, Y., Zheng, M., Lu, H., Feng, S., Ji, G., & Cao, J. Template synthesis of carbon nanofibers containing linear mesocage arrays, Nanoscale Res. Lett., 2010, 5(6), 913–916. May.

[9] Zelenski, C. M., & Dorhout, P. K. Template synthesis of near-monodisperse1 microscale nanofibers and nanotubules of MoS2, J. Am. Chem. Soc., 1998, 4(120), 734–742.

[10] Rolandi, M., & Rolandi, R. Self-assembled chitin nanofibers and applications, Adv. Colloid Interface Sci., 2014, 207, 216–222. May.

[11] Liao, H.-S., Lin, J., Liu, Y., Huang, P., Jin, A., & Chen, X. Self-assembly mechanisms of nanofibers from peptide amphiphiles in solution and on substrate surfaces, Nanoscale, 2016, 8(31), 14814–14820. Aug.

[12] Katsogiannis, K. A. G., Vladisavljević, G. T., & Georgiadou, S. Porous electrospun polycaprolactone (PCL) fibres by phase separation, Eur. Polym. J., 2015, 69, 284–295. Aug.

[13] Zhang, Z., Tu, W., Peijs, T., & Bastiaansen, C. W. M. Fabrication and properties of poly (tetra -fluoroethylene) nanofibres via sea-island spinning, Polymer (Guildf)., 2017, 109, 321–331. Jan.

[14] Hammami, M. A., Krifa, M., & Harzallah, O. Centrifugal force spinning of PA6 nanofibers – processability and morphology of solution-spun fibers, J. Text. Inst., 2014, 105(6), 637–647. Jun.

[15] Sarkar, K., et al. Electrospinning to forcespinning™, Mater. Today, 2010, 13(11), 12–14. Nov.

[16] Formhals, A., "Process and apparatus for preparing artificial threads," US50028330A, 1930.

[17] Jirsak, O., Sanetrnik, F., Lukas, D., Kotek, V., Martinova, L., & Chaloupek, J., "A method of nanofibers production from a polymer solution using electrostatic spinning and a device for carrying out the method," WO/2005/024101, 18-Mar-2005.

[18] Yalcinkaya, F., Yalcinkaya, B., & Jirsak, O. Analysis of the effects of rotating roller speed on a roller electrospinning system, Text. Res. J., 2017, 87(8), 913–928.

[19] Yener, F., & Jirsak, O. Comparison between the needle and roller electrospinning of polyvinylbutyral, J. Nanomater., 2012, 2012, 1–6.

[20] Yener, F., Yalcinkaya, B., & Jirsak, O. On the measured current in needle- and needleless electrospinning, J. Nanosci. Nanotechnol., 2013, 13(7), 4672–4679. Jul.

[21] Yalcinkaya, B., Jirsak, O., Callioglu Cengiz, F., & Yener, F. "The effect of PVA concentration on the spinning parameters of roller electrospinning," in 18th International Conference on Structure and Structural Mechanics of Textile Fabrics (STRUTEX), 2011, pp. 345–348.

[22] Yalcinkaya, B., Cengiz Callioglu, F., & Yener, F. Measurement and analysis of jet current and jet life in roller electrospinning of polyurethane, Text. Res. J., 2014, 84(16), 1–9, 1720–1728.

[23] Yalcinkaya, F., Yalcinkaya, B., & Jirsak, O. Influence of salts on electrospinning of aqueous and nonaqueous polymer solutions, J. Nanomater., 2015, 134251.

[24] Wang, L., Zhang, C., Gao, F., & Pan, G. Needleless electrospinning for scaled-up production of ultrafine chitosan hybrid nanofibers used for air filtration, RSC Adv., 2016, 6(107), 105988–105995. Nov.

[25] Li, J., Gao, F., Liu, L. Q., & Zhang, Z. Needleless electro-spun nanofibers used for filtration of small particles, Express Polym. Lett., 2013, 7(8), 683–689.

[26] Chattopadhyay, S., Hatton, T. A., & Rutledge, G. C. Aerosol filtration using electrospun cellulose acetate fibers, J. Mater. Sci., 2015, 51(1), 204–217. Jan.

[27] Ungur, G., & Hrůza, J. Modified polyurethane nanofibers as antibacterial filters for air and water purification, RSC Adv., 2017, 7(78), 49177–49187. Oct.

[28] Erickson, A. E. et al. High-throughput and high-yield fabrication of uniaxially-aligned chitosan-based nanofibers by centrifugal electrospinning, Carbohydr. Polym., 2015, 134, 467–474. Dec.

[29] Alamein, M. A., Stephens, S., Liu, Q., Skabo, S., & Warnke, P. H. Mass production of nanofibrous extracellular matrix with controlled 3D morphology for large-scale soft tissue regeneration, Tissue Eng. Part C Methods, 2013, 19(6), 458–472. Jun.
[30] Yalcinkaya, B., Yalcinkaya, F., & Chaloupek, J. Thin film nanofibrous composite membrane for dead-end seawater desalination, J. Nanomater., 2016, 2016, 1–12. Aug.
[31] Esmaeilzadeh, I., Mottaghitalab, V., Tousifar, B., Afzali, A., & Lamani, M. A feasibility study on semi industrial nozzleless electrospinning of cellulose nanofiber, Int. J. Ind. Chem., 2015, 6(3), 193–211.
[32] Fatma, Y., Siekierka, A., Bryjak, M., & Maryska, J. Preparation of various nanofibrous composite membranes using wire electrospinning for oil-water separation, IOP Conf. Ser. Mater. Sci. Eng., 2017, 254(10), 102011. Oct.
[33] Jiříček, T., et al. Flux enhancement in membrane distillation using nanofiber membranes, J. Nanomater., 2016, 2016, 1–7. Jun.
[34] Jiříček, T., Komárek, M., Chaloupek, J., & Lederer, T. Maximising flux in direct contact membrane distillation using nanofibre membranes, Desalin. Water Treat., 2017, 73, 249–255.
[35] Li, X., Zhang, Y., Li, H., Chen, H., Ding, Y., & Yang, W Effect of oriented fiber membrane fabricated via needleless melt electrospinning on water filtration efficiency, Desalination, 2014, 344, 266–273. Jul.
[36] Lu, W., Ma, M., Xu, H., Zhang, B., Cao, X., & Guo, Y. Gelatin nanofibers prepared by spiral-electrospinning and cross-linked by vapor and liquid-phase glutaraldehyde, Mater. Lett., 2015, 140, 1–4. Feb.
[37] Sasithorn, N., & Martinová, L. Fabrication of silk nanofibres with needle and roller electrospinning methods, J. Nanomater., 2014, 2014, 1–9. Sep.
[38] Mašek, J. et al. Multi-layered nanofibrous mucoadhesive films for buccal and sublingual administration of drug-delivery and vaccination nanoparticles - important step towards effective mucosal vaccines, J. Control. Release, 2017, 249, 183–195. Mar.
[39] Maryšková, M., Ardao, I., García-González, C. A., Martinová, L., Rotková, J., & Ševců, A. Polyamide 6/chitosan nanofibers as support for the immobilization of Trametes versicolor laccase for the elimination of endocrine disrupting chemicals, Enzyme Microb. Technol., 2016, 89, 31–38. Jul.
[40] Cengiz-Callioglu, F. Dextran nanofiber production by needleless electrospinning process, E-POLYMERS, 2014, 14(1), 5–13. Jan.
[41] Ozturk, M. K., Kalinova, K., Nergis, B., & Candan, C. Comparison of resonance frequency of a nanofibrous membrane and a homogeneous membrane structure, Text. Res. J., 2013, 83(20), 2204–2210.
[42] Ulrich, T., & Kalinova, K., "Helmholtz resonators systems treated with the nanofibrous membrane," in Proceedings of ICSV24, 2017, p. 278.
[43] Li, R. et al. A facile approach to superhydrophobic and superoleophilic graphene/polymer aerogels, J. Mater. Chem. A, 2014, 2(9), 3057.
[44] Yalcinkaya, F., Komarek, M., Lubasova, D., Sanetrnik, F., & Maryska, J. Preparation of antibacterial nanofibre/nanoparticle covered composite yarns, J. Nanomater., 2016, 2016, 1–7.
[45] Yalcinkaya, F., Komarek, M., Lubasova, D., Sanetrnik, F., & Maryska, J. Producing antibacterial textile material by weaving PVB/CuO nanocomposite fiber covered yarn, Nanocon 2014, 6th Int. Conf., 2015, OCTOBER, 2014.
[46] Niu, H., Gao, W., Lin, T., Wang, X., & Kong, L. Composite yarns fabricated from continuous needleless electrospun nanofibers, Polym. Eng. Sci., 2014, 54(7), 1495–1502. Jul.
[47] Chansaengsri, K., Onlaor, K., Tunhoo, B., & Thiwawong, T. Production of polyvinylidene fluoride nanofibers by free surface electrospinning from wire electrode, Mater. Today Proc., 2017, 4(5), 6085–6090. Jan.

[48] Mulder, M. *Basic Principles of Membrane Technology*, Springer Netherlands, Springer, Dordrecht, 1996.
[49] Baker, R. W. *Membrane Technology and Applications*, John Wiley & Sons, Ltd, , Chichester, UK, 2004.
[50] Yoon, K., Hsiao, B. S., & Chu, B. High flux nanofiltration membranes based on interfacially polymerized polyamide barrier layer on polyacrylonitrile nanofibrous scaffolds, J. Membr. Sci., 2009, 326(2), 484–492. Jan.
[51] Ma, Z., Kotaki, M., & Ramakrishna, S. Electrospun cellulose nanofiber as affinity membrane, J. Membr. Sci., 2005, 265(1–2), 115–123. Nov.
[52] Gopal, R., Kaur, S., Ma, Z., Chan, C., Ramakrishna, S., & Matsuura, T. Electrospun nanofibrous filtration membrane, J. Membr. Sci., 2006, 281(1–2), 581–586. Sep.
[53] Gopal, R. et al. Electrospun nanofibrous polysulfone membranes as pre-filters: Particulate removal, J. Membr. Sci, 2007, 289(1–2), 210–219. Feb.
[54] Aussawasathien, D., Teerawattananon, C., & Vongachariya, A. Separation of micron to sub-micron particles from water: Electrospun nylon-6 nanofibrous membranes as pre-filters, J. Membr. Sci., 2008, 315(1–2), 11–19. May.
[55] Kaur, S., Rana, D., Matsuura, T., Sundarrajan, S., & Ramakrishna, S. Preparation and characterization of surface modified electrospun membranes for higher filtration flux, J. Membr. Sci., 2012, 390–391, 235–242. Feb.
[56] Li, M., Wang, D., Xiao, R., Sun, G., Zhao, Q., & Li, H. A novel high flux poly(trimethylene terephthalate) nanofiber membrane for microfiltration media, Sep. Purif. Technol., 2013, 116, 199–205. Sep.
[57] Gautam, A. K., Lai, C., Fong, H., & Menkhaus, T. J. Electrospun polyimide nanofiber membranes for high flux and low fouling microfiltration applications, J. Membr. Sci., 2014, 466, 142–150. Sep.
[58] Wang, X., Fang, D., Yoon, K., Hsiao, B. S., & Chu, B. High performance ultrafiltration composite membranes based on poly(vinyl alcohol) hydrogel coating on crosslinked nanofibrous poly(vinyl alcohol) scaffold, J. Membr. Sci., 2006, 278(1–2), 261–268. Jul.
[59] Yoon, K., Kim, K., Wang, X., Fang, D., Hsiao, B. S., & Chu, B. High flux ultrafiltration membranes based on electrospun nanofibrous PAN scaffolds and chitosan coating, Polymer (Guildf)., 2006, 47(7), 2434–2441. Mar.
[60] Fan, Z., Wang, Z., Duan, M., Wang, J., & Wang, S. Preparation and characterization of polyaniline/polysulfone nanocomposite ultrafiltration membrane, J. Membr. Sci., 2008, 310(1–2), 402–408. Mar.
[61] Ritcharoen, W., Supaphol, P., & Pavasant, P. Development of polyelectrolyte multilayer-coated electrospun cellulose acetate fiber mat as composite membranes, Eur. Polym. J., 2008, 44(12), 3963–3968. Dec.
[62] Yung, L. et al. Fabrication of thin-film nanofibrous composite membranes by interfacial polymerization using ionic liquids as additives, J. Membr. Sci, 2010, 365(1–2), 52–58. Dec.
[63] Kaur, S., Sundarrajan, S., Rana, D., Matsuura, T., & Ramakrishna, S. Influence of electrospun fiber size on the separation efficiency of thin film nanofiltration composite membrane, J. Membr. Sci., 2012, 392–393, 101–111. Mar.
[64] Kaur, S., Barhate, R., Sundarrajan, S., Matsuura, T., & Ramakrishna, S. Hot pressing of electrospun membrane composite and its influence on separation performance on thin film composite nanofiltration membrane, Desalination, 2011, 279(1–3), 201–209. Sep.
[65] Bui, N.-N., Lind, M. L., Hoek, E. M. V., & McCutcheon, J. R. Electrospun nanofiber supported thin film composite membranes for engineered osmosis, J. Membr. Sci., 2011, 385–386, 10–19. Dec.

[66] Huang, L., Arena, J. T., Manickam, S. S., Jiang, X., Willis, B. G., & McCutcheon, J. R. Improved mechanical properties and hydrophilicity of electrospun nanofiber membranes for filtration applications by dopamine modification, J. Membr. Sci., 2014, 460, 241–249. Jun.
[67] Choi, S.-S., Lee, Y. S., Joo, C. W., Lee, S. G., Park, J. K., & Han, K.-S. Electrospun PVDF nanofiber web as polymer electrolyte or separator, Electrochim. Acta, 2004, 50(2–3), 339–343. Nov.
[68] Satinderpal, K., Zuwei, M., Renuga, G., Gurdev, S., Seeram, R., & Matsuura, T., "Plasma-induced graft copolymerization of poly(methacrylic acid) on electrospun poly(vinylidene fluoride) nanofiber membrane," 2007.
[69] Kim, G. M., Lach, R., Michler, G. H., Pötschke, P., & Albrecht, K. Relationships between phase morphology and deformation mechanisms in polymer nanocomposite nanofibres prepared by an electrospinning process, Nanotechnology, 2006, 17(4), 963–972. Feb.
[70] Rahul Sen,Bin Zhao,Daniel Perea,Mikhail E. Itkis,Hui Hu,James Love,Elena Bekyarova, and, and Robert C. Haddon, "Preparation of Single-Walled Carbon Nanotube Reinforced Polystyrene and Polyurethane Nanofibers and Membranes by Electrospinning" Nano Letters 2004, 4 (3), 459–464 DOI: 10.1021/nl035135s
[71] Wang, X., Zhang, K., Zhu, M., Hsiao, B. S., & Chu, B. Enhanced mechanical performance of self-bundled electrospun fiber yarns via post-treatments, Macromol. Rapid Commun., 2008, 29(10), 826–831. May.
[72] Yoon, K., Hsiao, B. S., & Chu, B. Formation of functional polyethersulfone electrospun membrane for water purification by mixed solvent and oxidation processes, Polymer (Guildf)., 2009, 50(13), 2893–2899. Jun.
[73] Huang, L., Manickam, S. S., & McCutcheon, J. R. Increasing strength of electrospun nanofiber membranes for water filtration using solvent vapor, J. Membr. Sci., 2013, 436, 213–220. Jun.
[74] Huang, L., & McCutcheon, J. R. Hydrophilic nylon 6,6 nanofibers supported thin film composite membranes for engineered osmosis, J. Membr. Sci., 2014, 457, 162–169. May.
[75] Razavi, S. M. J., Neisiany, R. E., Ayatollahi, M. R., Ramakrishna, S., Khorasani, S. N., & Berto, F. Fracture assessment of polyacrylonitrile nanofiber-reinforced epoxy adhesive, Theor. Appl. Fract. Mech., Elsevier, 26-Jul-2017.
[76] Kanafchian, M., Valizadeh, M., & Haghi, A. K. A study on the effects of laminating temperature on the polymeric nanofiber web, Korean J. Chem. Eng., 2011, 28(2), 445–448.
[77] Mohammadian, M., & Haghi, A. K. Study on the production of a new generation of electrospun nanofiber webs, Bulg. Chem. Commun., 2014, 46(3), 530–534.
[78] Yao, M. et al. Effect of heat-press conditions on electrospun membranes for desalination by direct contact membrane distillation, Desalination, 2016, 378, 80–91. Jan.
[79] Yoon, B., & Lee, S. Designing waterproof breathable materials based on electrospun nanofibers and assessing the performance characteristics, Fibers Polym., 2011, 12(1), 57–64. Feb.
[80] Yalcinkaya, B., Yalcinkaya, F., & Chaloupek, J. Optimisation of thin film composite nanofiltration membranes based on laminated nanofibrous and nonwoven supporting material, Desalin. Water Treat., 2017, 59, 19–30.
[81] Yalcinkaya, F. Mechanically enhanced electrospun nanofibers for wastewater treatment, 22, Nov, 2017.
[82] Yalcinkaya, F., Yalcinkaya, B., Hruza, J., & Hrabak, P. Effect of nanofibrous membrane structures on the treatment of wastewater microfiltration, Sci. Adv. Mater., 2016, 9(5), 747–757. May.

Index

https://doi.org/10.1515/9783110581393-013

polymer-free 11, 12

www.ingramcontent.com/pod-product-compliance
Lightning Source LLC
Chambersburg PA
CBHW061157240926
51846CB00006B/1041
* 9 7 8 3 1 1 0 5 6 1 8 0 7 *